Transformations in E-Business Technologies and Commerce:

Emerging Impacts

In Lee
Western Illinois University, USA

BUSINESS SCIENCE
Reference

Managing Director:	Lindsay Johnston
Senior Editorial Director:	Heather Probst
Book Production Manager:	Sean Woznicki
Development Manager:	Joel Gamon
Development Editor:	Myla Harty
Acquisitions Editor:	Erika Gallagher
Typesetters:	Michael Brehm
Print Coordinator:	Jamie Snavely
Cover Design:	Nick Newcomer, Greg Snader

Published in the United States of America by
Business Science Reference (an imprint of IGI Global)
701 E. Chocolate Avenue
Hershey PA 17033
Tel: 717-533-8845
Fax: 717-533-8661
E-mail: cust@igi-global.com
Web site: http://www.igi-global.com

Library of Congress Cataloging-in-Publication Data

Transformations in e-business technologies and commerce : emerging impacts / In Lee, editor.
p. cm.
Includes bibliographical references and index.
ISBN 978-1-61350-462-8 (hardcover) -- ISBN 978-1-61350-464-2 (perpetual access) -- ISBN 978-1-61350-463-5 (ebook) 1. Electronic commerce. 2. Technological innovations--Management. I. Lee, In, 1958-
HF5548.32.T724 2011
381'.142--dc23
2011043041

British Cataloguing in Publication Data
A Cataloguing in Publication record for this book is available from the British Library.

All work contributed to this book is new, previously-unpublished material. The views expressed in this book are those of the authors, but not necessarily of the publisher.

Table of Contents

Detailed Table of Contents

Online retailing is a rapidly expanding business. There are three major purchasing channels for online shoppers to choose when they conduct online purchasing: (1) shopping in online stores, (2) bidding in online auctions, and (3) direct purchasing in online auctions. Each channel has its strengths and weaknesses, and suits different kinds of online shoppers. This article explores the relationships between online shoppers' lifestyles and their preferences for online purchasing channels. Knowledge about online shoppers' preferences for online purchasing channels helps online retailers place right products in the right channel for the right customers and consequently generate stable revenues. Major findings are: (1) the online shoppers with extravert lifestyle prefer online auctions, (2) the online shoppers with indulgent lifestyle prefer online stores, and (3) the online shoppers with shrewd lifestyle have a more balanced preference for all three purchasing channels. Theoretical and practical implications are discussed.

In this paper, the authors the inventory control practices of a small-to-medium retailer to identify common challenges this type of organization experiences with respect to automated data capture (ADC) and the implementation of an enterprise wide information system. The study explores a single case of a hardware store in a regional town in New South Wales, Australia. Four semi-structured interviews were conducted with employees, focusing on issues related to inventory control including delivery discrepancies, checking and sorting of orders, locating stock and goods, loss prevention, customer purchasing and point of sale processing and replenishment. Flowcharts illustrate the current processes of the retailer with an understanding of how ADC technologies like bar code and radio-frequency identification (RFID) impact the retailer. The findings promote an evolutionary approach toward the use of automated data

Web-enabled customer relationship management, or e-CRM, holds the promise to simultaneously achieve the conflicting goals of strategic flexibility and operational efficiency in developing and executing innovative e-commerce marketing strategies. A new multi-disciplinary field involving such diverse areas as marketing, information systems, and consumer behavior, e-CRM seeks to strengthen customer relationship via a portfolio of trust-building practices with the support of powerful information technologies. While most literature in e-CRM has focused on how firms formulate and implement e-CRM initiatives, there is a dearth of literature on viewing the overall quality of an e-commerce firm's e-CRM practices from the consumer point of view. This research proposes such a consumer-oriented concept, e-CRM value, based on existing e-CRM research literature and examines the effect of e-CRM value on website loyalty. It was found that customers' perception of e-CRM value has a positive causal effect on their website loyalty. The implications of this finding for the theory-building of e-CRM and marketing practices of e-commerce firms are discussed.

This case study explores the attitudes and perceptions of franchise owner-operators in regard to their acceptance of advanced Internet innovation and technologies. Because these franchisees serve a dual role as both the decision maker and the end user of new technology and innovation, they provide a dichotomy of perspectives that yield insights into many aspects of business leadership, customer service, and operational proficiency. Findings suggest five key attributes of Internet use and adoption by franchisees. The data in this exploratory case study also reveals three areas of disparity regarding franchisee behavior worthy of consideration by practitioners and academics: (1) The inward focus of franchisees, (2) project costs are considered superficially, and (3) the micro SME as change agent. This research is significant and accretive by capturing the voice of franchisees in regard to Internet acceptance and by providing a strategic look at how the franchise micro SME is significantly different than non-franchised and larger SMEs.

Security is an increasingly important issue for business, and with it, the need for authentication. The use of biometric systems for personal authentication is a response to the rising issue of authentication and security. As with any new technology, user acceptance is often hard to measure, thus, in this work, a study of user acceptance of a biometrics authentication system in e-commerce, such as online banking within the Saudi society, was conducted. The study examined whether Saudis are practically willing to accept this technology by involving 306 participants in a large scale laboratory experiment that actively tested a biometric authentication system in combination of a survey. The Technology Acceptance Model

This article shows how the technical hype of 1990s has been transformed into the e-business organizations at the beginning of the 21st century. The authors took an interpretive stance in this study, grounded theory, and investigated the ontology of virtual organization by the metaphorical analysis. The metaphorical analysis adopted in this study provides the analytical power to conceptualize the social structure of virtual organization in the context of structuration theory with the process of grounded theory. As the e-business structuration indicates in this study, virtualization, the metaphor of virtual organization is interpreted as flexible domination, where the rational relationships in power distribution with other e-business characterized temporal (virtual) bond with low switching costs.

Current m-commerce business models show that m-commerce depends on complex networks of business relationships, which often comprise telecommunications service providers, mobile device makers, banking industry, Internet search engine providers, and various third-party value-adding companies. Due to the nature of m-commerce, the key to success in m-commerce lies predominantly in managing a network of alliances. This paper answers research questions, such as why do companies team up for m-commerce? What are the key challenges facing the alliances? How can companies address the challenges? What does the future hold for the study of strategic alliances including m-commerce alliances? This paper extends strategic alliance theories to the study of m-commerce alliances that are formulated in various cultural and national backgrounds. The authors examine both strategic and operational strategies for m-commerce alliances and discuss a wide range of issues in the formulation and implementation of m-commerce alliance strategy.

This conceptual paper revisits and reviews early milestone literature of influential authors, who discuss the impact of the Internet on major forces that shape a company's marketing strategy to assess whether their futurist predictions and conceptual claims turned out to be true or were later falsified by empirical research or market experience. In particular, the impact of the Internet on customers, design of products/services, pricing, marketing communications, intermediaries, publics, and social/cultural environment is explored. Throughout the paper, related under-researched and unsettled areas are discussed and future research avenues are suggested.

From a channel conflict perspective, this paper proposes a Relational Model identifying key determinants of manufacturers' intentions to establish an online direct sales channel (ODSC). The paper also develops a Strategic Framework as a guideline for manufacturers to formulate their Internet channel strategies. Unlike most other studies on channel conflicts that focus on channel management after the manufacturer has already established its ODSC, this study provides a comprehensive strategic analysis of channel conflicts based on a pre-ODSC context, which prevents unnecessary losses from channel conflicts before they occur.

Internet-based affiliate marketing programs have emerged as one of the fastest-growing methods for online retailers to acquire customers and increase sales by tapping into the power of independent web sites to reach a large, diverse audience of potential customers. However, while these programs have proven effective in increasing website traffic and sales, illegal or inappropriate activities on the part of affiliates could negatively impact a retailer's brand in the eyes of customers. This study is an exploratory analysis of governance mechanisms (formal contracts, partner selection, incentives and monitoring) in one-to-many affiliate programs in Spain. Agency theory and transaction cost analysis provide the theoretical background. The conclusion is that there is a significant lack of transparency in the guidance and restrictions communicated to affiliates, and a lack of systematic monitoring of affiliate behavior, which increases the risk of opportunism or misconduct. General recommendations for managers of affiliate programs are considered.

The first decade of the e-commerce era saw an increase in activity in the software development industries as new firms were created and existing firms made acquisitions. Many firms pursued a growth strategy and this growth required capital that was often obtained through an initial public offering (IPO) of equity. Software firm cost structures are very different from traditional physical goods firms because their marginal costs are much lower, but what is not known is whether this affects their financing strategies. In this study we compare software firm and traditional firm IPOs using data from 780 IPOs offered during the late dot-com era (1998-2002) to identify differences in firm and offer characteristics, investment risk, initial returns, and underwriting activity. We find that the characteristics and performance of software firm IPOs are significantly different from IPOs offered by traditional firms during this time period providing supporting for our conclusion that firm cost structure should be considered when analyzing IPOs and other strategic issues.

This study investigates whether brand name has an impact on personal customers' adoption of Internet banking (IB) in Hong Kong. Based on the Technology Acceptance Model (TAM) with proper adaptation to the online environment, we propose a model comprising an external factor, namely brand name, and three constructs of technology adoption, namely perceived credibility, perceived usefulness and perceived ease of use to assess customers' intention to use IB. In addition, we investigate the difference in the brand name's effect on customers' adoption of IB between banks with larger market shares and banks with smaller market shares. We collected data through a questionnaire survey and analyzed the data using structural equation modelling (SEM) to investigate the inter-relationships between the constructs and assess the overall explanatory power of the model. The results provided support for the model and confirmed its robustness in predicting the impact of brand name on personal customers' adoption of IB in Hong Kong.

Previous studies explored the effects of individual cognitive and affective reactions on online shopping and those of individual differences on usage of information systems. However, few studies evaluated how individual differences affect online shopping. This paper draws on the theory of reasoned action (TRA) and the technology acceptance model (TAM) as it attempts to gain understanding of individual difference impacts on online shopping attitudes. The model was tested using data collected from a Web survey, and 171 questionnaires were collected and analyzed. Results show that online shopping experience, level of Internet usage, cognitive absorption, personal innovativeness, and computer self-efficacy positively influence online shopping attitudes, and that perceived usefulness is a significant mediator, but only between online shopping attitudes and two individual differences variables: the level of Internet usage and cognitive absorption.

This research addresses difficulties in measuring e-commerce success by implementing the DeLone and McLean (D&M) model of IS success (1992, 2003) in an e-commerce environment. This research considers the influence of quality on e-commerce success by measuring the information quality and system quality attributes of an e-commerce system and the intention to use, user satisfaction and intention to transact from a sample of respondents. This research provides an empirical e-commerce application of the updated IS success model proposed by DeLone and McLean (2003). This paper found significant relationships between Information Quality and System Quality and three success dimensions: intention to use, user satisfaction and intention to transact. It found the following constructs to be most important in predicting success: ease of understanding, personalisation and reliability. In particular, that reliability is more important than usability where transactions are concerned and security though important, is not the most important factor.

Preface

EMERGING WEB 2.0-BASED BASED BUSINESS MODELS

Recently, the Web paradigm shifted from the business-centered to user-centered one. This paradigm shift has become known as "Web 2.0", coined by Tim O'Reilly in 2004 (O'Reilly, 2007. Web 2.0 refers to a website that has evolved beyond Web 1.0 and features user-created-content and online communities that allow website visitors to interact dynamically with the site. While technological distinctions between Web1.0 and Web 2.0 are often not clear in some areas, the social and technological environment for positive user participation and interactivity epitomizes Web 2.0. Web 2.0 is continuously evolving and offers new business models and support business processes, knowledge management, customer relationship management, and partner relationship management.

Despite the lack of a comprehensive framework for the use of Web 2.0 applications, businesses are eager to use them. Managers are already active users of the various Web 2.0 technologies such as social networks and blogs. Managers are realizing that Web 2.0 affects a wide spectrum of business activities from building product or brand awareness to after-sales services. The main purpose of this study is to explore emerging Web 2.0-based business models. This study proceeds as follows: Section 2 presents a literature review on Web 2.0. Section 3 explores emerging Web 2.0-based business models. Finally, Section 4 concludes with managerial implications.

LITERATURE REVIEW

Web 2.0 refers to the multitude of new ways that the Internet is used as a platform for developing and hosting software applications and developing and exchanging digital contents by the businesses and users. Due to the easy publication and editing of online content, Web 2.0 has already had great impacts on the ways that people interact and businesses operate. A global survey conducted by McKinsey in 2007 finds that the popularity of Web 2.0-based applications is growing among businesses (McKinsey Survey on Internet Technologies, 2007). While most companies surveyed have so far integrated a limited number of these applications into their business strategies, the large majority have indicated that Web 2.0 integration is important for maintaining the company's market position, providing a competitive edge, and addressing customer demand. Other studies on Web 2.0 adoption indicate that the benefits of the Web 2.0 applications come from knowledge management initiatives (Cayzer, 2004; Wagner, 2004), project management efforts (Miller, 2006), and social networks that connect employees (Middleton,

2008). Businesses can leverage Web 2.0 technologies in order to dynamically cooperate with customers and partners in efforts to generate new design innovations (Brown, 2008).

McAfee (2006) coined the term "enterprise 2.0" to describe the application of Web 2.0 to the enterprise utilizing wikis and social networking software to support and enhance the continuously changing and emergent collaborative structures of knowledge work across the enterprise. Organizations need to invest in Web 2.0 technologies differently from the way they invested in information technology (IT) projects in the past. Organizations will have to find new ways of management to respect the freedom, openness, and sociality inherent to Web 2.0 technologies (De Hertogh & Viaene, 2010).

The most important contribution of Web 2.0 is not in the software but in the information provided on the web sites (van Iwaarden, van der Wiele, Williams, & Eldridge, 2010). The tremendous increase in user-generated content (UGC) on the Internet has important consequences for Web users, as well as companies. While most studies argue that Web 2.0 is a healthy phenomenon, as it is promoting free expression and democracy and is becoming the new source of consumer influence and empowerment, critics argue that Web 2.0 also promotes low quality amateur journalism, threatens intellectual property rights, and blurs the boundaries between fact and fiction. Objections about UGC have also to do with issues of privacy and the lack of responsibility for online publications (Constantinides, 2010). Web 2.0 also emphasizes communication via the e-social networking, occurring on so-called e-community platforms. Therefore, interaction between users is the utmost importance. Interaction is easily conceivable for private as well as for commercial purposes (Kollmann & Lomberg, 2010).

Web 2.0 has contributed to an unprecedented growth of information volume, new forms of networking, customer empowerment, and new business models (Constantinides & Fountain, 2008). The following discusses major Web 2.0 applications including social networking sites, blogs, folksonomies, wikis, and integrated services.

Social Networking Sites

Social networking sites (SNS) allow individuals to form or maintain online social connections and share their skills, talents, knowledge, and/or preferences with other members. While every social networking site requires its members to create a profile, each site has different purposes and targets specific user populations. For example, Facebook focuses on friend networks, LinkedIn focuses business networks, and MySpace focuses on special interest topics such as movies and hobbies. Recently, these SNS have expanded their business models and have begun competing with each other.

Blogs

Blogs (short for weblogs) are online journals that are characterized by short entries and regular updates. Blogs are inherently flexible and can be used for a variety of purposes, ranging from personal opinions to knowledge management initiatives and customer relation tools (Ives & Watlington, 2005). One of the most useful features of blogs is the functionality that allows readers to comment on each entry (Kolbitsch & Maurer, 2006; Rosenbloom, 2004). The collective comments and links on blogs form a clustered online network termed the blogosphere (Schmidt, 2007). A variety of public blogging services are available to individuals and firms. For example, Twitter is the most popular social networking microblogging service which thrives on constant change and updates. Tweets are text-based posts of up to 140 characters displayed on the author's profile page and delivered to the author's subscribers, known as followers. Authors can restrict delivery to those in their circle of friends or allow open access. Recently, blogs have demonstrated the sheer power of information sharing and dissemination in areas of politics, natural phenomena, and celebrity activities.

Collectively Arranged Metadata: Folksonomies and Tags

Collectively arranged metadata are the result of user participation in the classification of digital objects. Collectively arranged metadata become more useful as more users participate in the creation. The process of individually assigning metadata about objects such as URLs, images, videos, and texts is called folksonomy or tagging. The process of assigning tags or labels to websites is also often referred to as social bookmarking. The primary benefit is that users will find information more easily and accurately. Folksonomies have become part of social software applications such as photograph annotation and bookmarking, and have become an important alternative to search engines or other instruments for surfing the web. An empirical analysis of the complex dynamics of tagging systems has shown that consensus around stable distributions and shared vocabularies emerge, even in the absence of a centrally controlled vocabulary (Halpin, Robu, & Shepherd, 2007). Some popular tagging sites include *Delicious*: a social bookmarking system, *Digg*: a story sharing community in which submissions are voted upon by users, and *Flickr*: a photo publishing/sharing site.

Wikis

Wikis are easy-to-use, browser-operated platforms that enable collaborative publication on the Internet (Ebersbach & Glaser, 2005). Wikis also embody a specific mindset towards collective intelligence. They allow many individual participants to contribute to an online discussion, usually via centrally managed content management systems. Wikis are designed to make it easy to correct mistakes and track changes. In contrast to blogs, the content of wikis tends to be more unbiased, as the author allows the readers to co-edit the original content. Through multiple revisions of a document by a group of co-editors, the content becomes more credible (Kolbitsch & Maurer, 2006). A number of validity checks are implemented to the contributions made to a given wiki topic. One of the most successful applications of wikis is Wikipedia, a popular online encyclopedia for which any member can contribute and edit contents.

Integrated Services

Integrated services utilize Web services. A number of web applications are based on Web services and service integration among businesses and users. Mashups are aggregations of services from different online sources to create a new service. One example includes pulling store locations from a database and displaying them on Google maps to show where the stores are located. Salesforce.com is an example of companies hosting and integrating corporate mashups. A podcast is a series of digital media files (either audio or video) that are released episodically and often downloaded through web feed. They are often distributed through an aggregator, such as an iPod. Many businesses are leveraging podcasting to their customers. A web feed is a data format used for allowing people to subscribe to online distribution of news, blogs, podcasts, or other information. Content distributors syndicate a web feed, thereby allowing users to subscribe to it.

E-Commerce Business Models

A business model is a framework of how an organization generates revenue and involves a series of planned activities or business processes. The literature on business models has been abundant. Timmers (1999) proposes his definition of business models that is applicable for e-commerce environments: "A business model is defined as the organization of product, service, and information flows, and the sources of revenues and benefits for suppliers and customers." Afuah and Tucci (2003) suggest that a business model is "the method by which a firm builds and uses its resources to offer its customers better value than its competitors and to make money doing so." They note that business models are designed to make money for long term. Rappa (2004) stated that "a business model is the method of doing business by which a company can sustain itself (that is, generate revenue); the business model spells out how a company makes money by specifying where it is positioned in the value chain.

In summary, all of these definitions share a common idea: it concerns what to produce and sell, and how to sell and to whom. The business model also addresses the value proposition of the product or service and earnings logic behind the profit making. The business model is a representation of management thinking and practice that helps companies to see, understand, and run their activities in a distinct and specific way (Chararbaghi, Fendt, & Willis, 2003). Based on that management thinking and practice, each firm is likely to develop a unique business model. Afuah and Tucci (2003) suggest the following eight components of a business model:

1. **Customer value:** The description of target customer value. How can the firm deliver value to customers? Will the firm offer differentiated or lower-cost products/services?
2. **Scope:** Target the right market segments with products or services that have the appropriate value mix to customers.
3. **Pricing:** The development of proper pricing strategies.
4. **Revenue sources:** The determination of all revenue sources.
5. **Connected activities:** The determination and timing of activities that underpin customer value.
6. **Implementation:** The determination of organizational structure, systems, people, and environment needed to carry out the activities and deliver customer value.
7. **Capabilities:** The existing capabilities and the desired capabilities needed to execute the value-adding activities, and the capability gaps needed to be filled.
8. **Sustainability:** The strategies that help the firm sustain competitive advantages and make it difficult for competitors to imitate the sustainability.

The term, "e-commerce business model" has been widely used by researchers and practitioners to loosely describe a unique aspect of a particular electronic commerce business. E-commerce business models use the web to carry out their activities and generate revenue. E-commerce business models are important for companies to survive in the global economy and are a fertile ground for innovation (Amit & Zott, 2000). E-commerce business models constantly evolve to adapt to changing consumer demands. For example, Amazon.com was launched as an online bookseller in 1995, but today it sells practically everything. Priceline.com started as a "name-your-own-price" site hosting travel bidding services, but later expanded its business to fixed-price products as a regular online travel agency.

A number of researchers have presented high-level e-commerce business models that have been influential in shaping strategy development and implementation of many online businesses. Their list of the e-commerce business models is not exhaustive. As shown below, their differing views on e-commerce business models are complementary with each other. Weill and Vitale (2001) identify the following eight e-commerce business models.

1. **Content Provider:** Provides content (information, digital products, and services). Examples include America Online (AOL) and accuweather.com.
2. **Direct to Customer:** Provides goods or services directly to the customer, often bypassing traditional channel members. An example is Dell.
3. **Full Service Provider:** Provides a full range of services in one domain (e.g. financial, health, application services). Examples include E*TRADE and Scottrade.
4. **Intermediary:** Brings together buyers and sellers by concentrating information. Examples include Expedia.com and eBay.
5. **Shared Infrastructure:** Brings together multiple competitors to cooperate by sharing common IT infrastructure. An example is CourseSmart, which is a venture supported by the leading publishers in North American higher education.
6. **Value Net Integrator:** Coordinates activities across the value net by gathering, synthesizing, and distributing information. An example is Cisco, which designs, manufactures, and sells networking equipment. Cisco utilizes resellers to install and support the configurator to reduce errors and collect market information.
7. **Virtual Community:** Creates and facilitates an online community of people with a common interest enabling interaction and service provision. Examples include Facebook and LinkedIn.
8. **Whole of Enterprise:** Provides a firm-wide single point of contact, consolidating all services/business models provided by a large multi-unit organization.

Rappa (2004) presents a collection of nine e-commerce business models:

1. **Brokerage:** Brokers are market-makers: they bring buyers and sellers together and facilitate transactions. Brokers play a catalyst role in business-to-business (B2B), business-to-consumer (B2C), or consumer-to-consumer (C2C) markets. Examples include marketplace exchanges (Orbitz, ChemConnect), auction brokers (eBay), and transaction brokers (PayPal, Escrow.com).
2. **Advertising:** The web advertising model is an extension of the traditional media broadcast model. The advertising website provides content and services (such as email, instant messaging, blogs) mixed with advertising messages. Examples include portals (Yahoo!), classifieds (Craigslist), and content-targeted advertising (Google).
3. **Infomediary:** Data about consumers and their consumption habits are valuable, especially when that information is carefully analyzed and used to develop targeted marketing campaigns. Examples include advertising networks (DoubleClick) and metamediaries (Edmunds).
4. **Merchant:** Wholesalers and retailers of goods and services. Sales may be made based on list prices or through auction. Examples include online merchants (Amazon.com) and bricks-and-clicks (Barnes & Noble).
5. **Manufacturer (Direct):** The power of the web allows a manufacturer to reach consumers directly and thereby compress the distribution channel. Dell Computer is a well-known direct manufacturer.
6. **Affiliate:** In contrast to the generalized portal, which seeks to drive a high volume of traffic to one site, the affiliate model provides purchase opportunities wherever people may be surfing. The af-

filiate model is inherently well-suited to the web. Examples include banner exchanges and revenue sharing programs (Barnes & Noble, Amazon.com, Target).

7. **Community:** An online community is an electronically supported social network. It can be seen as a group of people who have regular social interaction, independent of time and space, because of a common interest such as a problem, task, or feeling exchange. Examples of online communities are Internet forums, where users can gather to share information.
8. **Subscription:** Users are charged a periodic (daily, monthly, or annual) fee to subscribe to a service. Subscription and advertising models are frequently combined. Examples include content services (Netflix), internet services providers (America Online), and treasure hunting games (Geocaching).
9. **Utility:** The utility or "on-demand" model is based on metering usage, or a "pay as you go" approach. Unlike subscriber services, metered services are based on actual usage rates. Examples include metered usage (utility computing by Hewlett Packard).

As web technologies advances, new e-commerce business models continue to emerge. From the previous discussion, field studies, and examples, it is easy to conclude that at present many businesses explore and utilize Web 2.0. However, Web 2.0-based business models have not been fully investigated. In order to help businesses to utilize Web 2.0, the next section explores a basic classification of the Web 2.0-based business models.

3. EMERGING WEB 2.0-BASED BUSINESS MODELS

Web 2.0 is moving beyond the early diffusion stage and best practices emerge. Many new pure-play Web 2.0 business models arise by leveraging Web 2.0 technologies such as wikis and blogs. In addition, bricks-and-clicks organizations try to leverage Web 2.0 technologies to improve their business processes. Organizations must understand the impacts of Web 2.0 technologies on their existing business model in order to be competitive in this fast-paced environment. In this section, we investigate the emergence of new Web 2.0-based business models in which the use of Web 2.0 is the primary driver of revenue and corporate existence. Based on the analysis of popular Web 2.0-based organizations, we identified the following six Web 2.0-based business models.

Broad Online Community

A broad online community is an electronically supported social network of a wide range of user groups. It can be seen as a group of people who have regular social interaction but without any specific idiosyncratic group characteristics. The broad online community allows individuals to form or maintain online social connections and share their skills, talents, knowledge, and/or preferences with other members. The viability of this community model is based on user loyalty, since users need to invest both time and emotion to the community. Revenue can come from the sale of products, information services and/or advertising. A large community may expect revenue from subscriptions for premium services. Examples include Twitter, Facebook, Bebo, and Friendster.

Focused Online Community

A focused online community is a niche/specialty online community dedicated to people with a common interests and needs such as professions or hobbies. One of the major distinctions is in professional and private users (e.g., business social network services as LinkedIn and private social networking services as Blackplanet targeted at African American users). Users add their profile and portfolio to the community and become part of a common interest such as a problem, task, hobby, or business. Revenue can be based on the sale of specialty advertising, premium services, specialty-related products and services or contributions. Shopping communities bring like-minded people together to discuss, share, and shop. Using the wisdom of crowds, users communicate and aggregate information about products, prices, and promotions. An example of a focused online community is an Internet car forum, such as NASIOC (North American Subaru Impreza Owners Club, forums.nasioc.com). In this forum, members of the community can gather to share information such as car pricing, purchasing tips, technical details/troubleshooting, and events the community may hold in their region. They can also post products for sale or trade with other members. Like most focused online communities, forums.nasioc.com is funded through banner advertisements, as well as sponsors who sell their products on the forum.

Social Shopping

Social shopping, also called social commerce, brings buyers and sellers together and facilitates transactions by providing a method of e-commerce where shoppers' social interactions are emphasized in the shopping experience (Wikipedia, 2010). Social shopping attempts use technology to mimic the social interactions found in physical malls and stores. Social shopping can largely be divided into two categories: (1) Group shopping sites and (2) Social shopping marketplaces. Group shopping sites group individual consumers to purchase products and services together from merchants at discount prices. Examples include Groupon, Gilt City, LivingSocial, and BuyWithMe. Shopping marketplaces bring social shopping sites, merchants, and consumers together to connect and transact. The marketplace brings together independent buyers and sellers and creates a forum for them to conduct business transactions. Examples include Sttorenvy and Jasmere. Social shopping also encourages people to exchange information about products and services. The revenue sources include sales commission and advertising. Social shopping sites develop Android and iPhone-based mobile apps to provide location-based services.

The success of the social shopping sites depends on the customer satisfaction, customer loyalty, partner management, and quality policy, including refund policy. As the number of the social shopping sites grows rapidly, the competition intensifies and social shoppers' complaints increase about overbooking, poor product/service quality, and stockout. Participating companies usually pay a high rate of commission to social shopping sites. Many social shoppers are bargain hunters.

Content Intermediary

Content intermediaries are businesses that function as a third party between content generators and content users. Sometimes, users serve as both content generators and content users. Content intermediaries aggregate content and deliver it to users. In addition to written content, content generators often contribute to the sites with video, audio, or other types of rich media. Product reviews, comments, recommendations, and news or information posted on the sites represent high quality market information and an unbiased customer voice. Three types of content intermediaries are: 1) blog sites, 2) collective intelligence sites, and 3) content aggregating/sharing sites.

Blogs are online journals that are characterized by short entries and regular updates. Blogs are inherently flexible and can be used for a variety of purposes, ranging from personal opinions of the contributor to knowledge management initiatives and customer relation tools. Personal or public blog sites are used to host blogs and the posted messages can be distributed to other sites or readers via RSS. Examples include political blog sites, such as HuffingtonPost.com, consumer electronics blog sites, such as engadget.com, and entertainment blog sites, such as perezhilton.com.

As users contribute new content to the web, the web of connections and associations among users grows stronger as a result of their collective activities. Collective intelligence is formed out of massive user participation and collaboration via the web. Innovative web business models such as social bookmarking and online encyclopedias take advantage of the network effects: the more people participate in generating and refining content, the more useful they become to the users. The ease of content generation and the speed of content sharing are critical to the success of collective intelligence sites. Another innovative collective intelligence business is an online reputation system. Online reputation systems are based on intelligence of crowds. People provide opinions of the products, services, or users they have experienced in the form of scores, rankings, and comments. The reputation systems collect and publish reputation scores and comments to a community or general public. While most reputation systems are product review systems, the objects of the reputation system used by eBay are transaction participants who provide ratings of the transaction party after they conduct a transaction. eBay's reputation system is designed to reward good behavior and punish bad behavior. Examples of collective intelligence sites include online encyclopedias, such as Wikipedia, social bookmarking/tagging sites, such as del.icio.us and Digg, and online reputation systems, such as Epinions, Bizrate, and Yelp.

Content gathering/sharing sites gather web content (and/or sometimes applications) from users. The value of the sites depends on the quality and quantity of the content contributed by the users. This content is in the form of video, audio, music, images, and text. Compared to the blog sites, the journalistic function is limited. Examples include video-sharing sites, such as YouTube, event sharing sites, such as Upcoming, and photo publishing/sharing sites, such as Flickr.

Virtual World

Virtual worlds are persistent virtual reality spaces (Schroeder, 2008). Virtual world enables users to interact with each other without geographical barriers. In virtual worlds or games, it is possible for users to interact using avatars. Avatars are the representations of users in virtual worlds, often graphically displayed as 3-dimensional characters and completely customized according to the user's preferences. These worlds are available 24/7 – users can explore, socialize, and solve collaborative challenges. Virtual world technologies have affected education, information, and gaming industries. From a marketing perspective, virtual worlds create opportunities for a new form of commerce – virtual or v-commerce (Nasco, Boostrom Jr., & Coker, 2010). V-commerce is an alternative and/or supplement to traditional forms of commerce. Virtual commerce is a viable type of commerce in which products are created with infinite virtual resources and exchanged for real world money.

One of the more successful virtual worlds is 3-dimensional Second Life, created by Linden Lab in 2003. Avatars are central to the way that users interact with people and objects in virtual spaces. Second Life centers on socializing within communities, the sale and resale of goods, and the advancement of its virtual economy. Virtual characters known as Residents run businesses, own land, travel, and buy and sell goods and services with the Linden Dollar. Virtual worlds represent a significant Web 2.0 business model due to its business potential. The worldwide virtual world economy is valued at approximately $1.8 billion (Dibbell, 2007). Other virtual worlds include Smallworlds, Zwinktopia, ActiveWorlds, and Twinity.

Shared Web 2.0 Services

Shared Web 2.0 Services facilitate the growth of Web 2.0 populations by providing sharable services in the form of software and hardware. Using the shared Web 2.0 services, users/groups can develop their own Web 2.0 applications at minimum costs and technical skills. For example, users/groups may set up their own wiki sites using wiki development tools provided by wiki hosting sites such as Wikispaces and Wetpaint. WordPress is a popular blog hosting site available to individuals and firms. The revenue sources include advertising, premium services, and subscription fees. Examples of shared Web 2.0 services include social networking host services for private specialty social networking, mashup platform vendors, Widgets, wiki host services, Google apps for social networking and collaborations, blog host services, and collaborative host services.

4. CONCLUSION

In this study, we discussed major Web 2.0 applications including social networking sites, blogs, folksonomies, wikis, and integrated services. We also identified six Web 2.0-based business models: 1) broad online communities; 2) focused online communities; 3) social shopping; 4) content intermediaries; 5) virtual worlds, and 6) shared Web 2.0 services. These Web 2.0-based business models continue to grow in size and number through the interaction with individual users and business customers.

Web 2.0 has not only engendered Web 2.0-based businesses, but has also affected the other existing businesses. Existing businesses must understand the impacts of Web 2.0 technologies on their existing business models in order to be competitive in this fast-paced environment. Managers should clearly identify and prioritize their business model components that require Web 2.0 technologies to improve organizational performance. For instance, certain organizations may decide to incorporate Web 2.0 only as part of their information sharing and collaboration initiatives with partners whereas others may use it as a means to network with their customers. The overall business strategy will help determine Web 2.0 initiatives and implement them.

In Lee
Western Illinois University, USA

REFERENCES

Afuah, A., & Tucci, C. (2003). *Internet Business Models and Strategies*. Boston: McGraw-Hill/Irwin.

Amit, R., & Zott, C. (2001). Value creation in e-business. *Strategic Management Journal*, *22*(6-7), 493–520. doi:10.1002/smj.187

Cayzer, S. (2004). Semantic blogging and decentralized knowledge management. *Communications of the ACM, 47*(12), 47–52. doi:10.1145/1035134.1035164

Chararbaghi, K., Fendt, C., & Willis, R. (2003). Meaning, legitimacy and impact of business models in fast-moving environments. *Management Decision, 41*, 372–382. doi:10.1108/00251740310468013

Constantinides, E. (2010). Connecting small and medium enterprises to the new consumer: The Web 2.0 as marketing tool, Global Perspectives on Small and Medium Enterprises and Strategic Information Systems: International Approaches, (Bharati, Lee, & Chaudhury edited. Business Science Reference), IGI Global, 1-21.

Constantinides E., & Fountain S. (2008). Web 2.0: Conceptual Foundations and Marketing Issues, Journal of Direct, Data and Digital Marketing Practice, 9(3).

De Hertogh, S., & Viaene, S. (2010). Grounding principles for governing Web 2.0 investments. Encyclopedia of E-Business Development and Management in the Global Economy, (Lee edited. Business Science Reference), IGI Global, 1225-1234.

Dibbell, J. (2007). The life of the Chinese gold farmer. The New York Times. Retrieved March 25, 2009, from http://www.nytimes.com/2007/06/17/magazine/17lootfarmers-t.html?_r=1&pagewanted=all

Ebersbach, A., & Glaser, M. (2005). Wiki. *Informatik-Spektrum, 28*(2), 131–135. doi:10.1007/s00287-005-0480-7

Halpin, H., Robu, V., & Shepherd, H. (2007). The complex dynamics of collaborative tagging. Proceedings of the 16th International Conference on the World Wide Web (WWW'07), Banff, Canada, ACM Press, 211-220.

Ives, B., & Watlington, A. (2005). Using blogs for personal KM and community building. *Knowledge Management Review, 8*(3), 12–15.

Kolbitsch, J., & Maurer, H. (2006). The transformation of the Web: How emerging communities shape the information we consume. *Journal of Universal Computer Science, 12*(2), 187–213.

Kollmann, T., & Lomberg, C. 2010. Web 1.0, Web 2.0 and Web 3.0: The development of e-business. Encyclopedia of E-Business Development and Management in the Global Economy, (Lee edited. Business Science Reference), IGI Global, 1235-1242.

McAfee, A. (2006). Enterprise 2.0: The dawn of emergent collaboration. *MIT Sloan Management Review, 47*(3), 21–28.

McKinsey. (2007). How businesses are using Web 2.0: A McKinsey global survey. Retrieved June 1, 2010, from http://www.finextra.com/Finextra-downloads/featuredocs/hobu07.pdf

Middleton, C. (2008). The social side of business. Computer Weekly, 26-27.

Miller, D. I. (2006). Collaborating through wikis. CNet.

Nasco, S. A., Boostrom, R. E., Jr., & Coker, K. K. (2010). Virtual Commerce. Encyclopedia of E-Business Development and Management in the Global Economy, (Lee edited. Business Science Reference), IGI Global, 1192-1198.

O'Reilly, T. (2007). What is Web 2.0: Design patterns and business models for the next generation of software. *Communications & Strategies*, *65*, 17–37.

Rappa, M. (2004). Business Models on the Web. Retrieved: 25/07/2006, 2006, from http://digitalenterprise.org/models/models.html

Rosenbloom, B. (2004). *Marketing Channels* (7th ed.). Mason, OH: Thomson South-Western.

Schmidt, J. (2007). Blogging Practices: An Analytical Framework. *Journal of Computer-Mediated Communication*, *12*(4), 1409–1427. doi:10.1111/j.1083-6101.2007.00379.x

Schroeder, R. (2008). Defining virtual worlds and environments. *Journal of Virtual Worlds Research*, *1*(1), 1–3.

Timmers, P. (1999). *Electronic Commerce: Strategies and Models for Business-to-Business Trading*. Chichester, England: John Wiley.

van Iwaarden, J., van der Wiele, T., Williams, R., & Eldridge, S. (2010). Web 2.0: The era of user generated content on web sites. Encyclopedia of E-Business Development and Management in the Global Economy, (Lee edited. Business Science Reference), IGI Global, 1208-1214.

Wagner, C. (2004). Wiki: A technology for conversational knowledge management and group collaboration. CAIS, (13), 265-289.

Weill, P., & Vitale, M. R. (2001). *Place to Space: Migrating to eBusiness Models*. Boston: Harvard Business School Press.

Chapter 1
To Bid or to Buy?
Online Shoppers' Preferences for Online Purchasing Channels

Fan-Chen Tseng
Kainan University, Taiwan

ABSTRACT

Online retailing is a rapidly expanding business. There are three major purchasing channels for online shoppers to choose when they conduct online purchasing: (1) shopping in online stores, (2) bidding in online auctions, and (3) direct purchasing in online auctions. Each channel has its strengths and weaknesses, and suits different kinds of online shoppers. This article explores the relationships between online shoppers' lifestyles and their preferences for online purchasing channels. Knowledge about online shoppers' preferences for online purchasing channels helps online retailers place right products in the right channel for the right customers and consequently generate stable revenues. Major findings are: (1) the online shoppers with extravert lifestyle prefer online auctions, (2) the online shoppers with indulgent lifestyle prefer online stores, and (3) the online shoppers with shrewd lifestyle have a more balanced preference for all three purchasing channels. Theoretical and practical implications are discussed.

INTRODUCTION

With the prevalence of the Internet, more and more people are making purchases online (Carrie, 2004; Chou, 2006; Yang, 2007). There are three major purchasing channels for online shoppers when they conduct online purchasing:

DOI: 10.4018/978-1-61350-462-8.ch001

1. *shopping in online stores*: the shopper just pays the list price (sometimes plus a shipment fee) to acquire a product.
2. *bidding in online auctions*: the shopper (bidder) must engage in bidding to offer a price he/she is willing to pay for something. When more than one bidder is competing for the same item, the bidders have to raise the

price for an item by a series of rising bids. The highest bidder wins the item.

3. *direct purchasing in online auctions*: this is an alternative way for buying an item in online auctions. The auctioneer can set a "Buy It Now" price for an item. The bidder can purchase this item instantly at the "Buy It Now" price without engaging in bidding with other bidders.

Each channel has its own features and appeals to different segments of consumers. According to the report of Taiwan's *Institute for Information Industry* (Yang, 2007), 39.8% of online shoppers prefer shopping in online stores, 35.8% of online shoppers prefer direct purchasing in online auctions, and 22.5% of online shoppers prefer bidding in online auctions.

The purpose of this study, therefore, is to understand online shoppers' preferences for online purchasing channels. Specifically, this study aims to answer the following questions:

1. Do different kinds of online shoppers prefer different types of online purchasing channels?
2. Do different kinds of online shoppers often purchase different types of products online?
3. Do online shoppers prefer different online purchasing channels when purchasing different types of products?
4. Are these purchasing channels substitutable, and how much are they substitutable?

Understanding these issues can have both theoretical and practical significance. Theoretically, researchers have studied online shoppers in online stores (e.g., Rohm & Swaminathan, 2004; Childers et al., 2001) and in online auctions (e.g., Dholakia, 2005; Bapna et al., 2004; Ariely & Simonson, 2003) separately, but not jointly. Recognizing the fact that online stores and online auctions are both purchasing channels for online shopping, this study examines these two channels jointly to fill in the blank of existing knowledge. Practically, with the knowledge of online shoppers' preferences for online purchasing channels, firms or individuals selling items online can put right products in right channels to satisfy the right buyers to generate sizable revenues. Besides, network effects are prominent in online auctions and online store services (Tseng, Teng, & Chiang, 2007a, 2007b; Tseng, 2008). This means that if an online purchasing channel can fulfill the transactions satisfactorily, more and more buyers and sellers will gather to that channel, further enhancing its popularity and growth. This self-reinforcement process is an instance of the network effect.

This study explores the above-mentioned questions based on lifestyles of online shoppers because the lifestyle has been a powerful construct for analyzing consumer behavior (Hawkins et al., 2004; Solomon, 2007) and it has been proposed that lifestyle is more effective than personality and social economy in understanding consumers' purchase decisions (Lesser & Marie, 1986).

The rest of this paper is structured as follows. Related works are reviewed and hypotheses of this study are proposed. Preliminary interviews are then conducted to obtain a general understanding of online shoppers' preferences in online purchasing channels, followed by a survey to test the hypotheses about the association of online shoppers' lifestyles, preferred purchasing channels and purchased products. The general discussion section (1) summarizes the main findings, (2) compares them with the findings of earlier related studies, (3) enumerates practical applications of our findings, and (4) discusses the limitations of this study and future research directions.

RELATED WORKS AND HYPOTHESES

Berkman and Gilson (1974) suggested that lifestyle is unified patterns of behavior, which both determine and are determined by consumption.

Lifestyle integrates the strengths of demographics with the richness and dimensionality of psychological factors. It refers to daily behaviorally oriented facets of people as well as their feelings, attitudes, and opinions. (Wang et al., 2006)

With the stable growth of e-commerce, many researchers have extended the study of lifestyles to online shoppers. Notably, Swinyard and Smith (2003) argued that a key to the understanding of online shoppers is to understand their *Internet-related lifestyles*. They developed 38 items regarding online shoppers' Internets psychographics and extracted six factors as satisfiers and dissatisfiers for online shopping. Yang (2004) identified three lifestyle segments (experiencers, traditionalists, and self-indulgents) among Internet users in Taiwan and found the attitudes towards Internet advertising were different among these segments. Wang et al. (2006) found that online shoppers with different lifestyles (principle, active, and excited) assign different weights on the nine objectives of online shopping (Keeney, 1999). Specifically, active lifestyle consumers regarded privacy as the most critical value above all others. On the other hand, principle lifestyle and excited lifestyle consumers regarded equally privacy, safety, and product quality above all other objectives. However, different online purchasing channels may score differently in these objectives, and consumers have to choose one specific channel to complete their purchases. This study thus posits that consumers with different lifestyle will prefer different purchasing channels.

Although previous research works have studied lifestyles in online environments, two issues remain to be resolved. First, the lifestyles were not specifically Internet-related (e.g., Wang et al., 2006; Wu, 2006; Yang, 2004). Second, the studies focused on online stores without covering online auctions (e.g., Swinyard & Smith, 2003). To overcome these shortcomings, this study proposes its own lifestyles for online shoppers as described below (the process for deriving these lifestyles will be elaborated in subsequent sections).

1. The *extravert shoppers* are interested in social interactions, fashion-pursuing and make purchases casually.
2. The *shrewd shoppers* are in the practice of gathering information before shopping and make planned purchases, and will hunt for desired products decisively.
3. The *self-indulgent shoppers* are more susceptible to marketing promotions and enjoy spending money.

Hence, this study proposes the following hypothesis for online shoppers' preferences for online purchasing channels:

H1: Online shoppers with different lifestyles (extravert, shrewd, and self-indulgent) prefer different online purchasing channels.

The *means-end* chain (MEC) theory argues that an individual's decision to purchase a product is contingent on how the consumer relates the product to herself (Gutman, 1982). That is, product attributes are the means with which consumers obtain their desired ends (values), and consumers holding various values choose various categories of products to achieve their values. Some researchers (e.g., Brunso & Grunert, 1998; Olson & Reynolds, 1984) also suggested that the lifestyle framework should be integrated with the means-end chain to investigate consumer behavior. Recently, Brunso et al. (2004) proposed a framework that defined lifestyle as an intervening system of cognitive structures linking situation-specific product perceptions to more abstract cognitive categories and eventually to personal values. Hence, the lifestyle represents a system of individual differences in the use of knowledge structures that intervene between abstract goal states (personal values) and situation-specific product perceptions and behaviors. Therefore, this study postulates online shoppers with different lifestyles will chiefly buy different categories of products online.

H2: Online shoppers with different lifestyles (extravert, shrewd, and self-indulgent) often purchase different categories of products online.

The study of Korgaonkar et al. (2006) proposed three e-retailer types (prestigious department store e-tailers, discount e-tailers, and non-store-based e-tailer) and three product types (search, experience, and credence products). They found significant interaction effects of the online stores and product type. That is, for different product categories, consumers prefer to shop from different types of e-tailers since they have different perception of search costs, complexity, and risks toward each type of e-tailer. Also, according to the report of Yang (2007), the most-often purchased product categories, combining all three online purchasing channels (online store, direct purchase in online auctions, and bidding in online auctions) is computer hardware/software, followed by cosmetics and health-care products, followed by fashion apparels and accessories. In contrast, the most-often purchased product categories, considering only online auctions, is fashion apparels and accessories, followed by computer hardware/software, followed by cosmetics and health-care products. Clearly, the most-often purchased product categories are not identical for all three online purchasing channels.

This study therefore infers that different online purchasing channels have different features and may be appropriate for selling/buying different categories of products, and online shoppers choose different channels for purchasing different categories of products.

H3: Online shoppers prefer different purchasing channels for purchasing different categories of products online.

Figure 1 presents the conceptual framework for the hypotheses concerning the associations among lifestyles, purchasing channels, and product categories.

PRELIMINARY INTERVIEWS

Preliminary interviews were first conducted with online shoppers to obtain a general understanding of online shoppers' evaluation and preferences for online purchasing channels. Two graduate students with experiences in online shopping served as research assistants to contact their classmates, friends, and relatives who were experienced online shoppers for the face-to-face interviews. Purposive sampling was used to select these "experienced" online shoppers satisfying one of the following criteria: (1) with over one year of experience in

Figure 1. The conceptual framework

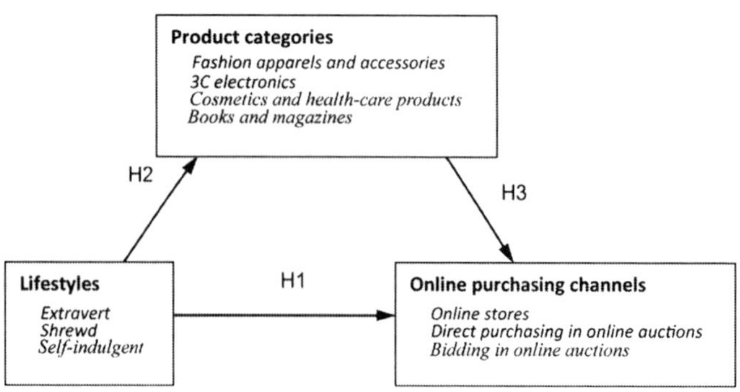

online shopping, or (2) with at least one transaction in every two months within latest one year.

Although only 14 online shoppers were selected, the sampling process tried to approximate the typical profile of the large-scale sample (with 5773 respondents) in a report of the semi-official organization III (Institute for Information Industry) (Yang, 2007). In the III report, the ratio of male to female is nearly 1:1, and most respondents (nearly 78%) were aged from 20 to 39. In jobs, most were students (44%) or salaried employees (38%). Most respondents (77%) had collegiate degrees or higher.

The 14 online shoppers in the preliminary interviews have the following demographics: in gender, seven males and seven females; in job, seven students and seven salaried employees; in age, from 19 to 39; in education, only one of them is a senior high-school student, others are collegiate students or holding collegiate or graduate degrees.

They were invited to talk about their opinions on the following three open-ended questions about online shopping:

1. Which of the three purchasing channels (online stores, direct purchase in online auctions, and bidding in online auctions) do you prefer? Why?
2. Under which conditions will you use alternative purchasing channels other than the one you use most often?
3. What categories of products do you often purchase in your preferred purchasing channels?

Preferred Purchasing Channels

Of the 14 online shoppers interviewed, two shoppers preferred online stores, eight shoppers preferred direct purchase in online auctions, and four shoppers preferred bidding in online auctions. Their evaluations for online purchasing channels are listed in Table 1. There are two categories of attributes for evaluation: common attributes that were mentioned for all three channels and unique attributes that are not shared by all three channels.

These 14 shoppers have different evaluations regarding the common attributes of the three channels, which is indicated by the numbers (1, 2, and 3) in Table 1 to show the rankings of the purchasing channels on these common attributes.

For example, shopping in online store is most convenient, followed by direct purchase in online auctions, and bidding in online auctions is least convenient. The inconvenience of bidding in online auctions comes from the need to interact

Table 1. Evaluations on online purchasing channels

		Shopping in online stores	Direct purchase in online auctions	Bidding in online auctions
Common attributes	Convenience	1	2	3
	Transaction security	1	2	3
	Cheaper prices	3	2	1
	Amusement	3	2	1
	Product quality	1	2	3
	Product description and website design	1	2	3
	After-sale services	1	2	3
Unique attributes		Allowing installment payments by credit card	1. With unique products 2. prices are negotiable	With unique products

with the auctioneer and the need to monitor other bidders' bidding prices.

On the attributes of security, quality, product descriptions and after-sale services, online stores rank highest, followed by direct purchase in online auctions. Bidding in online auctions ranks lowest on these attributes. The ranking results are reasonable because online stores are mostly operated by well-established companies and direct purchases in online auctions are run by smaller-sized firms, while online auctioneers are individuals with limited resources. Nevertheless, bidding in online auctions ranks highest on cheaper prices and amusement. Shoppers can usually find real bargains at prices lower than in other two purchasing channels. Besides, the bidding process, though time-consuming, is more interesting because bidders can interact with auctioneers and compete with other bidders.

Alternative Purchasing Channels

Shoppers will switch to alternative purchasing channels for two common reasons: (1) shoppers cannot find their desired products in the preferred channel, and (2) the prices in other channels are more favorable.

There are specific reasons for switching to alternative purchasing channels. Shoppers preferring online stores will try online auctions because of their friends' recommendations. Shoppers preferring direct purchase in online auctions will turn to online stores when they are eager to own desired products immediately. Shoppers preferring bidding in online auctions will go to online stores when they are eager to own desired products quickly or when they buy expensive items.

Product Categories in Three Purchasing Channels

There are product categories online shoppers will purchase using all the three purchasing channels, including fashion garments and accessories, cosmetics and health-care products, and 3C electronics.

There are product categories online shoppers will chiefly purchase in online stores, including books, expensive items such as jewelries, and products demanding assured qualities such as baby products. It is noted that there have been a few well-known online bookstores in Taiwan, so shoppers will usually visit these stores to find books.

There are product categories online shoppers will chiefly purchase in online auctions, including second-hand products, special items of limited quantities, and some special foods available only in online auctions.

Summary and Interpretation

From the interviews, we see that online shoppers have different preferences for online purchasing channels, because they perceive and pursue different benefits from different channels. Besides, online shoppers tend to use different purchasing channels to buy different categories of products. However, these channels are not mutually exclusive. Online shoppers may still try alternative channels for better prices or much-desired product items.

Three insights emerge from these interviews, which constitute the basis for the development of survey questions about online shoppers' lifestyles.

1. *Search*: online shoppers search various online purchasing channels (and the Web in general) for information about prices and products.
2. *Interaction*: online shoppers perceive online auctions not only as a purchasing channel, but also as a platform for *social interaction*. This means that bidders can talk or discuss with the auctioneer about the products or relevant knowledge about the products. When the bidder and auctioneer experienced several successful transactions, they may talk or even chat about other topics, somewhat

like friends talking with each other online. In addition, the process of competing with other bidders is an exciting experience for some online shoppers.

3. *Eagerness*: online shoppers usually perceive online auctions as a purchasing channel for cheaper prices. However, when they are in eagerness to own their much-desired products immediately, they will resort to online stores even this may cost them more money.

These three insights turn out to be three factors in online shoppers' lifestyles: (1) information search, (2) social interaction, and (3) decisiveness. The details will be given in the following sections.

STUDY SURVEY

The questionnaire contains three parts: Part I is for online shoppers' basic demographics; Part II is about the usage of online purchasing channels; and Part III is about the lifestyles of online shoppers, developed from preliminary interviews and a review of the relevant literatures (e.g., Swinyard & Smith, 2003; Yang, 2004; Rohm & Swaminathan, 2004).

Five online shoppers were invited to verify the descriptions and design of the questionnaire, and necessary corrections were made before the questionnaire was disseminated online.

The questionnaire was then posted on the website *YouthWant* for a convenient sampling. The website YouthWant is the largest community website for students and young people in Taiwan, offering various services: campus news, sponsorship for student societies, blogs, file sharing, and online surveys (the surveyors have to buy credits from YouthWant to post their surveys). There are several kinds of online surveys in YouthWant: academic surveys, marketing survey, and political polling. YouthWant requires the respondents fill in the questionnaire honestly to receive bonus credits. Otherwise, YouthWant may cancel the credit offering to the respondent, and the response will not be collected. In addition, the Internet Protocol (IP) addresses of all respondents were checked to prevent multiple responses.

RESULTS

Sample profile

A total of 172 valid samples were received. Of the 172 respondents, 102 (59.3%) were female and 70 (40.7%) were male. Their ages were mostly between 20 and 30 (58.7%) or between 30 and 40 (20.4%). Most of them (70.9%) had collegiate degrees or higher. And 76 of them (44.2%) had experience in online shopping for more than three years, 46 (26.7%) of them had experience in online shopping for one to three years, 50 (29.1%) of them had experience in online shopping for less than one year.

As for the preferences for online purchasing channels, 75 (43.6%) preferred online stores, 56 (32.6%) preferred direct purchasing in online auctions, and 41 (23.8%) preferred bidding in online auctions. This preference order is consistent with the report of Yang (2007). The four product categories most-often purchased online, considering all three purchasing channels, are fashion apparels and accessories (31.4%), followed by 3C electronics (26.2%), followed by books and magazines (17.4%), followed by cosmetics and health-care products (12.2%).

It is noted that the sample size of this study is rather limited, which may induce potential biases. Therefore, to evaluate the sample quality, this study compares its sample profile with that of a large-scale sample (with 5773 respondents) from the semi-official organization III (*Institute for Information Industry*) (Yang, 2007). It turns out that the distributions of these two samples are quite similar in age, education, and occupation.

Specifically, in this study, 58.7% of respondents are aged between 20 and 30, and in the III

Table 2. Lifestyle factor analysis for online shoppers

Factors	Lifestyle Items	Factor Eigenvalue	Factor Loading	Explained Variance (%)
F1 information search	When shopping online, I compare many product types and brands.	3.517	.852	15.99
	When shopping online, I visit many e-retailing websites.		.775	
	I can save money by searching for online sales promotions.		.753	
	I can find special items online that are unavailable in physical stores.		737	
	I will search the web for better and cheaper products.		.622	
F2 social interaction	I like to interact with sellers (auctioneers) to make friends with them.	2.613	.788	11.88
	I like to interact with sellers (auctioneers) to learn product-related knowledge.		.769	
	It is interesting to compete with other bidders.		.713	
	It is interesting to bargain with auctioneers over the price.		.698	
F3 casualness	I always follow my purchase plans.*	2.530	.872	11.50
	I usually make plans for shopping activities.*		.847	
	I often go shopping without any plans.		.777	
	* reverse coded statements			

study, 58.4% of respondents are aged between 20 and 29. In this study, 70.9% of respondents have collegiate degrees or higher, and in the III study, the proportion is 77.0%. In this study, 40.7% of respondents were salaried employees, and in the III study, the proportion is 44.0%.

Furthermore, the preferences for online purchasing channels in these two studies are consistent: online stores are the most preferred, followed by direct purchasing in online auctions, and followed by bidding in online auctions. Overall, although the sample size of this study is rather limited, it is workable.

Exploratory Factor Analysis For Online Shoppers' Lifestyles

An exploratory factor analysis was conducted on the 41 items for online shoppers' lifestyles, with principal-component analysis and Varimax rotation. Six factors with eigenvalues greater than one were abstracted, which explained 67.18% of the total variance. Table 2 shows the items for these six factors with factor loading greater than 0.5, resulting in 22 items.

Segmenting Online Shoppers Based On Their Lifestyles

Based on the results of lifestyle factor analysis, the factor scores were clustered to form online shopper lifestyle segments. A *K*-means clustering approach was used as recommended by researchers of internet consumer behavior (e.g., Bapna et al., 2004; Wu, 2006). Three distinctive clusters were discovered and named according to their prevailing factor loadings, as reported in Table 3.

The *extravert shoppers* are interested in social interactions, fashion pursuing and make purchases casually. The *shrewd shoppers* are in the practice of gathering information before shopping and make planned purchases, and will hunt for desired products decisively. The *self-indulgent*

Table 2. continued

Factors	Lifestyle Items	Factor Eigenvalue	Factor Loading	Explained Variance (%)
F4 Self indulgence	I will buy a product after receiving commercial emails.	2.422	.814	11.01
	I still want to buy things even when I cannot afford them.		.711	
	The sole purpose of making money is to spend it.		.648	
	Sometimes I feel like spending money on anything that catches my eyes.		.625	
F5 decisiveness	It pays to pay a higher price to win a bid for something I desire.	2.106	.802	9.57
	When I like something, I will buy it without any hesitation.		.738	
	It pays to pay more for something I desire.		.735	
F6 fashion pursuing	I often read fashion magazines or news.	1.591	.647	7.23
	I will buy a product that is a hot topic in online forums.		.600	
	I like to visit new scenic spots.		.574	

Table 3. Lifestyle descriptions of three online shopper segments

Factors	Clusters (mean of factor scores)			*F*	*p*(sig.)	Scheffe test
	1 (n=67) extra-vert shoppers	2 (n=14) shrewd shoppers	3 (n=91) indul-gent shoppers			
Information	-.14553	**1.00088**	-.04684	8.509	.000	2>1, 2>3
Interaction	**.32553**	-.00254	-.23929	6.555	.002	1>3
Casualness	**.26522**	-1.24789	-.00329	15.507	.000	1>2, 3>2
Indulgence	.08071	-1.22535	**.12909**	13.114	.000	1>2, 3>2
Decisiveness	.03961	**1.27499**	-.22531	16.182	.000	2>1, 2>3
Fashion	**.76026**	.52983	-.64126	74.419	.000	1>3, 2>3

shoppers are more susceptible to marketing promotions and enjoy spending money.

Online Shoppers' Preferences for Online Purchasing Channels

Online shoppers' preferred purchasing channels and their lifestyles are cross tabulated in Table 4. A chi-square test (with p-value=0.030) gives support to hypothesis H1 at a significant level of 0.05. That is, online shoppers' with different lifestyles prefer different online purchasing channels. To visualize the interaction between lifestyles and preferred purchasing channels, a bar chart is constructed from Table 4 and is shown in Figure 2.

It can be seen from Table 4 and Figure 2 that most self-indulgent shoppers (55%) prefer online stores. On the other hand, most extravert shoppers and shrewd shoppers prefer direct purchase in online auctions (38.8% and 42.9%, respectively). However, while 31.3% of extravert shoppers prefer bidding, only 21.4% of shrewd shoppers prefer bidding. Below is a tentative explanation.

Most self-indulgent shoppers (55.0%) prefer shopping at online stores. As indicated in preliminary interviews, online stores rank highest in convenience and product quality among three purchasing channels. These features are attractive for self-indulgent shoppers because they can enjoy spending money freely and possessing the desired

Table 4. Online shoppers' lifestyles and preferred purchasing channels

	Lifestyle segments			
Preferred purchasing channels	extravert shoppers	shrewd shoppers	self-indulgent shoppers	Sum
Online stores	20 (29.9%)	5 (35.7%)	**50 (55.0%)**	75 (43.6%)
Direct purchase in online auctions	**26 (38.8%)**	**6 (42.9%)**	24 (26.4%)	56 (32.6%)
Bidding in online auctions	**21 (31.3%)**	3 (21.4%)	17 (18.7%)	41 (23.8%)
Sum	67	14	91	**172**

Figure 2. The interaction between lifestyles and purchasing channels

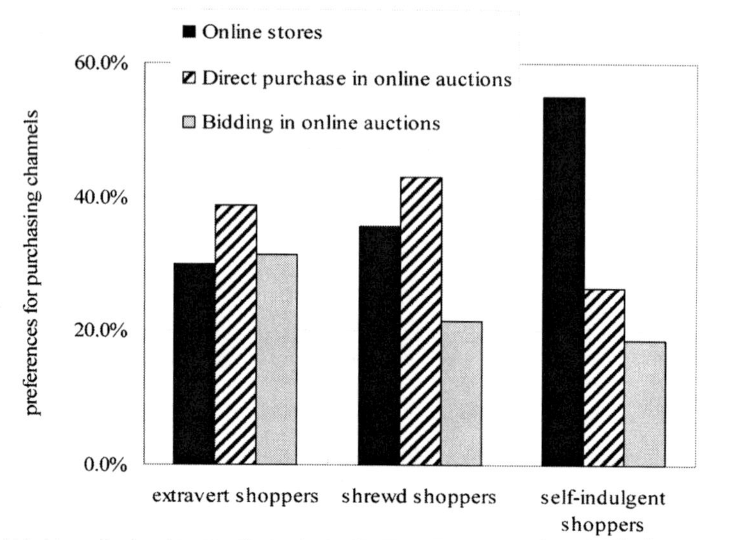

products quickly, without the haggle and hassle associated with online auctions.

The extravert shoppers are active in social interaction and fashion pursuing. Online auctions provide a platform for interacting with the auctioneers and other bidders, and serve as a channel for learning fashion trends. Therefore, the percentages of using direct purchase in online auctions (38.8%) and using bidding in online auctions (31.3%) are both higher than the percentage of using online stores (29.9%).

Compared with extravert and self-indulgent shoppers, the shrewd shoppers are more balanced in using online auctions and online stores. These shoppers are more rational and focused in their purchasing activities. They may search for bargains by direct purchase in online auctions, survey online stores to buy their targets of purchase, or compete with other bidders to capture their desired products.

Most-Often Purchased Product Categories and Online Shoppers' Lifestyles

Online shoppers' most-often purchased product categories and their lifestyles are cross-tabulated in Table 5. A chi-square test (with p-value=0.040) gives support to hypothesis H2. That is, online

Table 5. Online shoppers' lifestyles and most-often purchased products

	Online shopper lifestyle segments			
Most-often purchased product categories	extravert shoppers	shrewd shoppers	self-indulgent shoppers	Sum
fashions and accessories	**27 (40.30%)**	**5 (35.71%)**	22 (24.18%)	54 (31.40%)
3C electronics	14 (20.90%)	3 (21.43%))	**28 (30.77%)**	45 (26.16%)
books and magazines	7 (10.45%)	**4 (28.57%)**	**19 (20.88%)**	30 (17.44%)
cosmetics and health-care products	7 (10.45%)	1 (7.14%)	**13 (14.29%)**	21 (12.21%)
other product categories	12 (17.91%)	1 (7.14%)	9 (9.89%)	22 (12.79%)
Sum	67	14	91	**172**

Table 6. Cross table for product categories and purchasing channels

	Most-often purchased product categories					
Preferred purchasing channels	fashions and accessories	3C electronics	books and magazines	cosmetics and health-care products	other product categories	Sum
Online stores	16 (29.6%)	19 (42.2%)	**17 (56.7%)**	**16 (76.2%)**	7 (31.8%)	75 (43.6%)
Direct purchase in online auctions	**21 (38.9%)**	15 (33.3%)	8 (26.7%)	5 (23.8%)	7 (31.8%)	56 (32.6%)
Bidding in online auctions	**17 (31.5%)**	11 (24.5%)	5 (16.6%)	0 (0%)	8 (36.4%)	41 (23.8%)
Sum	54	45	30	21	22	**172**

shoppers' most-often purchased product categories are associated with their lifestyles.

Most extravert shoppers (40.30%) buy fashions and accessories most often, well above the overall percentage (31.40%) considering all segments of online shoppers. This reflects the fact that extravert shoppers are fashion pursuing and active in social interactions. Most shrewd shoppers (35.71%) also buy fashions and accessories, above the overall percentage. Besides, 28.57% of shrewd shoppers buy books and magazines most often, well above the overall percentage (17.44%). This reflects the fact that shrewd shoppers are eager for information. Self-indulgent shoppers buy three product categories more often than overall percentage: 3C electronics, books and magazines, cosmetics and health-care products. According to the report of Chou (2006), 3C electronics, and cosmetics and health-care products were two largest physical product categories in online retailing. The growth rate at 2006 was as high as 63% and 90%, respectively. Self-indulgent shoppers can easily find these products and buy them to spend their money.

Most-Often Purchased Product Categories and Online Purchasing Channels

The preferred purchasing channels and most-often purchased product categories are cross-tabulated in Table 6. A chi-square test (with p-value=0.013) gives support to hypothesis H3. That is, most-often purchased product categories and preferred purchasing channels are associated.

Below are observations and interpretations from Table 6:

1. Fashions and accessories are largely purchased in online auctions, either with direct purchase (38.9%) or with bidding (31.5%). Both percentages are higher than the overall percentages considering all product categories (32.6% and 23.8% respectively). These products are chiefly search goods (Zeithaml, 1981; Klein, 1998) which online shoppers can evaluate before purchase. Therefore, shoppers can just go to auction websites to find better prices for these products.
2. The proportions for purchasing 3C electronics in three channels (42.2%, 33.3%, and 24.5%, respectively) are closer to the overall percentages considering all product categories (43.6%, 32.6%, and 23.8%, respectively). As indicated in preliminary interviews, shoppers tend to use online stores for buying expensive items, and use online auctions for buying less expensive items. The 3C products are mostly commodities with standard specifications. However, price dispersion can be substantial for different brands or product quality. Therefore, online shoppers choose the most favorable channel for their purchase activities.
3. Books and magazines are mostly purchased in online stores, because there have been a few dominant online bookstores (such as www.books.com.tw) in Taiwan.
4. Cosmetics and health-care products are mostly purchased in online stores. This is because these products are mostly experience goods (Zeithaml, 1981; Klein, 1998) so consumers are more concerned about the product quality and after-sale services. As indicated in preliminary interviews, online stores rank higher than online auctions in these attributes.

Substitutability Among These Online Purchasing Channels

As indicated in preliminary interviews, online shoppers will switch to alternative purchasing channels to find better prices or special products. That is, these three purchasing channels are substitutable. To examine the substitutability of these purchasing channels, a cross analysis is conducted for online shoppers' preferred purchasing channels and alternative purchasing channels. The results are shown in Table 7. A chi-square test with p-value=0.000 indicates that online shoppers' preferred purchasing channels and alternative channels are associated.

Note the numbers in the diagonal line represent shoppers who will not consider using alternative channels. For example, for shoppers preferring online stores, 52% of them do not consider any other alternative channels.

The numbers off the diagonal line represent shoppers who will consider alternative purchasing channels. For example, for shoppers preferring

Table 7. Cross table for preferred channels and alternative purchasing channels

	Preferred purchasing channels			
Alternative purchasing channels	Online stores	Direct purchase in online auctions	Bidding in online auctions	Sum
Online stores	39 (52%)	**26 (46.4%)**	**7 (17.1%)**	72
Direct purchase in online auctions	**27 (36%)**	18 (32.2%)	**15 (36.6%)**	60
Bidding in online auctions	**9 (12%)**	**12 (21.4%)**	19 (46.3%)	40
Sum	75	56	41	**172**

Table 8. Summary of hypotheses

H1	Online shoppers with different lifestyles (extravert, shrewd, and self-indulgent) prefer different online purchasing channels	Supported by Table 4
H2	Online shoppers with different lifestyles (extravert, shrewd, and self-indulgent) often purchase different categories of products online	Supported by Table 5
H3	Online shoppers prefer different purchasing channels for purchasing different categories of products online	Supported by Table 6

online stores, 36% of them will consider direct purchase in online auctions as an alternative. On the other hand, for shoppers preferring direct purchase in online auctions, 46.4% of them will consider online stores as an alternative.

This study further defines the *mutual substitutability* between online stores and direct purchase in online auctions as *36% × 46.4%=16.7%*. Similarly, the *mutual substitutability* between direct purchase in online auctions and bidding in online auctions is 36.6% × 21.4%=7.8%, and the *mutual substitutability* between online stores and bidding in online auctions is 12% × 17.1%=2.1%.

Therefore, online stores and direct purchase in online auctions are most mutually substitutable, while online stores and bidding in online auctions are least mutually substitutable. This is consistent with the findings in preliminary interviews, where it can be seen from Table 1 that direct purchase in online auctions is a compromise between online stores and bidding in online auctions. Hence, direct purchase in online auctions is quite substitutable with the other two channels. In contrast, online stores and bidding in online auctions are hardly mutually substitutable.

GENERAL DISCUSSION

Main Findings

This study investigated the relationships among online shoppers' lifestyles, their most-often purchased products, and their preferred online purchasing channels. Table 8 summarizes the hypotheses findings.

Major findings in this study are:

1. Online shoppers with extravert lifestyle prefer online auctions.
2. Online shoppers with indulgent lifestyle prefer online stores.
3. Online shoppers with shrewd lifestyle have a more balanced preference for all three purchasing channels.
4. Online shoppers with different lifestyles often purchase different categories of products.
5. Online shoppers prefer different online purchasing channels for buying different categories of products.
6. The three online purchasing channels are substitutable, but with different degrees of substitutability.

Theoretical Implications

This study examined online stores and online auctions jointly, and explored online shoppers' preferences for the three popular online purchasing channels (online stores, direct purchase in online auctions and bidding in online auctions). Hence, this study has established a linkage between the research of online stores and the research of online auctions, and has broadened the scope of research on online shopping.

For example, the study of Rohm and Swaminathan (2004) on online grocery shopping identified a type of store-oriented shoppers who were more motivated by offline store features such as immediate possession and social interaction, and had lowest propensity to shop online. This study, by including online auctions, identified *extravert shoppers* who enjoy social interactions in online auctions.

On the other hand, the study on bidders' strategies identified two types of bidders (opportunists and sip-and-dippers) that were more eager to win bids but were more likely to pay higher prices in auctions (Bapna et al., 2004). In contrast, based on online shoppers' lifestyles, this study identified a type of *shrewd shoppers* characterized by information search and decisiveness to win. Most shrewd shoppers, however, do not engage themselves in bidding in online auctions. Rather, most shrewd shoppers prefer another online purchasing channel: *direct purchase* in online auctions. That is, they may negotiate with auctioneers to arrive at acceptable prices to buy products directly to avoid bidding wars with other bidders.

This study also enriches the research of online shoppers' lifestyles. For example, Swinyard and Smith (2003) used *Internet-related lifestyles* to find different segments of online shoppers and online non-shoppers, and Wang et al. (2006) found that online shoppers with different lifestyles assign different weights on the objectives of online shopping. This study further discovered the association between online shoppers' lifestyles and their preferred online purchasing channels.

Practical Implications

The findings in this study provide useful insights to online retailers. First, online retailers must pay attention to online shoppers' lifestyles, because online shoppers' preferred purchasing channels and frequently purchased products are both associated with their lifestyles. Second, online shoppers tend to use different purchasing channels to purchase different categories of products. Third, online retailers must also recognize the fact that these three purchasing channels are mutually substitutable and shoppers can switch among them. That is, online stores are not only competing with other online stores but also with online auctions, and vice versa.

This study offers the following suggestions to online retailers about the lifestyle segments of online shoppers:

1. The *self-indulgent shoppers* like to spend money freely and prefer online stores. Managers of online stores must grasp this segment of shoppers and facilitate the shopping process from product search and product descriptions to final transactions.
2. The *extravert shoppers* are interested in social interactions and prefer online auctions. Managers of online auctions must enhance the process of interaction between bidders and auctioneers. Furthermore, online auction websites can establish virtual communities to gather bidders to satisfy their needs of social interactions.
3. The *shrewd shoppers* are information seekers and planned buyers, and have a more balanced preference for online stores and online auctions. Managers of any kind of purchasing channel must provide enough information about products, and price their products reasonably to prompt shrewd shoppers to make purchases.

Limitations and Future Research Directions

The first stage of this study utilized in-depth interviews to explore the shopping behavior of online shoppers. Although this study attempted to find as many experienced shoppers as possible, only a few could be contacted owing to limited time and resources. The second stage used convenience sampling and the sample size is rather limited, potentially creating bias owing to the non-randomness. Besides, this study focused on only three main constructs: lifestyles, purchasing channels, and product categories, leaving many other factors unconsidered. Another important issue is culturally related. Since this study was conducted in Taiwan, it might not be generalized to other countries straightforwardly.

However, limitations of this study provide the foundation for continued research to improve our understanding of the factors related to online shoppers' preferences and buying behaviors. For instance, needs and wants could be factors as strong as lifestyles. Income levels could be another relevant factor on the preferences of shoppers and the types of products purchased. Shoppers' trust/distrust on vendors could also be another factor affecting the choice of online purchasing channels. Fear of fraud might influence the type of products purchased as well as the type of purchasing channels adopted.

Some more topics for further research could be conducted following this study. For instance, this study found that some shoppers would use alternative purchasing channels while some would not. That is, some shoppers are more loyal to their preferred channels while some are not. Future research should explore the factors that influence online shoppers' loyalty to their preferred channels. In addition, this study found that direct purchase in online auctions is gaining acceptance of online shoppers. Future research can study the development of this purchasing channel.

REFERENCES

Ariely, D., & Simonson, I. (2003). Buying, Bidding. Playing, or Competing? Value Assessment and Decision Dynamics in Online Auctions. *Journal of Consumer Psychology, 13*(1&2), 113–123.

Bapna, R., Goes, P., Gupta, A., & Yiwei Jin, Y. (2004). User Heterogeneity and its Impact on Electronic Auction Market Design: An Empirical Exploration. *Management Information Systems Quarterly, 28*(1), 21–43.

Berkman, H. W., & Gilson, C. C. (1974). Consumer life styles and market segmentation. *Journal of the Academy of Marketing Science, 2*(1), 189–200. doi:10.1007/BF02729512

Brunso, K., & Grunert, K. G. (1998). Cross-cultural similarities and differences in shopping for food. *Journal of Business Research, 42*(2), 145–150. doi:10.1016/S0148-2963(97)00114-8

Brunso, K., Scholderer, J., & Grunert, K. G. (2004). Closing the gap between values and behavior—A means-end theory of lifestyle. *Journal of Business Research, 57*(6), 665–670. doi:10.1016/S0148-2963(02)00310-7

Carrie, A. J. (2004). *US eCommerce overview: 2004 to 2010: A six-year forecast of US online retail sales*. Business View Research Documents. Forrester Research.

Childers, T. L., Carr, C. L., Peck, J., & Carson, S. (2001). Hedonic and utilitarian motivations for online retail shopping behavior. *Journal of Retailing, 77*(4), 511–535. doi:10.1016/S0022-4359(01)00056-2

Chou, W.-T. (2006). A forecast of the trends of Taiwan's online shopping development. Institute for Information Industry, Nov. 2, 2006. (in Chinese)

Dholakia, U. M. (2005). The usefulness of bidders' reputation ratings to sellers in online auctions. *Journal of Interactive Marketing, 19*(1), 31–40. doi:10.1002/dir.20029

Gutman, J. (1982). A means-end chain model based on consumer categorization processes. *Journal of Marketing, 46*(Spring), 60–72. doi:10.2307/3203341

Hawkins, D. I., Best, R. J., & Coney, K. A. (2004). *Consumer Behavior: Building Marketing Strategy* (9th ed.). Richard D. Irwin, Inc.

Keeney, R. L. (1999). The value of Internet commerce to the customer. *Management Science, 45*(4), 533–542. doi:10.1287/mnsc.45.4.533

Klein, L. R. (1998). Evaluating the potential of interactive media through a new lens: search versus experience goods. *Journal of Business Research, 41*(3), 195–203. doi:10.1016/S0148-2963(97)00062-3

Korgaonkar, P., Silverblatt, R., & Girard, T. (2006). Online retailing, product classifications, and consumer preferences. *Internet Research, 16*(3), 267–288. doi:10.1108/10662240610673691

Lesser, J. A., & Marie, A. H. (1986). The generalizability of psychographic market segments across geographic location. *Journal of Marketing, 50*(1), 18–27. doi:10.2307/1251275

Olson, J. C., & Reynolds, T. J. (1984). Understanding consumers' cognitive structures: Implications for advertising strategy. In Percy, L., & Woodside, A. G. (Eds.), *Advertising and consumer psychology* (pp. 77–90). Lexington, MA: Lexington Books.

Rohm, A. J., & Swaminathan, V. (2004). A typology of online shoppers based on shopping motivations. *Journal of Business Research, 57*(7), 748–757. doi:10.1016/S0148-2963(02)00351-X

Solomon, M. (2007). *Consumer Behavior* (7th ed.). Upper Saddle River, New Jersey: Prentice-Hall, Inc.

Swinyard, W. R., & Smith, S. M. (2003). Why people (don't) shop online: A lifestyle study of the internet consumer. *Psychology and Marketing, 20*(7), 567–597. doi:10.1002/mar.10087

Tseng, F.-C. (2008). Network effects and multiple adoption: Two fundamental factors in the competition of e-commerce. *Proceedings of the 4th IEEE International Conference on Management of Innovation and Technology*, 498-502.

Tseng, F.-C., Teng, C.-I., & Chiang, M.-H. (2007a). Delivering superior customer perceived value in the context of network effects. *International Journal of E-Business Research, 3*(1), 41–50.

Tseng, F.-C., Teng, C.-I., & Chiang, M.-H. (2007b). Superior customer value and network size in markets characterized by network effects. In Lee, I. (Ed.), *E-Business Models, Services and Communications* (pp. 84–96). Hershey, Pennsylvania: IGI Global.

Wang, E. T.-G., Yeh, H.-Y., & Jiang, J. J. (2006). The relative weights of Internet shopping fundamental objectives: Effect of lifestyle differences. *Psychology and Marketing, 23*(5), 356–367. doi:10.1002/mar.20116

Wu, S.-I. (2006). A comparison of the behavior of different customer clusters towards Internet bookstores. *Information & Management, 43*(8), 986–1001. doi:10.1016/j.im.2006.09.002

Yang, C.-H. (2007). Analysis of the consumer behavior of Taiwan's Internet users. Institute for Information Industry. Oct. 8, 2007. (in Chinese)

Yang, K. (2004). A comparison of attitudes towards Internet advertising among lifestyle segments in Taiwan. *Journal of Marketing Communications, 10*(3), 195–212. doi:10.1080/1352726042000181657

Zeithaml, V. A. (1981). How consumer evaluation processes differ between goods and services. In Donnelly, J. H., & George, W. R. (Eds.), *Marketing of Services* (pp. 186–190). American Marketing Association.

APPENDIX: QUESTIONNAIRE

There are three parts in this questionnaire. Part I (questions 1 to 10) is for the basic demographics and the preference of online purchasing channels. Part II has three branches: (1) questions 11 to 13 are for shoppers with a preference for online stores; (2) questions 14 to 16 are for shoppers with a preference for direct purchasing in online auctions; and (3) questions 17 to 19 are for shoppers with a preference for bidding in online auctions. Part III (questions 20 to 60) is for the lifestyles of online shoppers. The flow control for the questionnaire is depicted below (Figure 3).

The measurements in Part I and Part II are categorical with single choice, and the measurements in Part III are 7-point Likert scale ranging from 1 (strongly disagree) to 7 (strongly agree). The questions are given below.

Figure 3. The flow control of the questionnaire

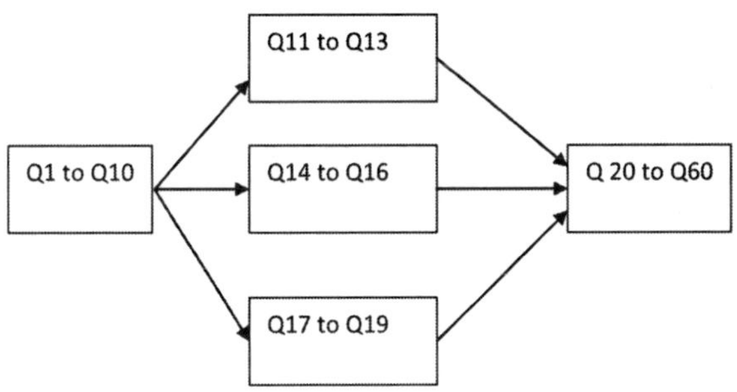

Part I

Q1	Your Gender: Male, Female
Q2	Your Age: <20, 20-30, 31-40, 41-50, >50
Q3	Your Education: ≦junior high, senior high, undergraduate, graduate
Q4	Your Occupation: student, salaried employee, entrepreneur, retire, jobless
Q5	Monthly income: ≦5000, 5001-10000, 10001-20000, …
Q6	Experience in online shopping (in years): <1, [1, 2), [2, 3), ≧3
Q7	Visiting online shopping websites recently: <once/week, once/week, 2-6 times/week, once/day, 2 or more times/day
Q8	How many purchases in online shopping (last 3 months): ≦3, 4-6, 7-10, 11-15, ≧16
Q9	How much spending in online shopping (last 3 months): ≦999, 1000-2999, 3000-4999, …
Q10	Most preferred online purchasing channel: online stores << *Please go to Q11* >> direct purchasing in online auctions << *Please go to Q14* >> bidding in online auctions << *Please go to Q17* >>

Part II

Q11	The most important reason for you to choose this channel: convenience, security, low price, payment flexibility, a wide assortment of goods, quality of goods, amusement, others
Q12	The most often purchased product types in this channel: fashion apparels and accessories, cosmetics and health-care products, books and magazines, 3C electronics
Q13	Your first-priority alternative online purchasing channel: direct purchasing in online auctions bidding in online auctions *<< Please go to Q20 >>*
Q14	The most important reason for you to choose this channel: convenience, security, low price, payment flexibility, a wide assortment of goods, quality of goods, amusement, others
Q15	The most often purchased product types in this channel: fashion apparels and accessories, cosmetics and health-care products, books and magazines, 3C electronics
Q16	Your first-priority alternative online purchasing channel: online stores bidding in online auctions *<< Please go to Q20 >>*
Q17	The most important reason for you to choose this channel: convenience, security, low price, payment flexibility, a wide assortment of goods, quality of goods, amusement, others
Q18	The most often purchased product types in this channel: fashion apparels and accessories, cosmetics and health-care products, books and magazines, 3C electronics
Q19	Your first-priority alternative online purchasing channel: direct purchasing in online auctions online stores *<< Please go to Q20 >>*

Part III (the answers are 7-point Likert scale, from 1 (strongly disagree) to 7 (strongly agree))

Q20	I like to visit new scenic spots.
Q21	Profit is more important than security when making investments.
Q22	All brands are similar.
Q23	I will buy a product that is a hot topic in online forums.
Q24	I will search the web for better and cheaper products.
Q25	I like to interact with sellers (auctioneers) to learn product-related knowledge.
Q26	Even though I like something very much, I would rather not buy it if it were not cheap enough.
Q27	It is interesting to compete with other bidders.
Q28	I often read fashion magazines or news.
Q29	I still want to buy things even when I can't afford them.

Part III (cont'd)

Q30	I am often persuade by advertisement.
Q31	I usually make plans for shopping activities.* (*reverse coded)
Q32	I will buy a product after receiving commercial emails.
Q33	Online shopping is an amusement to me.
Q34	I like trendy things.
Q35	The sole purpose of making money is to spend it.
Q36	As for jobs, more salary is more important than higher security.
Q37	The purpose of bidding is to get something cheap.
Q38	When buying something, the reputation of a brand is very important.
Q39	I take note of new happenings in the society.
Q40	It pays to pay more for something I desire.
Q41	I like to interact with sellers (auctioneers) to make friends with them.
Q42	When shopping online, I visit many e-retailing websites.
Q43	I will buy something that is advertised on the website.
Q44	I always follow my purchase plans.* (* reverse coded)
Q45	Sometimes I feel like spending money on anything that catches my eyes.
Q46	I can find special items online which are unavailable in physical stores.
Q47	The products of unfamiliar brand names are usually inferior in quality.
Q48	In general, I steer clear of risky things.
Q49	It does not hurt to be trendy if I like it.
Q50	It is interesting to bargain with auctioneers over the price.
Q51	Sometime I feel like buying something because of advertisements in mass media.
Q52	I often go shopping without any plans.
Q53	It pays to pay a higher price to win a bid for something I desire.
Q54	When I like something, I will buy it without any hesitation.
Q55	When shopping online, I compare many product types and brands.
Q56	I can save money by searching for online sales promotions.
Q57	I love to buy fashionable and trendy products.
Q58	I often buy name brand items.
Q59	I am often influenced by others' recommendations in my purchase decisions.
Q60	I'll be the first to admit that I'm an impulsive buyer.

This work was previously published in International Journal of E-Business Research Volume 6, Issue 2, edited by In Lee, pp. 1-20, copyright 2010 by IGI Publishing (an imprint of IGI Global).

Chapter 2
Overcoming Visibility Issues in a Small-to-Medium Retailer Using Automatic Identification and Data Capture Technology:
An Evolutionary Approach

Dane Hamilton
University of Wollongong, Australia

Katina Michael
University of Wollongong, Australia

Samuel Fosso Wamba
University of Wollongong, Australia

ABSTRACT

In this paper, the authors the inventory control practices of a small-to-medium retailer to identify common challenges this type of organization experiences with respect to automated data capture (ADC) and the implementation of an enterprise wide information system. The study explores a single case of a hardware store in a regional town in New South Wales, Australia. Four semi-structured interviews were conducted with employees, focusing on issues related to inventory control including delivery discrepancies, checking and sorting of orders, locating stock and goods, loss prevention, customer purchasing and point of sale processing and replenishment. Flowcharts illustrate the current processes of the retailer with an understanding of how ADC technologies like bar code and radio-frequency identification (RFID) impact the retailer. The findings promote an evolutionary approach toward the use of automated data capture technology by adopting barcode technology and subsequently introducing the complementary RFID technology.

DOI: 10.4018/978-1-61350-462-8.ch002

1 INTRODUCTION

The aim of this paper is to review the inventory control practices of a small-to-medium retailer and to show how automatic identification and data capture (AIDC) technology can overcome operational challenges. Inventory control is the activity that organizes the availability of products to customers through the coordination of the purchasing, manufacturing and distribution functions (Wild, 1997). The aim of inventory control is to maximize customer service, profit and efficiency of purchasing and production, whilst at the same time minimizing inventory investment. Inventory control practices are significant to all organizations, especially small-to-medium retail enterprises that are more susceptible to inventory control issues (Gunasekaran, Forker, & Kobu, 2000; Zipkin, 2000). Traditionally, to improve inventory control organizations have semi-automated their processes through the implementation of barcode systems. The implementation of barcode technology has translated into operational improvements in the form of superior data accuracy, efficiency, consistency and inventory and asset management within the organization (Ellram, Londe, & Weber, 1999; Zebra Technologies, 2007). However, in more recent times the widespread acceptance of electronic commerce has created a more competitive environment prompting organizations to seek ways to improve operations through the use of information technology.

RFID is an emerging technology that is yet to be proven in retail organizations. RFID can be defined as an electronic tagging technology that allows an object, place or person to be automatically identified using an electromagnetic challenge/response exchange (Want, 2004). Unlike barcodes RFID supports non-line of sight scanning and can record data in real-time (Jones, Clarke-Hill, Shears, Comfort, & Hillier, 2004; Sarma, 2004). RFID systems can function well in harsh conditions, dissimilar to barcode systems (Finkenzeller, 2003; Michael & McCathie, 2005). For these reasons, numerous authors have heralded RFID as a technology with the potential to provide benefits to retail organizations, especially in relation to inventory control (Garfinkel & Rosenberg, 2005; Heinrich, 2005; Symbol Technologies, 2004; Tellkamp, Angerer, Fleisch, & Corsten, 2005). Potential benefits aside, technical issues, the high cost of the technology, lack of standardization and privacy concerns have negatively influenced its uptake, rendering it impractical for many organizations, especially those that are small-to-medium in size. This paper examines the inventory control practices of a small-to-medium retailer, as well as discussing the potential for RFID to improve inventory visibility.

2 LITERATURE REVIEW

2.1 Automatic Identification and Barcoding

Inventory systems are used by organizations to maintain control over their materials and products. Traditionally, to improve inventory control and inventory systems, organizations have semi-automated their processes through the implementation of barcode systems. Bar coding is a form of automatic identification that collects data through keyless entry. The barcode is essentially a binary code that is comprised of a pattern of bars and gaps arranged in a parallel configuration (Palmer, 1995). The information stored by barcodes is captured in 'line of sight' by optical readers, one barcode at a time. The first bar code system was developed in the United States of America during the late 1940's by Sylvania/GTE applied Research Lab. This system was used in the rail industry, aimed at controlling freight cars. The system utilized a white light optical scanning system to illuminate horizontal bars of reflective red, white and blue on non-reflecting black background (Collins & Whipple, 1994). Since its inception the barcode has been used in many different industries, including the military, health and insurance industries. Commercial use of the barcode came about during the early 1970's and became widespread during the 1980's and 90's. Barcodes have become the most pervasive automatic identification technology over the past few decades as they have many advantages over other technologies. Barcodes are affordable, easy to use and have increased reliability and accuracy over manual collection techniques and other technologies (McCathie & Michael, 2005). Barcodes have also become highly standardized, especially in retail where the UPC/EAN standard is used by more than a million companies in 141 nations. These advantages of barcode technology have translated into operational improvements in the form of superior data accuracy, efficiency, consistency and inventory and asset management within the organization (Ellram, Londe, & Weber, 1999).

2.1.1 Recent Trends in Retail Operations and Inventory Control

Over the past decade operations and processes carried out by organizations have changed, which has had implications for inventory control. For instance, many Japanese firms have been able to operate with lower inventories than their western equivalents, sparking the trend of keeping inventories as lean as possible (Zipkin, 2000). Practices such as just-in-time (JIT) inventory management and build-to-order (BTO) are outcomes of this new wave. Wild (1997) and Ellram, Londe and Weber (1999) believe that inventory control and inventory management have developed rapidly due to the increasing focus on the customer, the more competitive environment and the wider use of information and communication technology. Zipkin (2000) agrees that advances in technology have had a major influence on inventory control and that technology-enabled innovations in inventory control have the potential to restructure entire industries. Organizations are continually seeking ways to develop a competitive advantage through information technology.

2.2 RFID Technology Basics

RFID can be defined as an electronic tagging technology that allows objects to be automatically identified at a distance without direct line of sight, using an electromagnetic challenge/response exchange (Want, 2004). An RFID system primarily consists of RFID tags (transponders) and RFID readers, but can be extended to include antennas, radio characteristics and the computer network used to connect RFID readers (Finkenzeller, 2003). Figure 1 illustrates the configuration, components and interactions present in an RFID system.

RFID readers contain radio frequency modules that emit pulses of radio energy that are detected by tags and responded to with information, such as the tag's serial number. RFID tags are the labels that are attached to the object to be identified. RFID tags consist of an antenna, a small silicon chip that contains a radio receiver, a radio modulator, control logic, memory and a power system (Garfinkel & Rosenberg, 2005). RFID tags are classified as being passive, semi-passive or active based on their composition. Passive tags are the most common tags, solely powered by the radio frequency signal that is used to transport information, whereas active tags are equipped with onboard batteries (Nemeth, Toth, & Hartvanyi, 2006). Semi-passive tags are passive tags that have had a battery added to boost signal range (Angeles, 2005).

Figure 1. Components of an RFID system

2.2.1 Advantages of RFID

The characteristics of RFID technology described previously differentiate them from other automatic identification technologies. One of the main physical advantages of RFID technology is that tags, unlike barcodes do not require line of sight to be read and multiple tags can be read simultaneously (Jones, Clarke-Hill, Shears, Comfort, & Hillier, 2004). A field test carried out by UK retailer Marks and Spencer tagged 3.5 million bins and recorded that it took just 3 minutes to read 25 trays when it used to take 17.4 minutes using barcodes, an 83% reduction in reading time for each bin. Dissimilar to barcodes, RFID systems are not affected by harsh manufacturing environments (Finkenzeller, 2003). RFID systems are unaffected by dust, moisture, oils, coolants, cuttings, gases. In addition to this RFID tags can operate in extreme temperatures and last for longer periods, in some cases longer than the items they are attached to (Michael & McCathie, 2005). RFID tags and systems are also characterized by having a greater data density and data quantity than traditional automatic identification technologies in the form of barcodes. This allows RFID tags to carry unique serial numbers more easily than a barcode, which would require a long symbol or a two-dimensional variant, which is difficult to scan and fit into available space. A final major advantage of RFID systems is that they capture data in real time. Capturing data in real time allows organizations to improve data quality, as the information captured is more timely and accurate. All these advantages detailed have the potential to improve operations within organizations, like retailers.

2.2.2 Challenges of RFID

Studies and literature reveal that due to RFID's novelty in commercial and manufacturing applications, a number of challenges have created concerns about the feasibility of its implementation. The majority of problems that have been encountered when implementing RFID are technical and hardware issues (Albano & Engels, 2002). Issues have been raised such as the reliability of RFID tag reads. It has been discovered that when a tag is oriented perpendicular to a reader it is difficult to read it. Michael and McCathie (2005) state that radio waves can be absorbed by moisture in the immediate environment. Radio waves can be hidden, distorted or reflected by metal and the noise from electric motors and that fluorescent lights can also interfere with RFID communications. In the past the cost of RFID technology has had an impact on its uptake as tags were considered to be too expensive, especially for item-level tagging. However the cost per tag continues to fall. The absence of global standards is another major problem of RFID as they are still developing through the formation of the EPC global network. To date systems have utilized multiple standards restricting interoperability (Lefebvre, Lefebvre, Bendavid, Wamba, & Boeck, 2006). Organizations that implement RFID systems must also review the information technology infrastructure they have in place, as a copious amount of real-time data is captured by these systems. The final challenge of RFID relates to privacy concerns. As RFID is used to track items, privacy activists are concerned about the use of technology on retail items such as clothes which could allow retailers to send and receive information after items have been purchased. Privacy concerns, like all the other issues identified in this section, require more attention for RFID to become widespread in the commercial and manufacturing industries.

2.3 RFID and Inventory Control

According to Viale (1996), "[i]nventory is a very expensive asset that can be replaced with a less expensive asset called 'information'. In order to do this, the information must be timely, accurate, reliable, and consistent. When this happens you carry less inventory, reduce cost and get products to customers faster." RFID's ability to provide real-time information can help provide retailers with the 'timely, accurate and reliable and consistent' information. This in turn helps organizations adjust to changes in business operations by moving from a *push* to a *pull* model, where inventory is pulled from the supply chain by customer activity. Mills-Harris, Soylemezoglu and Saygin (2005) observe that the "management of inventory should be robust yet sensitive to change with respect to varying demand patterns". The touted benefits of RFID include the promotion of tighter supply chain and inventory management controls resulting in attendant cost savings, reduced labor costs, customer service improvements, shrinkage reduction, an improved replenishment process and clearer targeting of customers and tracking of their purchasing behavior (Bhuptani & Moradpour, 2005; Fleisch & Tellkamp, 2003).

To date there has been substantial literature published examining RFID's impact on supply chain management, but scant literature exists documenting its impact on inventory control. A paper by Kärkkäinen (2003) found that RFID can provide significant benefits to an organization without supplier participation, based on an RFID trial at a major grocery retailer. However this paper was primarily concerned with increasing efficiency in the supply chain. Similarly, Albano and Engels (2002) and Bendavid, Wamba and Lefebvre (2006) have used practical trials to emphasize the benefits that RFID has for inventory control when implemented into the supply chain. Lee, Cheng and Leung (2004) use a simulation model and subsequent quantitative analysis to gauge supply chain performance. The paper reveals that RFID technology can improve inventory accuracy, shelf replenishment policy and inventory visibility, but its primary focus regards the supply chain like other papers previously mentioned. Likewise, Fleisch and Tellkamp (2003) use a simulation model to quantitatively analyze supply chain performance. Angeles (2005), Chuang and Shaw (2005) and Nemeth, Toth and Hartvanyi (2006) also investigate the impact of RFID in the supply chain but through less practical methods. Angeles (2005) outlines the main aspects of the technology and draws on example RFID implementation cases. Chuang and Shaw (2005) also identify the impact RFID has on the supply chain by considering current implementation cases using the role playing simulation game (*Beer Game*). However the findings of this study are more relevant to suppliers not retailers. Nemeth, Toth and Hartvanyi (2006) simply describe the benefits of RFID and its use in the supply chain.

All the papers discussed in this section are focused on RFID's impact on the supply chain. The only papers found specifically related to RFID and inventory control are aimed at detailed technical issues. A paper by Hu, Sundara, Chorma and Srinivasan (2005) consider the use of a bitmap data type to support RFID item tracking, while a paper by Abraham, Ahuja, Ghosh, and Pakanati (2002) presents the pros and cons of four different collision resolution protocols used by RFID-enabled inventory control systems.

2.4 RFID and Small-to-Medium Retailers

Small-to-medium retailers employ more than 5 and less than 200 people and are responsible for a significant proportion of the working population (ABS, 2002). Small retailers are characterized by little job specialization where the store owner is generally responsible for overall management and merchandising tasks. This characteristic of small-to-medium retailers has an impact on the level of formalization present in these organization's processes and operations. According to Gunasekaran, Forker, and Kobu (2000) small-to-medium enterprises, especially retailers have problems with unreliable inventory control systems, with poor stock tracing and cost control. These problems eventually lead to excess obsolete stock and eroding customer service levels (Gunasekaran, Forker, & Kobu, 2000). RFID is a technology that could potentially improve inventory control within small-to-medium retailers. However there is currently limited literature relating to RFID implementation in small-to-medium enterprises. Kärkkäinen (2003) and Jones, Clarke-Hill, Shears, Comfort, and Hillier (2004) document RFID trials and implementations that have been carried out in large established organizations. The implications RFID has for large retail organizations and their supply chains is demonstrated in a research study by Lefebvre, Lefebvre, Bendavid, Wamba, and Boeck (2006) when they write:

[t]he retail giant stands to save substantial amounts from supply chain optimization, Just-in-Time deliveries and the elimination of stock-outs since it will know exactly where all products are throughout its entire supply chain and could theoretically track any product from its supplier's supplier's location.

However, small-to-medium retailers are fundamentally different to larger established retailers and the effect RFID has on these organizations has not been widely studied. Studying the impact that RFID has on inventory control in a small-to-medium retailer is more relevant than examining the technology's impact on the supply chains of these organizations, as small-to-medium retailers do not possess the bargaining power that larger established retailers have over their suppliers. This paper analyzes the inventory control practices of a single small-to-medium retailer and assesses whether the implementation of barcode and RFID can be beneficial to the inventory control practices of these types of organizations.

3 METHODOLOGY

The research study documented in this paper involves a case study examining a single small-to-medium retailer. A case study method has been employed as it is ideal for investigating contemporary events and is able to take into account a wide variety of evidence (Yin, 1994). For this study data has been gathered through the collection of procedural documents, semi-structured interviews and a participant observation. This paper presents the data collected from the semi-structured interviews conducted with employees of the organization, as well as revealing the business process flows (through flowcharts) of the organization in order to determine whether RFID is a feasible automated data capture technology for small-to-medium retailers. An observational study was also conducted over a period of two weeks in 2007. A daily diary was kept by the participant and this data was analyzed together with full-length transcripts. A single small-to-medium hardware retailer is focused on in this paper in order to analyze and present inventory control practices.

3.1 Case Study

3.1.1 Background

The organization examined in this study is located on the south coast of New South Wales, approximately 128 kilometers from the centre of Sydney. The company employs ten staff including casuals and is classified as a small-to-medium hardware retailer. The current proprietors have operated the business since 2003.

3.1.2 Physical Layout

The premises of the retailer measures approximately 2000 square meters, with about 550 square meters of this area making up the internal shop floor. The shop floor is composed of four sheds, each with independent access. There are two small internal offices, one designed to deal with customer purchasing and Point of Sale (POS) transactions while the other is used by managers and bookkeepers for ordering, accounting and other administrative practices. The external perimeter of the organization is surrounded by an eight foot high barbed wire fence. Figure 2 shows the organization's physical layout.

Figure 2. The physical layout of the retailer's premises

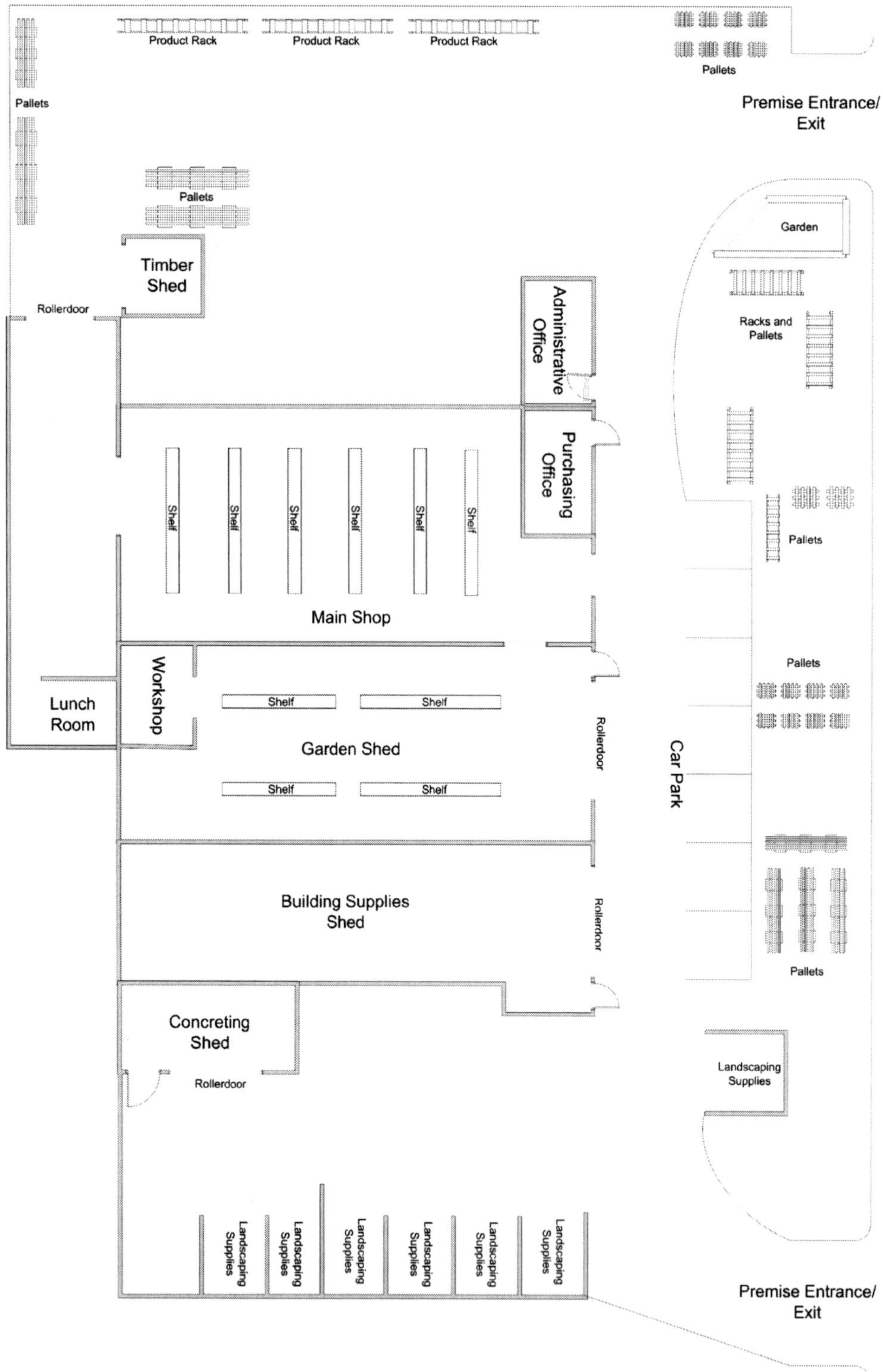

3.1.3 Stock and Inventory

The retailer currently possesses between $300,000 to $400,000 worth of inventory which is kept on the premises. The inventory held by the organization is estimated to consist of 5000 products lines, which are provided by 110 active suppliers. Products and other inventory are stored or displayed before purchase inside the store or outside within the confines of the premises. Items and stock within the store are positioned based on the type of product as well as the supplier. Most items kept inside the store are also shelved on racks that measure 2.1m in height. The shop floor is divided into five separate areas that include general hardware, timber, gardening, cement and building supplies. Products stored outside are generally unaffected by environmental and weather conditions such as landscaping supplies, cement blocks, treated pine sleepers and sheets of steel reinforcing. Stock is usually delivered to the store packaged at pallet, crate, and carton or item level.

3.1.4 Services

The retailer provides many services to its customers primarily through the selling of hardware and other building related supplies. The organization provides a delivery service to its customers if they purchase products that are too large to be transported or products that they wish to be delivered on a certain day. Products are delivered to customers in one of the three vehicles the organization owns. A flat top truck is used for steel deliveries, a tip truck is used for landscaping supplies and a utility vehicle is used for general deliveries. The organization also has a front-end loader that it uses to load landscaping supplies on vehicles. The organization offers accounts for customers that purchase products frequently.

3.1.5 Information Technology (IT) Infrastructure

The retailer currently has limited Information Technology (IT) infrastructure and does not utilize a server, as the current operations of the business do not require a large volume storage device. The organization utilizes two desktop computers in their administration office that are primarily used to manage customer accounts through the software package MYOB Premier Version 10. At the end of each month, the organization uses the MYOB software to generate invoices which are sent out to account holding customers, requesting that they pay for the items they have purchased. The organization has another desktop computer which is used by employees to search a program that acts as an index of paint colors provided by different paint suppliers. All computers within the organization are able to access the Internet.

3.2 Interviews

3.2.1 Interviewees

Insights into the current inventory control practices at the small-to-medium retailer are based on semi-structured interviews carried out on four employees of the organization. The roles and duties of these employees are documented in Table 1.

As can be seen from Table 1, employees of the organization have minimal job specialization, which reinforces Diamond and Pintel's (1996) observations of small businesses. The proprietor/manager and proprietor/part-time manager are responsible for the overall running of the business whereas the store manager is specifically responsible for shop maintenance and management. The delivery truck driver is primarily responsible for making outbound deliveries. The store manager and delivery truck driver are answerable to both of the proprietor/managers.

Table 1. Employee roles and duties

ID	Job Title	Duties
1	Proprietor/Manager	Stock replenishment, capital purchasing, staff management, delivery scheduling, customer service
2	Proprietor/Part-Time Manager	Stock replenishment, staff management, delivery scheduling, delivery truck driver, customer service
3	Store Manager	Stock control and replenishment (for some of the stock), customer service, staff management, shop maintenance
4	Delivery Truck Driver	Stock delivery, stock control and replenishment (for some of the stock), customer service

Figure 3. The Inventory Cycle

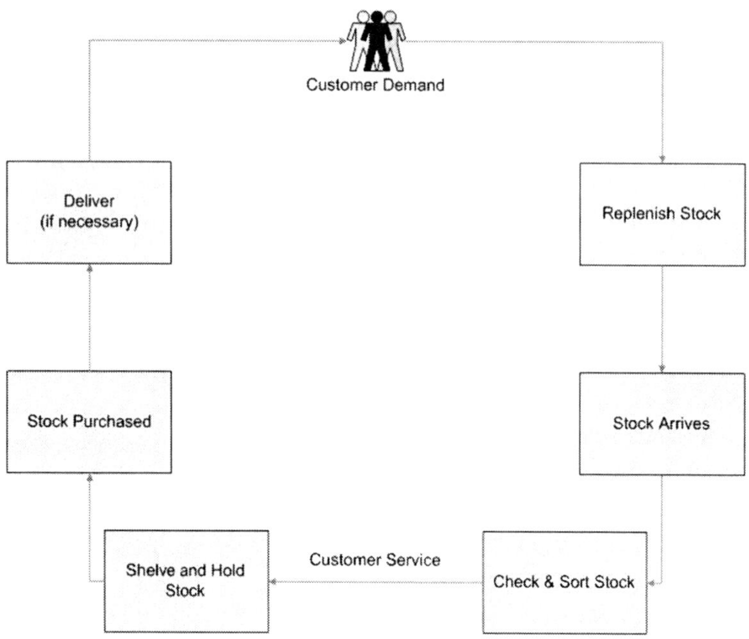

3.2.2 Interview Questions and the Inventory Cycle

Inventory control as defined by Wild (1997) involves "coordinating the purchasing, manufacturing and distribution functions to meet marketing needs". Coordinating these functions requires many discrete activities including ordering stock or materials and shelving or putting it in the correct position so that customer's have access to it. In this section, the inventory control process has been broken down so that the inventory practices of the small-to-medium retailer can be explored in greater detail. Figure 3 illustrates the Inventory Cycle. It should be noted that Tersine's (1998) Inventory Flow Cycle is focused on the flow of raw materials to their finished state, while this inventory control cycle has been developed based on a retailer that sells finished goods.

As can be seen in Figure 3, customer demand triggers the ordering or re-ordering of stock. Stock then arrives at the retailer, where it is checked and sorted before being shelved in the correct position. Stock is then purchased by a customer and delivered by the retailer if necessary.

The inventory cycle demonstrated in Figure 3 was considered when developing questions for the semi-structured interviews. The majority of the questions asked related to the six different processes that were identified in the inventory control cycle. There were a total of twenty-eight questions included in the original semi-structured interview protocol but additional probing sub-questions were asked where the respondent was able to expand their response due to their knowledge of operations. The questions covered the background of the company case, the role of the employee in the organization, questions related to the current mode of operation to gauge the current inventory control practices and set-up, and more speculative questions regarding the transition of the organization from a manual-based system to barcode and/or RFID. For instance the proprietor was asked:

1. Can you describe the process that you use to check that orders have been delivered with the correct contents?
2. Do you keep any sort of record of how much stock you carry, either in physical or electronic form?
3. How would you describe the theft prevention measures in your workplace?
4. What triggers your organization to reorder or order stock?
5. Are there any issues affecting your adoption of automated data capture technology?
6. Do you think that RFID could be used within your business to improve inventory control?

The interview transcripts were analyzed using a qualitative approach and the findings were presented using a modular narrative style based on the steps in the inventory control cycle. The following sections summarize the findings of the semi-structured interviews.

3.3 Participant Observation

A participant observation requires the researcher to become a direct participant in the social process being studied, by becoming a member of an organization. The participant observation was carried out over a two week period with the intention of recording observations relating to the inventory control practices used within the small-to-medium retailer. This study utilizes an overt participant observation as members of the organization were already aware of the researcher's presence due to interviews being carried out at an earlier date. The overt approach was perceived to have had minimal influence on the behavior of the organization's members as they were informed that the purpose of the study was to examine inventory control practices of the retailer, not their personal behaviors. During the participant observation annotations and issues were documented through the use of a diary. Field notes were recorded during each day, and were formalized at the end of the day.

3.4 Procedural Documentation

The small-to-medium retailer's procedural documents were used to complement the semi-structured interviews and participant observation. Documentary secondary data, such as an organization's communications, notes, and other policy and procedural documents have been examined. Knight (2002) states that official documents, like procedural documents can be treated as unproblematic statements of how things are or were. The procedural documents have been used as evidence to support the determination of the inventory control practices of the small-to-medium retailer. The interviews conducted, participant observation and the collection of procedural documents were combined to develop the business process flows of the organization. A narrative presentation is used to bring together participant observational data and interviewee responses.

4 CURRENT INVENTORY ISSUES

4.1 Delivery Discrepancies, Checking and Sorting Orders

Managing delivery discrepancies and checking and sorting orders is an initial inventory control process that occurs when stock is delivered to an organization. Delivery discrepancies refer to deliveries or orders that have been received with incorrect contents, with either the wrong type or amount of stock being delivered than was originally requested. The checking and sorting of orders or deliveries is carried out as a delivery arrives to determine if a delivery has been filled correctly. It was determined, based on the results of the interviews that the small-to-medium retailer receives incorrect deliveries on a weekly basis. The store manager estimated that every third order that would come into the hardware store was incorrect, and that "quite often" it would happen "all the time". Despite the store manager's claims there are no exact figures on how many deliveries are made incorrectly as there is no monitoring of this within the organization. The lack of monitoring was mirrored by inconsistencies in the frequencies of incorrect deliveries reported by all interviewees. The process used to check and sort orders and deliveries, involves obtaining an invoice attached to the goods and then using this invoice to check if all items in the delivery match the amount and type of product printed on the invoice. If there are any discrepancies with the delivery, such as incorrect contents or quantities of contents then staff contact the supplier to rectify these issues.

The retailer not only receives deliveries, but also offers a delivery service for customers of its own. Interviewees stated that incorrect outbound deliveries do not happen very often, but once again the organization does not monitor how many of these incorrect outbound deliveries occur. It should also be noted that in the past that employees of the retailer have been informed by customers that deliveries have been made incorrectly, although the organization maintains otherwise. The store manager stated that the organization could only prove that it has made deliveries correctly through "paper work and hear say". The proprietor/manager pointed out that an incorrect delivery may take place once every quarter and that if customers complain or claim a delivery has been made incorrectly, "we (the business) take the customers word and rectify."

4.2 Locating Stock and Goods

Locating stock and products is significant to all retailers, especially the small-to-medium retailer due to the geographical size of the organization's premises. The proprietor/manager and the store manager affirmed that they have had trouble locating orders that have been delivered, mainly due to them being put in the wrong place. Deliveries made to the hardware store generally come in cartons (boxes), pallets and crates, with cartons (boxes) being the most common form of packaging for deliveries according to the interviewees. It was also confirmed that the type of packaging can impact how long it takes staff to locate stock. The proprietor/part-time manager stated that packaging affected identifying individual items within orders, while the store manager was clear that if pallets and boxes were delivered to the wrong section of the store then they would have difficulties locating the order.

After a delivery has been located it is sorted, priced and placed on the shelves ready for purchase. Interviewees stated that they not only had trouble locating deliveries but that they also had trouble locating stock once it has been shelved. According to the store manager and the delivery truck driver the reason for being unable to find stock is that it has been shelved incorrectly by casual staff. The proprietor/manager believed that a lack of communication between staff was responsible for being unable to locate stock. Staff also believed that customers misplace stock around the store regularly, most likely on a daily basis.

4.3 Loss Prevention

Loss prevention refers to reducing inventory cost through operational errors, and both customer and employee theft. Interviews revealed that the loss prevention measures of the small to medium retailer were almost non-existent which, according to the interviewees, could be improved by restricting the number of entrances/exits to the shop and changing the direction of the aisles so that staff could see down them and monitor customers physically. The interviewees suspect that theft occurs within the store, but have no way to prove it as they cannot remember anyone being caught for theft. Additionally, there is no way to calculate how much stock the organization holds at any given point in time. The delivery truck driver commented "[w]ell you don't know [that theft is occurring], but you just surmise that there is, my feeling that there's stuff, just little things walk out the door that you don't know about." The proprietor/part time manager echoed this sentiment, "[w]ell we can never directly tell [that the organization is subject to theft] really, we've had our suspicions that we have." The proprietor/manager suspects that theft occurs on a weekly basis and along with other interviewees share the belief that smaller items are the main target of thieves.

4.4 Customer Purchasing and Point of Sale Processing

The small-to-medium hardware store does not have an electronic or automated point of sale system, rather using manual processes to identify and price items at point of sale. Staff determine the price of items based on the price tag attached to them or by referencing a price book (an index containing prices for products) for items that do not have a price tag (Figure 4A). Due to the manual system used by the retailer, employees have experienced some difficulties in pricing items at point of sale. The store manager stated that the organization held a lot of old stock dating back to when the shop was managed by the previous owner:

a lot of times they will bring in an old item which has no price on it and we can't find a price for it so we have to ring the company to find out how much those things were two or three years ago and work out a price from there.

Other reasons given for not being able to identify and price stock at point of sale included stock being put on the shelf without being priced, price tags falling off, supplier representative placing stock on the shelf and stock having two conflicting price tags.

The retailer offers regular customers accounts where the purchases of the account holder are recorded by staff on a formalized account sheet (Figure 4B). At the end of each day the account sheet is given to a bookkeeper to enter the details or purchases into the accounting program Mind Your Own Business (MYOB) Premier Version 10. Other than recording the purchases made by account holders electronically, the organization does not record or hold any other information about customer purchasing. The proprietor/manager and proprietor/part-time manager agreed that the business would be able to target customers based on their purchasing preferences if they were able to retain customer purchasing information.

4.5 Replenishment and Supplier Ordering

Replenishment refers to the process that is used by the organization to ensure that stock and products are available to customers. The ordering or replenishment process currently employed at the small-to-medium organization involves physically checking the shelves to see if a particular item of stock is depleted (Figure 5A). The store manager shed further light on the ordering process by stating "[w]ell on a particular hook we could

Figure 4. A and B: Sample product pricelist and daily account sheets in the manual-based system

4A

CEMENT 2

20KG BAGS			RETAIL				
GP CEMENT		$	7.30			3	4-1300
OFF WHITE CEMENT		$	9.90			3	4-1300
CONCRETE MIX		$	6.20	$	6.35	3	4-1300
RAPID SET CONCRETE		$	7.20	$	7.35	3	4-1300
SAND & CEMENT		$	8.20	$	6.35	3	4-1300
RAPID SAND & CEMENT		$	13.90	$	9.40	3	4-1300
HYDRATED LIME 20 Kg		$	10.60	$	8.60	3	4-1300
RENDERERS CLAY 20 Kg		$	9.90	$	7.90	3	4-1300
SAND		$	6.20			3	4-1300
RIVER SAND		$	6.00			3	4-1300
SAND BAGS		$	1.50			3	4-1300
WHITE POLY BAGS		$	1.20			3	4-1300
MORTAR MIX		$	6.50			3	4-1300
RENDER MIX		$	14.40			3	4-1300
25KG OXIDE							
318 BLACK		$	130.60			3	4-1300
222 RED		$	112.50			3	4-1300
920 YELLOW		$	148.20			3	4-1300
SANDSTONE		$	180.15			3	4-1300
680 BROWN		$	205.50			3	4-1300
686 CHOCOLATE		$	191.40			3	4-1300
GN GREEN		$	591.80			3	4-1300
ARDIT Z8		$	57.60			3	4-1300
ARDIT PRIMER (ARDION 510)		$	82.50			3	4-1300
BYCOL 20 LTR		$	45.60			3	4-1300
CONDUIT							
FLEXIBLE CONDUIT	PER METER		4.05			1	4-1100
16MM	LENGTH		5.00			1	4-1100
20MM	LENGTH		6.80			1	4-1100
20MM ORANGE	LENGTH		8.70			1	4-1100
25MM	LENGTH		12.35			1	4-1100
25MM ORANGE	LENGTH		12.55			1	4-1100
32MM	LENGTH		14.60			1	4-1100
32MM ORANGE	LENGTH		15.10			1	4-1100
40MM	LENGTH		18.20			1	4-1100
40MM ORANGE	LENGTH		22.00			1	4-1100
50MM	LENGTH		27.50			1	4-1100
50MM ORANGE	LENGTH		30.00			1	4-1100
16 X 10 MM CLIP ON	LENGTH		19.50			1	4-1100
16 X 16 MM CLIP ON	LENGTH		23.00			1	4-1100
25 X 15 MM CLIP ON	LENGTH		28.75			1	4-1100
25 X 25 MM CLIP ON	LENGTH		34.05			1	4-1100
40 X 25 MM CLIP ON	LENGTH		43.05			1	4-1100
TILING							
Fixall	20KG		12.00			3	4-1300
Megalastic	20LTR		110.00			3	4-1300
Megalastic	5LTR		36.70			3	4-1300
Superfixall off White	20KG		18.00			3	4-1300
Megafix	20KG		41.30			3	4-1300
Conflex	25KG		41.30			3	4-1300
ASAFlex Liquid	15Ltr		63.70			3	4-1300
ASAFlex Powder	30Kg		29.40			3	4-1300
Multifix	20L		56.40			3	4-1300
Multifix	4L		20.80			3	4-1300
Multifix	1L		11.10			3	4-1300
Smooth White Grout	20KG		20.90			3	4-1300
Smooth White Grout	10KG		12.90			3	4-1300
Smooth White Grout	5KG		8.60			3	4-1300

29/06/2007 PRICELIST xls

4B

DATE: / 7 /07 PAGE:

A/C NAME	JOB ON	QTY	DESCRIPTION	EACH	TOTAL	ACCOUNT

Day Sheet xls

have one item that has a maximum and minimum, once it gets to that minimum we then reorder that product if we discover that it's down, (it) doesn't always happen". The proprietor/manager however stated that a lot of the time that the minimum and maximum are not used in the ordering process, "that's pretty much how we do it really just by keeping an eye on the stock you know physically with your eye [*laughs*]… and then we make a call if we need to put the purchase in". Based on the information provided in the interviews, replenishment seems to be a major issue, revealing that the organization sells out of some products on a weekly basis, while the majority of orders are made on a monthly basis. The proprietor/manager and store manager mentioned the cost of freight and its effect on the ordering cycle. The store manager reflected:

> *[s]ometimes we will hold five of one item and a customer will come in and want six… That happens quite often, almost every month, there's something like that that happens and it makes it difficult for us to order in that one item when companies charge us 50 or 60 dollars for freight. So that's another reason why we have to wait and put in one order a month or two orders a month for some companies, just to keep the cost down. Freight is a big problem nowadays.*

As can be seen most of the ordering is carried out fortnightly and monthly, however it was mentioned by the proprietor/manager that the bulk of stock in the store comes from one supplier, which is ordered and delivered on a weekly basis (Figure 5B).

In regards to how orders are actually placed by the business currently, it was ascertained that the bulk of orders are placed via facsimile, so

Figure 5. A and B: Sample product-specific Stocktake sheet and replenishment ordering sheet

5A

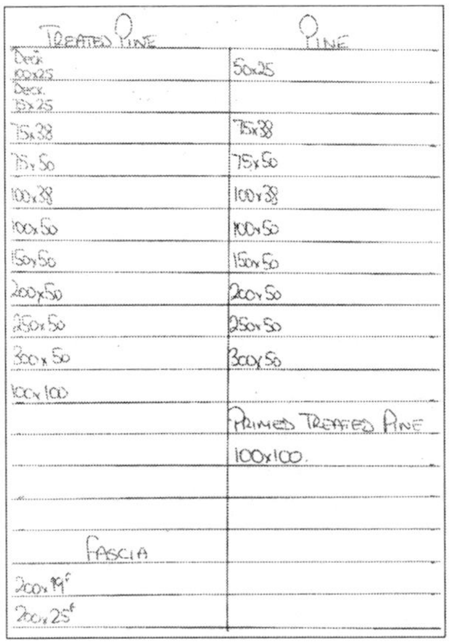

Treated Pine	Pine
Deck 100x25	50x25
Deck 75x25	
75x38	75x38
75x50	75x50
100x38	100x38
100x50	100x50
150x50	150x50
200x50	200x50
250x50	250x50
300x50	300x50
100x100	
	Primed Treated Pine
	100x100.
Fascia	
200x19f	
200x25f	

5B

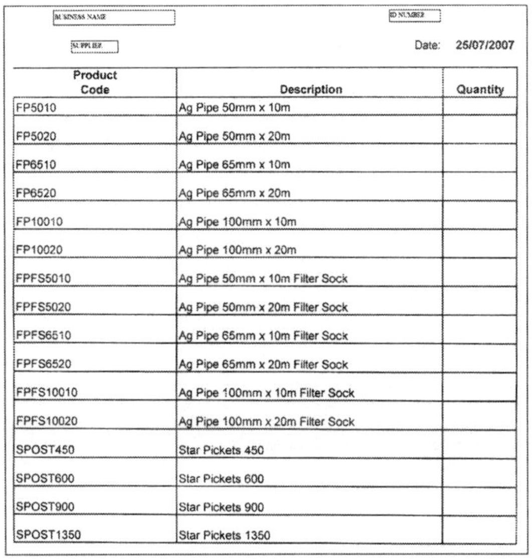

BUSINESS NAME

ID NUMBER

SUPPLIER

Date: 25/07/2007

Product Code	Description	Quantity
FP5010	Ag Pipe 50mm x 10m	
FP5020	Ag Pipe 50mm x 20m	
FP6510	Ag Pipe 65mm x 10m	
FP6520	Ag Pipe 65mm x 20m	
FP10010	Ag Pipe 100mm x 10m	
FP10020	Ag Pipe 100mm x 20m	
FPFS5010	Ag Pipe 50mm x 10m Filter Sock	
FPFS5020	Ag Pipe 50mm x 20m Filter Sock	
FPFS6510	Ag Pipe 65mm x 10m Filter Sock	
FPFS6520	Ag Pipe 65mm x 20m Filter Sock	
FPFS10010	Ag Pipe 100mm x 10m Filter Sock	
FPFS10020	Ag Pipe 100mm x 20m Filter Sock	
SPOST450	Star Pickets 450	
SPOST600	Star Pickets 600	
SPOST900	Star Pickets 900	
SPOST1350	Star Pickets 1350	

the business has a record of the order in case it is delivered incorrectly. A small number of suppliers, however, will ring and ask whether any stock is required, at which point stock is ordered over the phone if it is required.

5 KEY CHALLENGES OF THE SMALL-TO-MEDIUM RETAILER

Analyzing the inventory control practices of the small-to-medium retailer has revealed that the organization has many challenges in relation to how stock and products are managed. The memory locator system described by Muller (2003) is similar to that currently used within the retailer. Memory systems can be used successfully when the conditions are appropriate but the retailer observed in this study does not exhibit these conditions. Currently the retailer stocks a wide variety of products, with reasonable stock movement over a large area or shop floor. The inventory system that has been employed has contributed to the problems that the retailer is encountering. Currently within the store, the replenishment process could be improved so that it better meets the needs of the customers by making orders more frequently. Employees are also having difficulties locating stock due to it being shelved incorrectly and organizational problems such as a lack of communication between employees. The most significant problem that the organization has is that it does not keep a record of how much inventory it holds. Company-wide stocktakes are not carried out and neither is purchasing history kept by the proprietors, which prevents the organization from knowing the current inventory level, which stock is being purchased and how frequently. Not being able to locate stock efficiently as well as not knowing inventory levels has a negative effect on the organization. Muller (2003) declares that, "[i]f you cannot control the location of your product or raw material from a physical or recordkeeping standpoint then your inventory accuracy will suffer". Figure 6 illustrates the problems experienced by the organization and presented in the form of a "pain chain" (IBM Learning Services, 2001). The pain chain presents pains (i.e. challenges)

Figure 6. The pain chain (R = Reason for the pain)

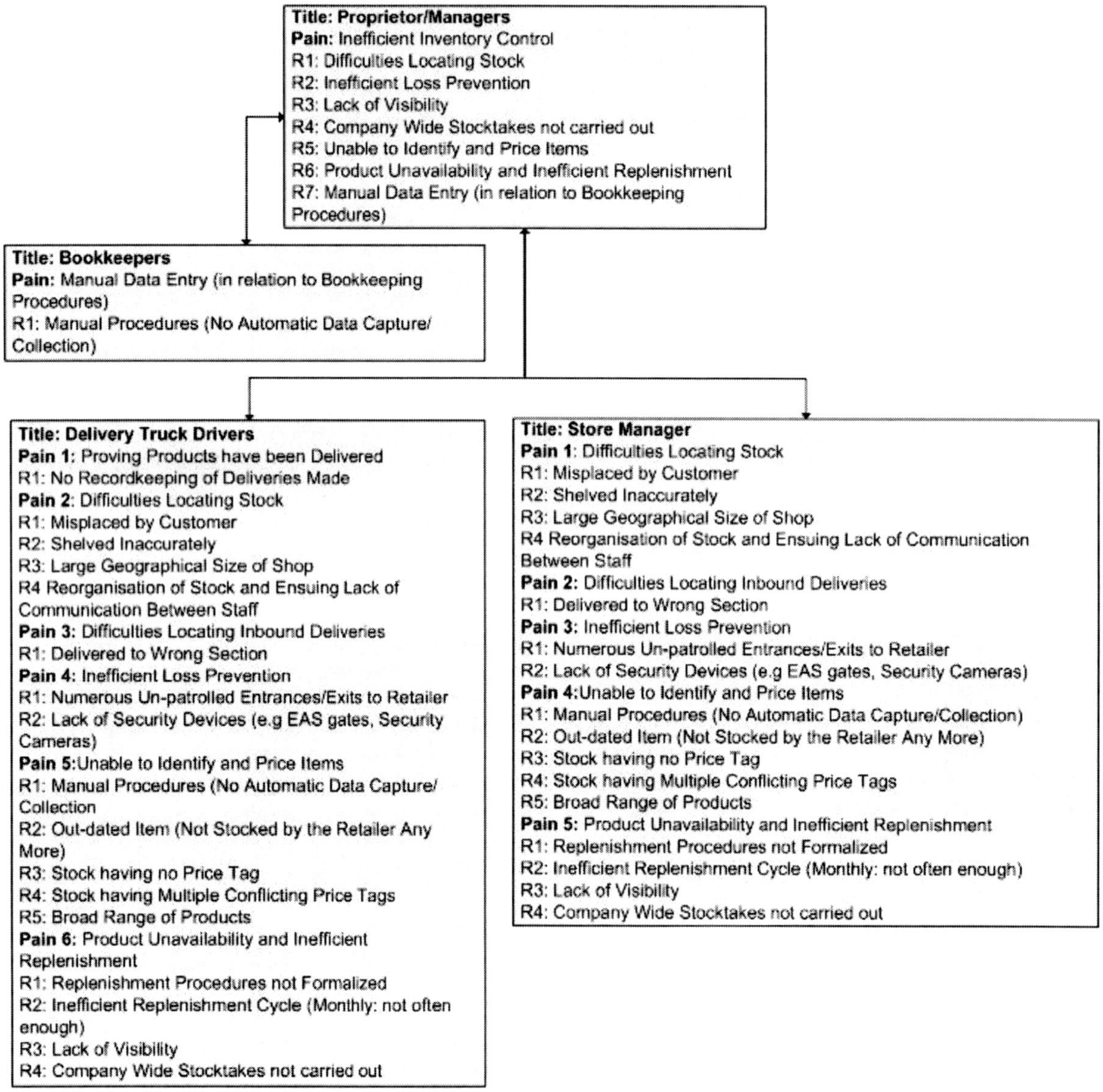

and how they are interlinked. For each pain depicted there may be more than one reason (R) for that particular challenge, and they are also presented. The pain chain was devised by multiple sources of evidence including the interviews, the participant observation, and existing procedural documentation.

6 COMPARISON OF PRESENT MODE OF OPERATION (PMO) AND FUTURE MODE OF OPERATION (FMO)

The purpose of this section is to illustrate the current inventory control processes used by the small-to-medium retailer and compare these to possible processes used by the retailer if an AIDC system, specifically RFID, was implemented in the organization. Flowcharts are used to depict the processes as they are valuable for analyzing inventory control systems and in being used to uncover dysfunctions in these systems. In addition, flowcharts are easier to comprehend than written procedures and allow a sequence of events to be examined without showing the interdependence between events (Muller, 2003).

Examining the current inventory control practices of the small-to-medium retailer has revealed that there are four inventory control processes in place which include, checking incoming orders/ deliveries, pricing and identifying stock, the ordering and replenishment process and outgoing delivery process. The four processes identified are represented in Figure 7 as part of the current inventory control practices, while Figure 8 illustrates RFID-enabled versions of the processes. The differences between the current and RFID-enabled processes are discussed in sections 6.1 through 6.4.

6.1 Checking Incoming Orders and Deliveries

The checking of incoming orders and deliveries process is shown in Figure 7. In this process deliveries are made, their contents are manually checked against the invoice provided by the supplier before being priced and shelved ready to be purchased. If a delivery is discovered to have incorrect contents then the manager will fill out a formalized credit request form to notify the supplier about the discrepancy related to the delivery. The supplier will then either resend the correct items or notify the business that they have credit on their account.

The checking of incoming orders and deliveries process shown in Figure 8 is identical to the process illustrated in Figure 7 except items delivered would have to be RFID tagged instead of labeled with a price tag. The RFID-enabled process assumes that the new system will allow the retailer to create a credit request electronically and email it to the supplier instead of completing this process manually. Based on the process depicted in Figure 8 it is apparent that if an RFID-based inventory control system was implemented in the small-to-medium retailer, it would have minimal affect on the checking orders/deliveries process. However if RFID technology was implemented throughout the supply chain of the retailer, then this process would be a lot more efficient, allowing contents of the delivery to be checked automatically and registered on the system as items would be already tagged when they are delivered.

6.2 Pricing and Identifying Stock at Point of Sale

Pricing and identifying stock at point of sale is a manual process, involving no form of automated data capture technology. Products to be purchased are brought to the front counter where staff identify the item by sight and either obtain the price of the item from its price tag, or if its not marked look the item up in the price book. Customers will then pay for the purchase by cash, check or through the Electronic Funds Transfer at Point of Sale (EFTPOS) facilities available. If a customer holds an account then information about the products they purchase, such as product name, price and quantity are written down on the formalized account sheet by staff. The account holder will receive a summary of purchases at the end of each month. This process is depicted in Figure 7.

If RFID or any other automatic identification and data capture technology was used in the small-to-medium retailer, difficulties experienced pricing and identifying items would be eliminated. The RFID tagged item would be brought to the counter, where staff would scan the item using a handheld barcode/RFID reader and accept payment from the customer. If the customer held an account then the item would be scanned and purchase details would be added to their account on the system. The implementation of RFID at point of sale would also help the retailer to effectively

Figure 7. The current inventory control practices

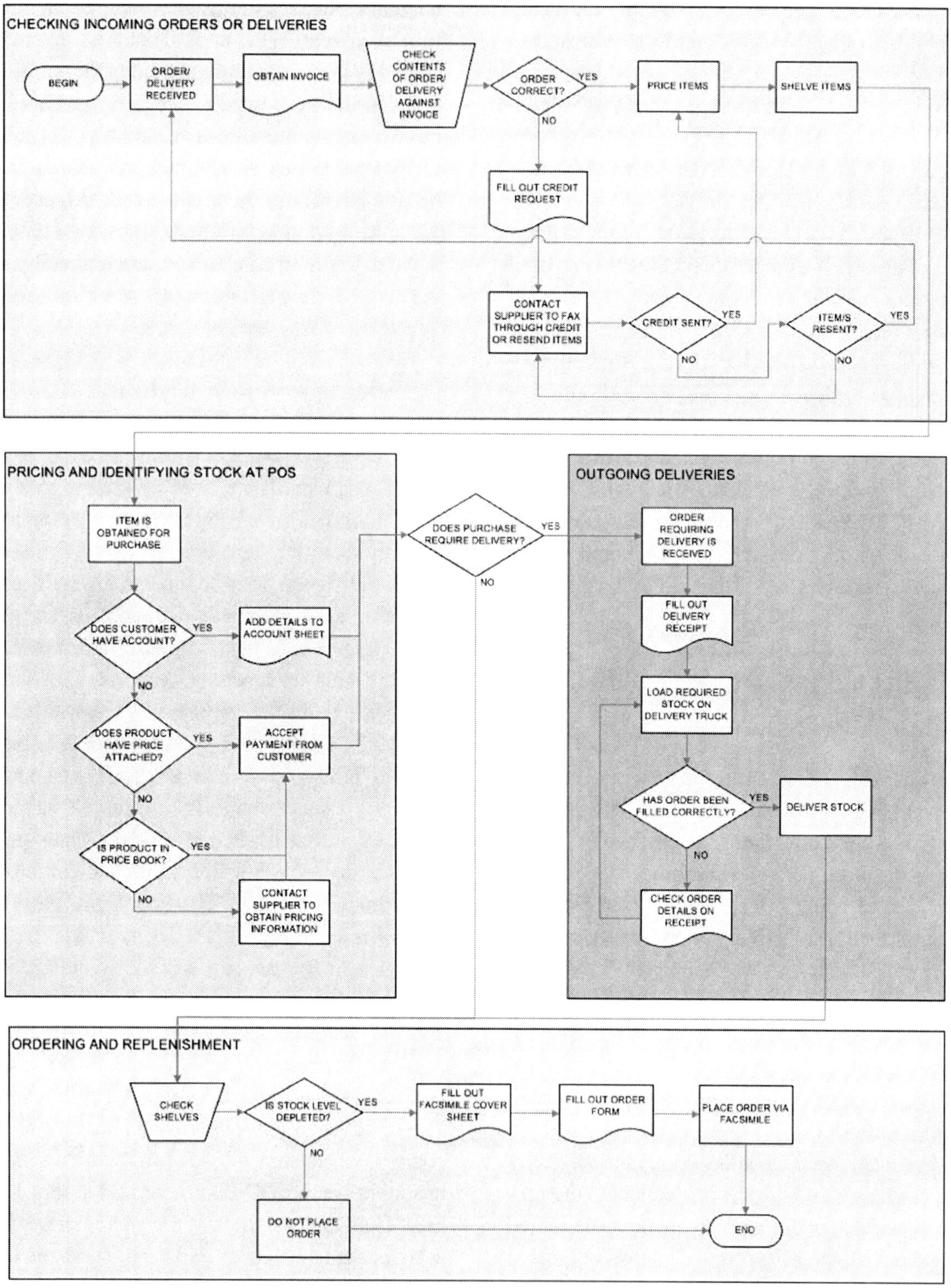

Figure 8. RFID-Enabled inventory control processes

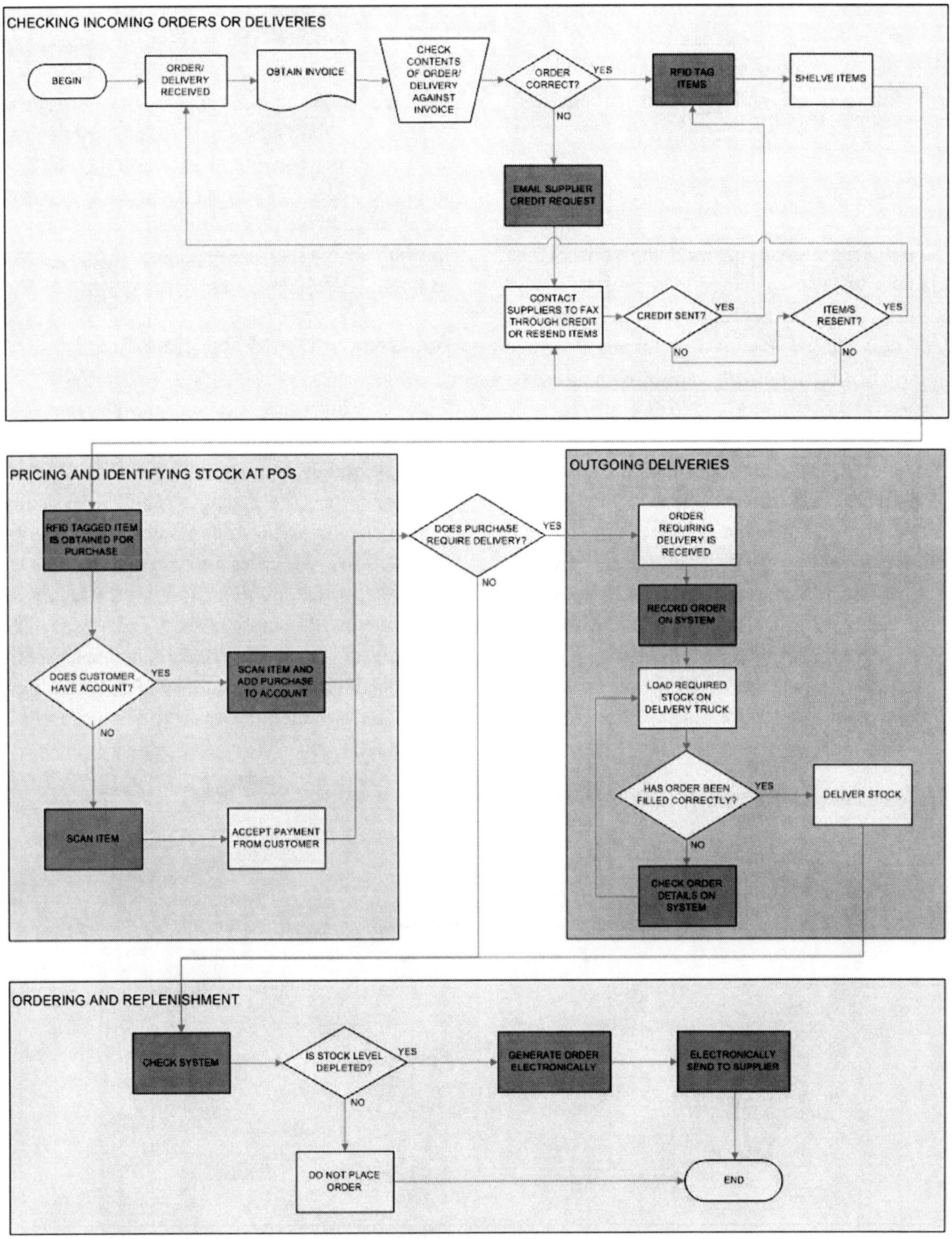

target customers. The RFID system would allow the organization to store information relating to customer purchasing patterns electronically which could be then used to develop catalogues based on popular products and items that are being sold frequently.

6.3 Outgoing Delivery Process

The outgoing delivery process shown in Figure 7 begins when a customer informs staff at the hardware store that they require an item or product to be delivered. The employee taking the order initially fills out a delivery receipt with information such as the customer's address, contact details and the description of the product they wish to be delivered. Once this information is recorded the delivery truck driver will then load all the products onto a truck and deliver the goods. The delivery receipt is usually only used as a reference for staff of the organization and is not given to the customer as a receipt or retained after for record keeping purposes after a delivery is made.

The implementation of RFID would have a minimal affect on the velocity of the outgoing deliveries process, rather formalizing the process and providing an accurate record of products that have been delivered. The small-to-medium retailer currently does not keep any record of the deliveries it has made and in the past has had customers complaining that some orders have been delivered with incorrect contents. RFID portals could be implemented at exit points of the retailer so when items of stock are placed on vehicles and moved through an exit they are registered on the system as having left the premises (Figure 9).

6.4 Ordering and Replenishment

Like other inventory processes used by the organization, the ordering and replenishment process is simple and is characterized by a minimal level of automation. The manager usually physically checks shelves and re-orders stock when quantities are down. An order will usually be made by facsimile, which involves filling out a facsimile cover sheet and in most cases an order form. The facsimile cover sheet is a document that has been developed by the organization and the order form is usually provided to the retailer by suppliers.

Figure 9. The retailer's utility vehicle and the proof of concept RFID portal

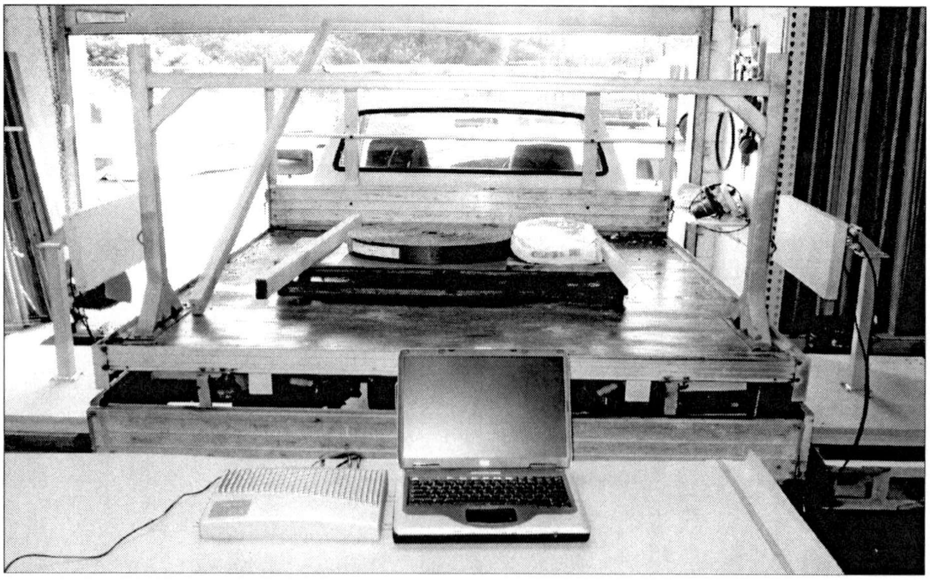

Some orders are made ad-hoc by managers if a supplier telephones the business but the majority of orders are made using the process that is encapsulated in Figure 7.

The ordering and replenishment process described in Figure 7 could be automated and made more accurate through the use of an RFID system (refer to Figure 8). RFID captures real-time information which could provide an accurate insight into the inventory levels of the organization in real-time. The managers of the retailer would be able to check the system in relation to the amount of stock held by the organization, and then automatically generate an order based on the quantity of items available. The order could then be sent electronically to the supplier. This application of RFID would have the greatest effect on the small-to-medium retailer as it would provide a level of visibility that the organization currently does not possess.

7 CRITICAL RESPONSE TO FUTURE MODE OF OPERATION (FMO)

The small-to-medium retailer explored in this study had minimal visibility and no way of monitoring the amount of inventory it held due to the manual-based system processes it employed. RFID technology could be used to automate inventory control within the organization helping to overcome the pains experienced by the organization. The ability of RFID technology to provide real-time information would be of major benefit. If an RFID system was implemented then the organization could potentially be provided with an accurate view of how much inventory they held, what items were being sold and may even helped to reduce the amount of stock carried. Being able to read more than one tag at a time and not requiring line of sight to identify products are two other characteristics that RFID possesses which could make it easier to locate stock and carry out stocktakes within the organization. The loss prevention measures could be enhanced by the implementation of RFID and the replenishment process could be formalized and automated (Huber & Michael, 2007). The adoption of RFID technology could also aid in streamlining supply chain operations. SAP AG (2002) and Harrington (2002) state that in order to improve supply chain management, and inherently improve inventory control, organizations must become adaptive to changes in the supply chain and customer capabilities by managing the visibility, velocity and variability of information. Managing visibility refers to the organization being able to sense changes to their supply chain and customer capabilities in real-time. Managing velocity is the next step to improving supply chain management that relates to organizations being able to plan and execute in real-time. The final aspect of adaptive supply chain management is managing variability which allows an organization to respond to change almost instantaneously. The first step to adopting adaptive supply chain management is significant to this study as it focuses on intra-enterprise integration and inventory control to facilitate visibility within the organization. The following phase leading to adaptive supply chain management involves collaborating and sharing information with other supply chain stakeholders. Once an organization progresses through these stages adaptive supply chain management is adopted where transactional, operational, and financial data is shared throughout the supply chain to enhance network competitiveness and optimize network profitability.

Although RFID does have the capacity to be an enabling technology that could transform how organizations operate, its feasibility must be considered by small-to-medium retailers. Technical and economic issues appear to be the main barriers preventing the feasible implementation of RFID in the small-to-medium retailer (Huber, Michael, & McCathie, 2006). From a technical perspective RFID requires an information technology infrastructure that can support the copious amount of real-time data being captured

(Jones, Clarke-Hill, Shears, Comfort, & Hillier, 2004). Employees within the organization must also have a reasonable knowledge of the technology and its system if they wish to successfully manage an RFID–based system internally. The organization documented in this paper has a basic IT infrastructure in place, without a server, which would be a minimum requirement if RFID was to be implemented. The interviews revealed that both the proprietor/manager and proprietor/part-time manager had minimal knowledge in relation to the technical aspects of a prospective AIDC system. The implementation of RFID technology must also be considered from an economic perspective. The proprietor/manager and proprietor/part-time manager both identified cost as being the major issue affecting their adoption of RFID technology. Even if the organization did implement RFID it would have to be at pallet or carton level as it would be economically unrealistic to tag at the item level as the cost of an RFID tag would be greater than the retail price of many items stocked.

Based on these reasons it is recommended that the small-to-medium retailer consider an evolutionary approach by first adopting barcode technology which is relatively simple, ubiquitous and simple to operate with a small-scale enterprise-wide information system, and then to introduce the complementary RFID technology in the future. The retailer should only look to invest in RFID once its suppliers begin to adopt the technology so they are able to experience the full benefits provided by RFID when implemented throughout the supply chain. Moving from manual inventory control processes to a barcode-based solution is a safer investment for the small-to-medium retailer, which would help to refine the inventory control processes used in the organization and minimize the technical complexity and economic burden on the organization. Figure 10 illustrates the evolutionary path that the organization should take to improve inventory control and to adopt adaptive supply chain management.

7.1 Research and Managerial Implications

There is a lack of empirical trials of RFID for small-to-medium retailers. While market leaders in the retail sector including, Tesco, McDonald's and Goodyear are trialing the technology there continues to be little evidence of the benefits of

Figure 10. The evolutionary path to RFID

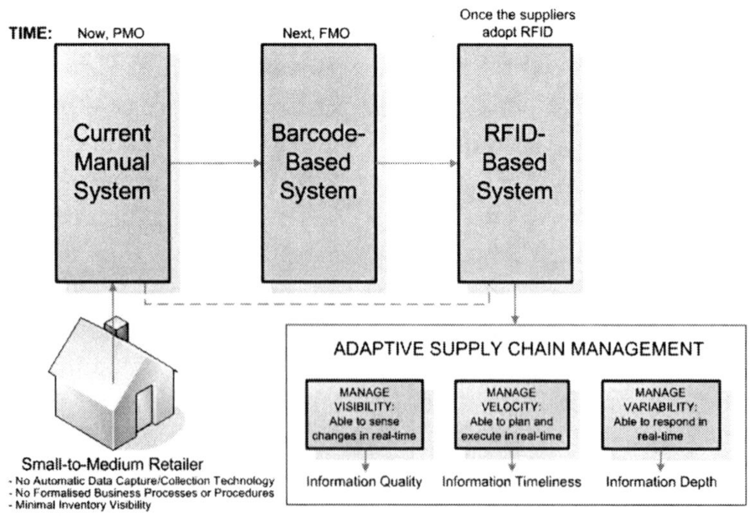

RFID in small-scale operations (DMD, 2006). There is also a need to test the applicability of RFID in retailers that stock a variety of metals, liquid substances and other materials such as concrete and timber. While item level tagging with smart shelves would realize the full potential of RFID, tiny items like nuts and screws would be cost-prohibitive to tag. Manual-based systems might work but they are labor intensive, time consuming, and error-prone. The question remains whether small-to-medium organizations like that studied in this paper will ever introduce RFID into their practices although it is without a doubt that barcodes will be introduced. Further research could also be conducted in studying several small-to-medium retailers and to consider their overall contribution to reducing the Bull-whip effect, an exaggeration of demand in an upward direction in a supply chain network.

The managerial implications of this paper have to do with an investment decision, not only for an AIDC system but for an enterprise-wide application such as enterprise resource planning (ERP) that will allow for internal and intra organizational processes to be automated, such as receiving, picking, and ordering. The online collection of data is paramount to the success of any business in the longer term. In the future a small-scale ERP would be able to take advantage of the benefits bestowed by both barcoding and RFID technology.

8 CONCLUSION

The inventory control issues and challenges experienced by a small-to-medium hardware retailer have been documented in this paper through the use of semi-structured interviews, participant observation and procedural documentation. Flowcharts were also utilized to illustrate and compare the manual business processes of the organization to RFID–enabled processes. Examining the literature has highlighted RFID as a revolutionary technology possessing significant benefits for inventory control and supply chain management. The literature portrays small-to-medium retailers as organizations that could benefit from the use RFID technology as they have experienced problems with inventory control in the past. This paper concludes that RFID can offer significant improvements for inventory control especially in relation to ordering and replenishment, but organizations must carefully consider its feasibility before committing to implementation. Small-to-medium retailers must consider the ability of their information technology infrastructure to handle an RFID-based system and assess whether staff have the technical aptitude to maintain and operate such a system internally. The organization must also consider the economic resources they have available to them as an RFID solution is more costly compared to other automated data capture technologies. For these reasons, it is difficult to foresee the retailer examined in this study moving from manual processes to an RFID-based system. Barcode technology can be viewed as a short-term alternative to RFID for this organization, which could overcome the visibility issues that they are experiencing. However the physical characteristics of RFID differentiate it from barcode technology, as RFID can provide retailers with an enhanced level of visibility and automation. Unless a technology that captures data in real-time, like RFID is put into practice by small-to-medium retailers, the adoption of adaptive supply chain management can be ruled out. This paper proposes an evolutionary approach to the implementation of automated data capture technologies for small-to-medium retailers, recommending that organizations incorporate RFID when technical and economic issues are overcome and once their suppliers begin to adopt the technology as it is perceived that the use of RFID throughout the supply chain has a more significant Return on Investment (ROI) than its use internally.

REFERENCES

Abraham, C., Ahuja, V., Ghosh, A. K., & Pakanati, P. (2002). *Inventory management using passive RFID tags: a survey*. Retrieved October 10, 2007, from http://www.it.iitb.ac.in/~it601/dep/paperlist/tag-identification-survey.pdf

ABS. (2002). *Small businesses in Australia 2001*. Retrieved October 10, 2008, from http://www.abs.gov.au/ausstats/abs@.nsf/e8ae5488b598839cca25682000131612/97452f3932f44031ca256c5b00027f19!OpenDocument

Albano, S., & Engels, D. (2002). *Auto-ID Center field trial: phase I summary.* Retrieved October 10, 2008, from http://autoid.mit.edu/whitepapers/MIT-AUTOID-TR006.PDF

Angeles, R. (2005). RFID technologies: supply-chain applications and implementation issues. *Information Systems Management, 22*(1), 51–65. doi:10.1201/1078/44912.22.1.20051201/85739.7

Bendavid, Y., Wamba, S. F., & Lefebvre, L. (2006). *Proof of concept of an RFID-enabled supply chain in a B2B e-commerce environment.* Paper presented at the 8th International Conference on Electronic Commerce: The new e-commerce: Innovations for Conquering Current Barriers, Obstacles and Limitations to Conducting Successful Business on the Internet, Fredericton, Canada.

Bhuptani, M., & Moradpour, S. (2005). *RFID Field Guide: Deploying Radio Frequency Identification Systems*. Upper Saddle River, NJ: Prentice Hall.

Chuang, M., & Shaw, W. (2005). *How RFID will impact supply chain networks?* Paper presented at the 2005 IEEE Engineering Management Conference, Newfoundland, Canada.

Collins, D. J., & Whipple, N. N. (1994). *Using Bar Code: Why It's Taking Over*. Duxbury, MA: Data Capture Institute.

Diamond, J., & Pintel, G. (1996). *Retailing* (6th ed.). Upper Saddle River, NJ: Prentice Hall.

DMD. (2006). RFID Application in Inventory Control. *The DecisionMakers' Direct: new insights, new tools, new solutions*. Retrieved November 21, 2008, from http://www.decisioncraft.com/dmdirect/pdf/rfidapplications.pdf

Ellram, L., Londe, B. L., & Weber, M. (1999). Retail logistics. *International Journal of Physical Distribution and Logistics Management, 29*(7), 477–494. doi:10.1108/09600039910291993

Finkenzeller, K. (2003). *RFID Handbook: Fundamentals and Applications in Contactless Smart Cards and Identification*. West Sussex, UK: John Wiley & Sons.

Fleisch, E., & Tellkamp, C. (2003). *The impact of inventory inaccuracy on retail supply chain performance: a simulation study.* Retrieved from http://autoid.mit.edu/whitepapers/STG-AUTOID-WH003.pdf

Garfinkel, S., & Rosenberg, B. (2005). *RFID: Applications, Security, and Privacy*. Upper Saddle River, NJ: Addison-Wesley.

Gunasekaran, A., Forker, L., & Kobu, B. (2000). Improving operations performance in a small company: a case study. *International Journal of Operations & Production Management, 20*(3), 316–335. doi:10.1108/01443570010308077

Harrington, L. (2002). *Winning supply chain strategies: adversity breeds creativity.* Retrieved October 10, 2007, from http://www.inboundlogistics.com/articles/features/1202_feature02.shtml

Heinrich, C. (2005). *RFID and Beyond: Growing Your Business Through Real World Awareness*. Indianapolis, IN: Wiley Publishing Incorporated.

Hu, Y., Sundara, S., Chorma, T., & Srinivasan, J. (2005). *Supporting RFID-based item tracking applications in Oracle DBMS using a bitmap datatype.* Paper presented at the 31st International Conference on Very Large Data Bases, Trondheim, Norway.

Huber, N., & Michael, K. (2007, July 9-11). *Minimizing Product Shrinkage across the Supply Chain using Radio Frequency Identification: a Case Study on a Major Australian Retailer.* Paper presented at the Sixth International Conference on Mobile Business, Toronto, ON, Canada.

Huber, N., Michael, K., & McCathie, L. (2006). *Barriers to RFID in the Supply Chain.* Paper presented at the RFID Eurasia, Turkey.

IBM Learning Services. (2001). *e-Business Selling Workshop: Opportunity Identification (Pain Chains).*

Jones, P., Clarke-Hill, C., Shears, P., Comfort, D., & Hillier, D. (2004). Radio frequency identification in the UK: opportunities and challenges. *International Journal of Retail & Distribution Management*, *32*(3), 164–171. doi:10.1108/09590550410524957

Kärkkäinen, M. (2003). Increasing efficiency in the supply chain for short shelf life goods using RFID tagging. *International Journal of Retail & Distribution Management*, *31*(10), 529–536. doi:10.1108/09590550310497058

Knight, P. (2002). *Small-Scale Research.* London: Sage.

Lee, Y., Cheng, F., & Leung, Y. (2004). *Exploring the impact of RFID on supply chain dynamics.* Paper presented at the 2004 Winter Simulation Conference, Yorktown Heights, NY.

Lefebvre, L., Lefebvre, E., Bendavid, Y., Wamba, S. F., & Boeck, H. (2006). *RFID as an enabler of B-to-B e-Commerce and its impact on business processes: a pilot study of a supply chain in the retail industry.* Paper presented at the 39th Annual Hawaii International Conference on System Sciences, Hawaii, USA.

McCathie, L., & Michael, K. (2005). *Is it the end of barcodes in supply chain management?* Paper presented at the Proceedings of the Collaborative Electronic Commerce Technology and Research Conference LatAm, Talca, Chile.

Michael, K., & McCathie, L. (2005). *The pros and cons of RFID in supply chain management.* Paper presented at the International Conference on Mobile Business, Sydney, Australia.

Mills-Harris, M., Soylemezoglu, A., & Saygin, C. (2005). *RFID data-based inventory management of time-sensitive materials.* Paper presented at the Industrial Electronics Society (IECON 2005), Raleigh, NC.

Muller, M. (2003). *Essentials of Inventory Management.* New York: American Management Association.

Nemeth, P., Toth, L., & Hartvanyi, T. (2006). *Adopting RFID in supply chains.* Paper presented at the IEEE International Conference on Mechatronics.

Palmer, R. C. (1995). *The Bar Code Book: Revised and Expanded.* Dublin, NH: Helmers Publishing, Inc.

SAP AG. (2002). *Adaptive supply chain networks.* Retrieved October 10, 2007, from http://www.sap.com/solutions/business-suite/scm/pdf/50056466.pdf

Sarma, S. (2004). Integrating RFID. *Queue*, *2*(7), 50–57. doi:10.1145/1035594.1035620

Symbol Technologies. (2004). *Business benefits from radio frequency identification (RFID).* Retrieved from http://www.motorola.com/staticfiles/Business/Product%20Lines/Symbol/Symbol%20Mobile%20Computers/_Documents/Static%20files/RFIDBenefits.pdf

Tellkamp, C., Angerer, A., Fleisch, E., & Corsten, D. (2005). *From pallet to shelf: improving data quality in retail supply chains using RFID.* Retrieved October 10, 2007, from http://www.autoidlabs.org/uploads/media/AUTOIDLABS-WP-BIZAPP-019.pdf

Tersine, R. (1998). *Principles of Inventory and Material Management* (4th ed.). Upper Saddle River, NJ: Prentice Hall.

Viale, D. (1996). *Basics of Inventory Management: From Warehouse to Distribution Center*. Menlo Park, CA: Crisp Publications Incorporated.

Want, R. (2004). The magic of RFID. *Queue, 2*(7), 40–48. doi:10.1145/1035594.1035619

Wild, T. (1997). *Best Practice in Inventory Management*. New York: John Wiley & Sons.

Yin, R. (1994). *Case Study Research: Design and Methods* (2nd ed.). London: Sage Publications.

Zebra Technologies. (2007). *Bar coding 101... What you need to know.* Retrieved October 8, 2007, from http://www.zebra.com/id/zebra/naen/documentlibrary/whitepapers/bar_coding_101.DownloadFile.File.tmp/WP1144Barcode101_R5_1.07.pdf?dvar1=White%20Paper&dvar2=Bar%0Coding%20101...What%20You%20Need%20to%20Know

Zipkin, P. (2000). *Foundations of Inventory Management*. Boston: McGraw-Hill.

This work was previously published in International Journal of E-Business Research Volume 6, Issue 2, edited by In Lee, pp. 21-44, copyright 2010 by IGI Publishing (an imprint of IGI Global).

Chapter 3
The Hispanic View of E-Mail, Popup, and Banner Advertising

Pradeep K. Korgaonkar
Florida Atlantic University, USA

Ronnie Silverblatt
Florida International University, USA

Enrique P. Becerra
Texas State University, USA

ABSTRACT

In this paper, the authors broadly investigate Hispanic consumer attitudes and their sociological behaviors toward online advertising. The study documents the correlation between Hispanic attitudes, behaviors, and three types of online advertising that have an effect on cultural distinctions. Additionally, the study compares the group's attitudes toward TV advertising vis-à-vis the three types of online advertising to compare the two medium's similarities and differences. Notably, Hispanics favor e-mail advertising and banner advertising over popup advertising, while Hispanics continue to enjoy advertising on television more than any of the three types of Web advertising that this paper will discuss.

INTRODUCTION

The Hispanic View of E-Mail, Popup, and Banner Advertising

In the last decade both the academic and business community has shown a growing interest in studying Hispanic consumers. This interest is the result of the substantial growth of the U.S. Hispanic population and their corresponding wealth. Marketers in particular are paying special attention to this distinct ethnic segment because the Hispanic market comprises over 46 million people and accounts for about 15 percent of the U.S. population (Sun & Johnson, 2009). Almost one in every seven Americans is of Hispanic origin. Moreover, as a population, it is growing about six times faster than the general population. The 2008 U.S. Census estimates the population of Hispanics to more than double by the year 2050 (U.S. Census Bureau, 2008). The size and dynamics of Hispanic minorities have redefined several aspects

DOI: 10.4018/978-1-61350-462-8.ch003

of the U.S. economy. According to the Interactive Advertising Bureau (2006), U.S. Hispanic Buying Power is increasing at a much faster rate than non-Hispanics Buying Power. The buying power of Hispanics is reported to be around $1 trillion in the marketplace (Sun & Johnson, 2009). Thus, the importance of the Hispanic market to advertisers is obvious.

Although a few studies in the past have investigated the media habits of this ethnic group, (e.g., Guernica, 1982), little is known about the beliefs, attitudes and behaviors of Hispanic consumers towards the three types of dominant Web advertising, namely, e-mail, pop-ups, and banner advertising.

Also, for Hispanic consumers, it is unknown how these three types of Web advertisements stack up against traditional television advertisements. The current study attempts to fill these gaps in the published literature.

Web advertising targeted to Hispanics is highly visible and monetarily significant. Web advertising targeting Hispanics accounts for more than 75 million dollars (Cox, 2004) up from $10 million spent in 2002. Most market research on Hispanics attitudes and beliefs towards marketing and/or advertising are general in nature and do not focus on specific advertising techniques. Research has shown many Hispanics use the Internet to learn more about brands and prices. A 2006 study by AOL reports that 55% of all Hispanics use the Internet; of the online Hispanics, 81% are mostly or partially acculturated, 77% use the Internet to learn about product brands, and 72% use the Internet to compare prices (www.imnewswatch.com).

Similar results were reported in a study quoted by Burns (2005). That study indicated Hispanic Internet usage equals or exceeds the non-Hispanic population in a variety of Internet activities. A recent study released by Yahoo! Telemundo and Experian Simmons Research (2007) suggested that online Hispanics are highly experienced and multifaceted on the Internet (Contacto Magazine, 2007). Summing up, Hispanics as a segment of the U.S. population and as a user group of the Internet offers a lucrative and growth market in the U.S. (Tellis & Ortiz 2008; U.S. Census Bureau, 2008).

Therefore, the study results are expected to be of interest to academics and practitioners alike. The next section reviews the pertinent literature and establishes the study hypotheses.

Literature and Hypotheses

Published literature in the field of Hispanic marketing has explored a variety of topics. Studies have investigated the effect of choice of language, i.e., English vs. Spanish, in advertising to Hispanics, brand loyalty, preference for shopping at neighborhood stores, and the propensity to purchase prestigious brands (e.g., Segal & Sosa, 1986; Hernandez & Newman, 1992). Advertising research has indicated that Hispanics have favorable attitudes toward advertising in general and towards direct marketing advertising in particular (Korgaonkar et al., 2000). Webster (1991) found that Hispanic consumers value advertising as an important source of information. Yankelovich et al. (1981) reported Hispanics rely significantly on commercial sources of information including advertising. It has also been suggested that Hispanics value brands most frequently advertised and believe in materialism (Bellenger & Valencia, 1982). Finally, studies by Fones (1981) and Webster (1992) suggested that advertising in radio, magazines, and brochures influence Hispanic consumers. More recently Korgaonkar et al. (2001) suggested Hispanics were ambivalent towards Web advertising.

The fall 2005 Simmons National Consumer Survey indicated more Hispanics found online advertising useful than non-Hispanics (Interactive Advertising Bureau, 2006). Although these studies are helpful as a preliminary attempt in understanding Hispanic consumers, we believe study results would be more useful if more specific forms of Web advertising such as e-mail, the pop-up, and the banner advertising were investigated instead of "general" advertising.

Although there is a significant business interest in understanding the different aspects of Web advertising, the literature pertaining to Hispanic Internet users is still scant. Most of the past studies have focused on only a single type of Web advertising and rarely had a Hispanic sample. In general, we know that the clicks thru rates are very low and intrusiveness is an issue. A few past studies focusing on Hispanic attitudes toward general advertising and Web advertising suggest that Hispanics are favorably disposed toward it (Korgaonkar et al., 2001; Webster, 1992; Deshpande et al., 1986; Yankelovich et al., 1981; Valencia, 1989). Nevertheless, studies of Hispanic attitudes regarding the three types of Web advertising are still lacking.

Drawing from past studies of non-Hispanic Internet users and a single type of Web advertising, we know individual types of Web advertisements have differential impacts on their audiences. For example, a few published studies suggest that e-mail advertising is effective if it is perceived to be useful (Martin et al., 2003), banner ads are considered ineffective because of their poor click thru rates (Cross, 1999), and popup ads are reported to yield more click thru rates, more favorable attitudes, and higher purchase intention than standard banner ads (Cho et al., 2001). These results are comparable to radio and TV advertisements having different impacts on their audiences although both are categorized as broadcast media. Similarly, we expect significant differences in the Hispanic consumers' attitudes between the three types of Web advertising. This leads us to hypothesize that:

H1: *Hispanic beliefs, attitudes, and behavior toward the three types of Web advertising will be significantly different from each other.*

Attitude and Behavior Consistency

Several past studies in advertising and marketing suggest a strong and positive relationship between consumer attitudes and behaviors. We posit a similar relationship between the Hispanic consumers' attitudes towards the three types of Web advertising and Hispanic consumers' behavior towards the three advertising techniques. The specific hypothesis to be tested is that

H2: *There will be a significant positive relationship between the attitudes towards each of the three types of Web advertising and behaviors of Hispanic consumers*

TV Advertising vs. E-Mail, Popup, and Banner Advertising

Compared to traditional TV advertising, Hispanic evaluation of the three advertising methods is expected to be less favorable. This result is expected because Hispanic culture is suggested to be conservative and traditional in its values as well as its behavior (Valdes & Sadane, 1995; Bellenger & Valencia, 1982). Additionally, a few studies in shopping context suggest that Hispanics are less prone to take risks and less likely to try new brands or ideas (e.g., Valdes, 1991; Cartagena, 2005). The Web and the three types of Web advertising are relatively new when compared to television. Accordingly, this leads us to suggest that Hispanics are less likely to embrace the new medium and its ads as readily as traditional TV ads. As noted earlier, we expect Hispanics to have significant differences in their beliefs and attitudes towards e-mail, popup, and banner advertising. We expect these differences to spillover to their evaluations of these three types of advertising vis-à-vis TV advertising. Consequently, this leads us to hypothesize that

H3: *Significant differences are expected between the evaluations of each of the three types of Web advertising vis-à-vis TV advertising.*

Methodology

Sample

The sample for the study consists of 660 consumers in the Southeast United States. The sample is drawn from a metropolitan area with a population of about five million people, 30 percent of which is of Hispanic ethnicity. The area has two major English and one major Spanish newspaper; it also has four major English and three major Spanish TV stations. As noted by many scholars, the ability to administer probabilistic sampling procedures in the investigation of ethnic studies is very difficult (Marin & Marin, 1995). Recognizing the difficulty of sampling Hispanics without a population listing, the surveys of adult Hispanics were conducted via personal interviews in several geographical areas to capture the diversity in socioeconomic as well as Hispanic background. Interviewers fluent in both English and Spanish contacted the subjects at home or work. The interviewers were instructed to sample respondents as randomly as possible and to administer the surveys on different days and at different times. Only those adult respondents who had surfed the Web were asked to respond to the survey. The interviewers explained the objectives of the study and gave the respondent an option of completing an English or Spanish language questionnaire. A total of 588 usable questionnaires were obtained from a sample of 660.

One sample group responded to banner advertising questions (n=267), a second group to popup advertising questions (n=287), and a third group to the e-mail advertising questions (n=106). The demographic profile of each respondent group is shown in Table 1. The sample of respondents for the banner advertising was slightly younger than the other two groups; the sample of the email advertising group were more likely to have family members speaking Spanish at home than the other two groups.

Measurement

To obtain a clearer understanding of consumers' Web advertising beliefs and attitudes, we examined published literature. Bauer and Greyser (1968) were the first to systematically examine attitude toward advertising; they assessed advertising beliefs as two clusters: economic and social effects. Succeeding studies have used variations of Bauer and Greyser's two-factor model (Muehling, 1987). More contemporary studies suggest several, not just two, facets of consumers' advertising beliefs.

Alwitt and Prabhaker (1992) found that consumers' perceptions of television advertising spans six dimensions, and in a (1994) follow-up study, they observed consumers' advertising evaluations comprise six dimensions. In his 1994 study Mittal describes ten consumer advertising perceptions and beliefs.

Pollay and Mittal (1993) as well as Korgaonkar et al. (2001) support a seven-factor belief model. Because the two studies have established the validity and reliability of these measures with both Hispanic and non-Hispanic consumers, we chose the seven-factor model they developed. In the two studies, the seven factors represent three personal uses of advertising: product information, social role and image, and hedonic/pleasure, along with four social effects of advertising: good for the economy, materialism, value corruption, and falsity/no sense. We describe below the seven belief factors as they relate to our study. The actual statements used to measure each belief as well as the overall attitude construct are shown in Table 2.

Seven Belief Factors

Product information. This personal use factor suggests that one of advertising's roles is to provide information. Web advertising information allows for greater marketplace efficiencies, where

Table 1. Demographic profile of the respondents

	Email advertising	Popup advertising	Banner advertising	Significance level
Gender				(.170) p<.2
Male	54	105	99	
Female	52	182	168	
Education				(.108) p<.15
High School Graduates	16	42	24	
Trade School Graduate	1	3	5	
Some College	62	101	138	
College Graduate	22	72	53	
Post Graduate/professional degree	5	24	20	
Age				(.021) p<.05
Under 20 years	19	27	27	
20 to 30 years	60	137	149	
31 to 40 years	13	42	32	
41 to 50 years	10	27	23	
51 to 60 years	3	6	7	
Over 60 years	1	3	2	
Family speaks Spanish at home				(.080) p<.10
All the time	65	94	93	
Mostly	0	8	16	
Sometimes	44	48	57	
Rarely	0	12	13	
Never	0	80	61	
Were you born in another country?				(.220) p<.25
Yes	51	111	110	
No	51	131	130	

consumers' needs and wants are better matched with advertisers' offerings.

In his 1996 study, Ducoffe (1996) found a substantial and significant correlation between informative and Web advertising value. "Product Information" is measured by: a) "_____ advertisements are a very valuable source of information about local sales;" b)"________advertisements tell me which brands have the features I am looking for;" and c) "________ advertisements help me keep up-to-date about products available in the marketplace." The coefficient alpha of the construct was 0.673.

Social role and image. Different forms of Web advertising, like other forms of mass media advertising, promote social and lifestyle messages through associated status, portrayal of ideal us-

Table 2. Measurements and significance tests of beliefs, attitudes, and behavior toward email, popup, banner advertising

Belief Construct/Measure *	**E-mail mean value (SD) N=106**	**Pop Ups mean value (SD) N=240**	**Banner mean value (SD) N=242**	**Significance level**	**Reliability coefficient alpha**
Product information					.673
______ advertisements are a very valuable source of information about local sales	1.6792 (.46898)	1.7355 (.92695)	2.2042 (2.10697)	p<.01	
______ advertisements tell me which brands have the features I am looking fo r	1.6792 (.46898)	1.5950 (.83072)	1.8500 (.88833)	p<.01	
______ advertisements help me keep up-to-date about products available in the marketplace	1.7736 (.42050)	1.7149 (.93642)	2.0417 (.96287)	p<.001	
Social role and image					.813
From _____ advertisements I learn what is in fashion and what I should buy for keeping a good social image	1.5755 (.49662)	1.5455 (.82994)	1.9333 (1.04488)	p<.001	
I like it when ___ advertisements show people like me using the brand I am using	1.7075 (.45705)	1.7149 (.94962)	1.9875 (.99992)	p<.01	
In _______ advertisements I often see and hear my own beliefs and values being portrayed	1.5849 (.49508)	1.4793 (.71845)	1.8083 (.97497)	p<.001	
Hedonic/Pleasure					.749
I find _______ advertisements a welcome break in viewing websites	1.5755 (.49662)	1.4339 (.75518)	1.6875 (.90919)	p<.01	
Some _____ advertisements make me feel good	1.6604 (.47583)	1.5950 (.84557)	1.9708 (.99538)	p<.001	
Value corruption					.504
_________ advertisements sometimes make people live in a world of fantasy	1.8396 (.36870)	2.0661 (1.12117)	2.2417 (1.09005)	p<.05	
A lot of ______ advertisements are based on ideas and values which are opposite to my own personal values	1.8962 (.30642)	2.2769 (1.09402)	2.5167 (2.69039)	p<.2	
___________ advertisements invade my privacy	1.8585 (.35020)	2.7603 (1.42605)	2.6042 (1.17398)	**p<.01**	
Falsity/No Sense					.239
One can put more trust in products advertised through _____ advertisements than in those not advertised on the Web.(reverse scale) [d]	1.6321 (.48453)	1.6033 (.78871)	1.8792 (.94070)	p<.01	
_________ advertisements advertise poor quality products	1.8679 (.34018)	2.4298 (1.07654)	2.5417 (2.82960)	p<.2	
With all the ________ advertising going on, I do not Quite know what to believe and what not to believe	1.8491 (.35969)	2.2934 (1.32039)	2.4000 (1.19902)	p<.1	
Good for the economy					**.755**
_________ advertisements improve people's standard of living	1.6226 (.48703)	1.4380 (.68040)	1.6750 (.79919)	p<.001	
_________ advertisements help the consumer buy the best brand for the price	1.6981 (..46126)	1.6446 (.78745)	1.9375 (.95519)	p<.001	

continued on following page

Table 2. Continued

Belief Construct/Measure *	E-mail mean value (SD) N=106	Pop Ups mean value (SD) N=240	Banner mean value (SD) N=242	Significance level	Reliability coefficient alpha
If there were no _____advertisements deciding what to buy would be difficult.	1.4811 (.50202)	1.4050 (.79499)	1.5208 (.74849)	p<.2	
Materialism					**.582**
________ advertisements make you buy things you do not really need	1.8585 (.35020)	2.3182 (1.20940)	2.4833 (1.14609)	p<.1	
________ advertisements increase dissatisfaction among consumers by showing products which some consumers can't afford	1.8491 (.35969)	2.2231 (1.01022)	2.2667 (.93050)	p<.1	
There have been times when I have bought something because of a _____ advertisement.	1.5755 (.49662)	1.4421 (.82414)	1.6667 (.89474)	p<.01	
Attitude					**.607**
Overall, do you consider _______ advertisements a good or bad thing?** [c]	2.5660 (1.02371)	2.0992 (2.21614)	2.4667 (.98468)	p<.05	
Overall, do you like or dislike ________ advertisements?*** [c]	2.2453 (1.11108)	1.5909 (.85092)	2.1042 (1.00708)	p<.001	
Behavior					.593
When you see a_______ advertisement how close attention do you play? ****	2.2264 (1.09790)	1.6570 (.87977)	.8458 (1.03353)	p<.01	
When you see a ________ advertisement how often do you click on the ad to find more information? ****	1.8868 (1.06298)	1.5041 (.88907)	1.6708 (.90788)	p<.01	
When you see a ________ advertisement how often do you get annoyed and leave the website? Reverse item ****	3.4057 (1.255568)	2.9711 (1.40951)	3.2958 (1.28395)	p<.05	

[a] *On a five-point scale of 1 = "strongly disagree" thru 5 = "strongly agree."
** On a five-point scale of 1 = "very bad" thru 5 = "very good"
***On a five-point scale of 1 = "strongly dislike" thru 5 = "strongly like"
****On a five-point scale of (1) Never (2) Occasionally (3) Sometimes, (4) Often, (5) Always
[b] Mean values are reported, not raw scores.

ers, social reaction to purchase, and brand image (Pollay &

Mittal, 1993; Korgaonkar et al., 2001). Moreover, the vivid, interactive messages in advertisements enhance its ability to convey these social messages. Additionally, it is expected that major companies will devote more money to brand advertising as measurements become more refined, fueling consumers' desire to pay premium prices for branded items. "Social role and image," also a personal factor, is measured by "From _____ advertisements, I learn what is in fashion and what I should buy for keeping a good social image" b) "I like it when ____ advertisements show people like me using the brand I am using" and c) "In _______ advertisements I often see and hear my own beliefs and values being portrayed." The coefficient alpha of the construct was 0.813.

Hedonic pleasure. Web advertising with its multimedia capabilities can be beautiful, sentimental, motivating, humorous and entertaining. Hoffman and Novak (1996) maintain that image and content-rich Web sites embedded with enjoyable, interactive presentations represent the future of Web advertising. To the extent that consumers enjoy viewing Web sites, there should be a positive impact on Web advertising beliefs. Thus, the creativity of the Internet advertising adds value to the advertised products and services. "Hedonic/ Pleasure," the third personal factor, is measured by: a) "I find ______advertisements a welcome break in viewing Websites." and b) "Some _____ advertisements make me feel good." The reliability alpha of the construct was 0.749.

Value corruption. Web advertisements can undermine the values parents try to impress upon their children. Web advertising has the power to mold users' values, and thus, has the potential to corrupt the values of its users. Value corruption is a major user concern prompting "ad blocker" software that prevents ads from appearing on a Web page (Bridis, 1998). "Value corruption," the first of four societal effects, is measured by: a) "________ advertising sometimes makes people live in a world of fantasy." b)" A lot of ________ advertisements are based on ideas and values which are opposite to my own personal values." c)" ________ advertisements invade my privacy." The reliability alpha of the construct was 0.504.

Falsity/no sense. In their rush to have presence on the Web, many firms that ordinarily employ sound research practices abandon their logic and develop ineffective Web sites (Nadilo, 1998). Some of the characteristics of these sites include half-truths, deceptive claims and intelligence-insulting prose. Like other forms of advertising, Web advertising often promises consumers magical results and fuels the falsity factor. Falsity/No sense scale with the alpha of 0.239,is measured by: a) "One can put more trust in products advertised through ______ advertisements than in those not advertised on the Web (a reverse scale)." b) "_______advertisements advertise poor quality products." c) "With all the ________ advertising going on, I do not quite know what to believe and what not to believe."

Good for the economy. Advertising saves consumers' time and makes it possible for them to choose from a wide variety of merchandise conveniently from almost any location. Additionally, the three types of Internet advertising facilitate the introduction of new products and services.

Particularly, when shopping for big-ticket items where price comparison in an information rich environment is fairly easy, and the potential for savings significant, the economic motivation to shop on the Web could be strong. "Good for the economy," another societal effect, is measured by: a) "_______ advertising improves people's standard of living." b) "________advertisements help the consumer buy the best brand for the price." and c) If there were no _______ advertisements deciding what to buy would be difficult." The coefficient alpha of the scale was 0.755.

Materialism. Materialism is defined by Pollay and Mittal (1993) as a set of belief structures that sees consumption as the route to most, if not all, satisfactions. Web users are faced with an enticing array of goods which may promote concerns about materialism. On-line storefronts and on-line malls displaying a wide variety of goods tempt consumers with their enticing display of wares. "Materialism," the fourth societal effect, is measured by: a) "____________ advertisements make you buy things you do not really need." b) "__________ advertisements increase dissatisfaction among consumers by showing products which some consumers can't afford." c) "There have been times when I have bought something because of a __________advertisement." The coefficient alpha of the scale was 0.582.

As previously stated, this study measures the seven dimensions of beliefs and attitudes toward Web advertising via adopting Pollay and Mittal (1993) and Korgonkar et al. (2001) measurement scale for Web advertising. Each of the seven scale dimensions (product information, social role and image, hedonic pleasure, value corruption, falsity/ no sense, good for the economy, and materialism) was operationalized in terms of three or four items. For each item, respondents were asked to indicate the extent to which they agreed or disagreed with;

each respondent was asked about one of the three types of Web advertisements under investigation. The item evaluations were elicited using a five-point scale with descriptive anchors ranging from "strongly agree" (coded 1) to "strongly disagree" (coded 5).

Attitude

The attitude construct was operationalized using two items utilized in past studies (Ducoffe, 1998; Korgaonkar et al., 2001). The two items are a) Overall, do you consider Web advertising a good or bad thing? Measured on a five-point scale of (1) "very bad" thru (5) "very good" and b) Overall, do you like or dislike Web advertising? measured on a five-point scale of (1)"strongly dislike" thru (5) "strongly like" The coefficient alpha of the scale was 0.607.

Web Advertising Behavior

Web advertising behavior is conceptualized as an activity variable of three dimensions. The three dimensions are: 1) When I see a _________ advertisement on my computer screen I pay close attention to it (attention), 2) When I see a _________ advertisement on my computer screen I click on the advertisement to find more information (interest), 3) When I see a ____________ advertisement on my computer screen I leave the Web site (avoidance, a reverse item). Each item is measured on a five-point scale of (1) Never (2) Occasionally, (3) Sometimes, (4) Often, (5)Always. The coefficient alpha of the scale was 0.593. Thus despite the difficulties involved in translating the meanings of the terms from English to Spanish the overall reliabilities for almost all the measures except Falsity/No Sense were fairly acceptable ranging from 0.504 to 0. 813.

Web vs. TV

The evaluation of the specific type of Web advertising in comparison to TV advertising was measured by asking the question: "Compared to _______ advertisements, how will you evaluate TV advertising? Is TV advertising more (1), the same (2), or less (3)" on a list of pertinent advertising related attributes.

The attributes drawn from previous advertising literature (e.g., Korgaonkar et al., 2001; Ducoffe, 1996) and were enjoyable, offensive, informative, deceptive, annoying, useful, good for the economy, intrusive and invades privacy.

Demographics

Additionally, demographic information such as gender, education, age, the extent of Spanish spoken at home, preference for speaking the Spanish versus English language and country of birth was collected.

Analysis and Results

Table 2 shows the specific items used to measure beliefs, attitudes and behavior toward each of the three types of Web advertising. Additionally, it shows differences, if any, between email, popup, and banner advertising with regards to the statements. As Table 1 indicates the three samples for the Web advertising were significantly different in terms of age ($p<.05$) and the extent of Spanish language spoken at home ($p<.10$).

Thus the results in Table 2 are reported after controlling for these variables.

The overall results are supportive of hypothesis one. As we examine the overall attitudes there were significant differences ($p<.05$) to the statement responses "Do you consider each one of the three types of Web advertisements good or bad?" The popup ads scored the least favorable response when compared to both the e-mail and banner ads. The pop-ups are the least favored because

Table 3. Attitudes and behavior consistency correlation coefficients for email, popup, and banner advertising

Email advertising (Sig. Level) 0.399 (p<.001)	Popup advertising (Sig. Level) 0.388 (p<.001)	Banner advertising (Sig. Level) .279 (p<.001)

of the inability of the Web surfers to anticipate, control them or delete them. The pop-ups are the most intrusive among the three types of web advertising. Consequently, perhaps they are also the most annoying. The responses to all three behavior statements regarding the three types of ads show that the pop-ups are significantly (p<.05 or better) the least useful. This is consistent with a study suggesting that the visitors to a Web site will adopt more negative attitudes because of intrusive ads such as pop-ups and pop-unders (McCoy et al., 2007).

The responses to the beliefs statements reveal a similar pattern. With the exception of two belief categories, five of the belief categories indicated responses to popup are the least favorable. The exceptions are the responses to the constructs of value corruption and falsity/no sense.

On these two constructs the e-mail advertising fairs the least popular method of Web advertising (p<.1 or better). These e-mail advertisements, although intrusive, could be deleted without having to read them. Additionally, once identified as spam by the Internet service provider, they are routinely deposited in a spam folder by most providers.

The responses to the statement, "When you see an ad how often do you click on it to find more information?" received the most positive response to e-mail ads and the least to the pop-ups, with banner ads in the middle (p<.01). Overall, the banner ads faired the best in comparison to the other two. These are less intrusive (p<.05), more pleasurable (p<.01), more informative (p<.01), more trustworthy (p<.1 or better) and better for the economy (p<.01 or better) when compared to e-mail and popup ads. However, as expected, the banner ads attract the least attention when compared to the other two (p<.01).

Table 3 shows the correlation coefficients between the attitude score of the summated two-item scale and the behavior score of the summated three item score with significance levels for each type of advertising.

The results are supportive of hypothesis two. All three coefficients are statistically significant at the 0.001 level. This again underscores the rationale for developing positive attitudes by studying the underlining beliefs towards each type of method of Web advertising. Table 4 shows the mean values for TV advertising vis-à-vis each of the three types of advertising for the eight attributes with tests for significance differences.

Overall, the television advertisements fare well in comparison to the Web based advertisements. The results show that Hispanic consumers in this study value television advertising for information, usefulness, and enjoyment aspects. The subjects also consider television ads to be less offensive, less deceptive, less annoying, and less invasive than the three types of Web advertising. This bodes well for traditional television advertising.

Turning to the differences among the three types of the Web ads, significant differences (p<.1 or better) were found for five out of eight constructs.

The television advertisements were considered to be the least offensive in comparison to the popup ads (p<.05), the most informative in comparison to the banner ads (p<.05), the most deceptive in comparison to the banner ads (p<.1), less annoying than the popup ads (p<.05), and more useful than the popup ads p<.1).

Table 4. Compared to ________ advertisements, how will you evaluate TV advertisements: Are TV advertisements more, the same or less enjoyable, informative ...etc?

	Email advertisements mean score*	Popup advertisements mean score*	Banner advertisements mean score*	Significance level
Enjoyable	1.2195	1.2643	1.2846	NS
Offensive	2.3902	2.5071	2.3408	p< 0.05
Informative	1.5772	1.4214	1.4120	p< 0.05
Deceptive	2.2114	2.1786	2.0861	p< 0.10
Annoying	2.4146	2.4714	2.3296	p< 0.05
Useful	1.6179	1.4646	1.5094	p< 0.10
Good for US economy	1.5935	1.5750	1.6854	NS
Invades my privacy	2.5935	2.5929	2.5468	NS

*Scale Values: 1 = More, 2 = About the same, 3 = Less

These results could be useful to advertisers using one or more of these Web advertising methods to reach Hispanic consumers. Advertisers should be aware of the differences between each of the three Web methods on various important attributes. Additionally, results indicate that television ads are still popular with the Hispanic consumers.

CONCLUSION

To conclude, U. S. Hispanic consumers' usage of the Web is on the rise and growing at a much faster rate than the U. S. non- Hispanic population. According to Leggatt (2007) the online spending of Hispanics is on the rise. However, usage does not necessarily translate into positive attitudes towards all types of Web advertising. This finding negates a suggestion by Ducoffe (1996) that media context influences the value of advertising. Specifically, the Hispanic consumers in this study favored e-mail advertising and banner advertising over popup advertising.

E-mail advertising was favored because of its attention getting value was perceived to be higher than banner ads and popup ads respectively. The banner ads are believed to be the most enjoyable by the Hispanics in our sample. Given that the banner ads are more prevalent than the email and pop ups there is usually a concern of overexposure leading to a potential backlash. The results of this study suggest that Hispanics continue to favor banner advertising. This is encouraging given that banner advertising is reported to affect purchasing patterns over the Web (Manchanda et al., 2006). The popup ads are considered the least effective of the three advertising methods. Compared to the other two, popup advertisements interrupt browsing activity. It forces the audience to physically as well as cognitively pay attention. While emails and banner ads could be potentially avoided it is not true for pop-ups.

Another important finding was a significant correlation between attitudes and behaviors towards all three types of Web advertising. This provides the nomological validity of the two scales. Finally, Hispanics continue to enjoy advertising on television more than any of the three types of Web advertising. Although the Internet shows great promise in expanding outreach designed to tap the Hispanic market, the results indicate that television advertising still captures the minds and hearts of this segment of the population.

A recent article in the Wall Street Journal (Steel, 2009) reported that Google and ad agency WPP are funding research on how traditional and digital ads work together to influence consumer behavior. This study is a small step in that direction. The study attempts to fill a void in the published literature by examining evaluations of three types of

Web advertising by the growing Hispanic segment of the U.S. population (Leggatt, 2007; Michman et al., 2003). The study results are limited to one geographic region and could be expanded to other Hispanic enclaves such as California and/or Texas in future studies to improve the generalized findings to a wider market of the Hispanic population.

REFERENCES

Alwitt, L. F., & Prabhaker, P. R. (1992). Functional and Belief Dimensions of Attitudes Toward Television Advertising: Implications for Copytesting. *Journal of Advertising Research, 32*(September/October), 30–42.

AOL. (2006). *16 Million Hispanics Use Internet.* Retrieved from http://www.imnewswatch.com/archives/2006/09/aol_16_million.html

Bauer, R. A., & Greyser, S. A. (1968). *Advertising in America: The Consumer View*. Boston: Harvard University Press.

Bellenger, D. N., & Valencia, H. (1982). Understanding the Hispanic Market. *Business Horizons, 25*(3), 47–50. doi:10.1016/0007-6813(82)90128-8

Bridis, T. (1998, August 27). New users battle Web Advertising. *Detroit News,* p. 6.

Burns, E. (2005). *Hispanic Online Users Adopt Broadband.* Retrieved from http://www.clickz.com/3521301

Cartagena, C. (2005). *Latin Boom! Everything You Need to Know To Grow Your Business in the U.S. Hispanic Market*. New York: Ballantine Books.

Cho, C. H., Lee, J. G., & Tharp, M. (2001). Different Forced-Exposure Levels to Banner Advertisements. *Journal of Advertising Research, 41*(4), 45–56.

Cox, J. (2004). Getting Connected. *Hispanic Business Magazine*. Retrieved December from http://www.hispanicbusiness.com/news/2004/11/22/getting_connected.htm

Cross, K. (1999). *Wither the Banner.* Retrieved from http://www.business2.com/b2/web/articles

Deshpande, R., Hoyer, W. D., & Donthu, N. (1986). The Intensity of Ethnic Affiliation: A Study of the Sociology of Hispanic Consumption. *The Journal of Consumer Research, 13*(September), 214–220. doi:10.1086/209061

Ducoffe, R. H. (1996). Advertising Value and Advertising on the Web. *Journal of Advertising Research, 36*(5), 21–35.

Fones, M. (1981, April 6). Two Worlds Together: Towns Along the US-Mexico Border Could Provide Marketing Edge. *Advertising Age*, 8.

Gordon, M. (1981). Models of Pluralism: The New American Dilemma. *The Annals of the American Academy of Political and Social Science, 4*(54), 178–188. doi:10.1177/000271628145400115

Guernica, A., & Kasperuk, I. (1982). *Reaching the Hispanic Market Effectively*. New York: McGraw-Hill.

Hernandez, S. A., & Newman, L. M. (1992). Choice of English vs. Spanish Language in Advertising to Hispanics. *Journal of Current Issues and Research in Advertising, 14*(2), 1–11.

Hoffman, D., & Novak, T. (1996). Marketing in Hypermedia Computer Mediated Environments: Conceptual Foundations. *Journal of Marketing, 80*(4), 50–68. doi:10.2307/1251841

Interactive Advertising Bureau. (2006). *Reach U.S. Hispanics Through Online Marketing.* Retrieved from http://www.iab.net/insights_research/iab_research/1666

Kaufman, C., & Hernandez, S. (1990, October/November). Barriers to Coupon Use: A View from the Bodega. *Journal of Advertising Research*, *30*, 18–25.

Korgaonkar, P. K., Karson, E. J., & Lund, D. (2000). Hispanics and Direct Marketing Advertising. *Journal of Consumer Marketing*, *17*(2), 137–157. doi:10.1108/07363760010317204

Korgaonkar, P. K., Silverblatt, R., & O'Leary, B. (2001). Web Advertising and Hispanics. *Journal of Consumer Marketing*, *18*(2), 134–150. doi:10.1108/07363760110386009

Leggatt, H. (2007). Online Spending Power of Hispanic Population Increasing. *Bizreport*. Retrieved from http://www.bizreport.com/2007/07/onlinespendingpowerofHispanicpopulationincreasing.html

Magazine, C. (2007). *Online U.S. Hispanics are Internet Trend Setters*. Retrieved from http://www.contactomagazine.com

Manchanda, P., Dube, J. P., Goh, K. Y., & Chintagunta, P. K. (2006). The Effect of Banner Advertising on Internet Purchasing. *JMR, Journal of Marketing Research*, *43*(February), 98–108. doi:10.1509/jmkr.43.1.98

Marin, G. (1989, December). Barbara VanOss Marin, Regina Otero-Sabogal, Fabio Sabogal, and Eliseo J. Perez-Stable, The Role of Acculturation in the Attitudes, Norms, and Expectancies of Hispanic Smolers. *Journal of Cross-Cultural Psychology*, *20*, 399–415. doi:10.1177/0022022189204005

Marin, G., & Marin, B. V. (1991). *Research with Hispanic Populations*. Thousand Oaks, CA: SAGE.

Martin, B. A. S., Van Durme, J., Raulas, M., & Merisavo, M. (2003). Email Advertising: Exploratory insights from Finland. *Journal of Advertising Research*, *43*(September), 293–300.

McCoy, S., Everard, A., Polak, P., & Galletta, D. F. (2007). The Effects of Online Advertising. *Communications of the ACM*, *50*(3), 84–88. doi:10.1145/1226736.1226740

Michman, R. D., Mazze, E. M., & Greco, A. J. (2003). *Lifestyle Marketing: Reaching the New American Consumer*. Westport, CT: Praeger Publishers.

Mittal, B. (1994). Public Assessment of TV Advertising: Faint Praise and Harsh Criticism. *Journal of Advertising Research*, *34*(February), 35–53.

Muehling, D. D. (1987). An Investigation of Factors Underlying Attitude-Toward-Advertising In-General. *Journal of Advertising*, *16*(1), 32–40.

Nadilo, R. (1998). On-Line Research Taps Consumers Who Spend. *Marketing Research*, *8*(June), 12.

Olmedo, E. L., & Padilla, A. M. (1978). Empirical and Construct Validation of a Measure of Acculturation for Mexican Americans. *The Journal of Social Psychology*, *105*, 179–187.

Pollay, R. W., & Mittal, B. (1993). Here's the Beef: Factors, Determinants and Segments of Consumer Criticism of Advertising. *Journal of Marketing*, *57*(July), 99–114. doi:10.2307/1251857

Rogers, J. L. (1989/1990, December/January). Consumer Response to Advertising Mail. *Journal of Advertising Research*, *29*, 18–24.

Roslow, P., & Roslow, S. (1980, June). How U.S. Latins Use Radio. *Journal of Advertising Research*, *20*, 19–22.

Rossman, M. (1994). Multicultural Marketing: Selling to a Diverse America. *American Management Association*, *1994*, 46–80.

Segal, M. N., & Sosa, L. (1983). Marketing to the Hispanic Community. *California Management Review*, *26*(1), 120–134.

Steel, E. (2009, March 18). WPP, Google to Fund Web-Ad Research. *Wall Street Journal*, B6.

Sun, T., & Johnson, A. (2009, March 4). *In down times, Hispanic Market is booming.* Retrieved from http://www.msnbc.msn.com/id/29470143

Tellis, E., & Ortiz, V. (2008). *Generations of Exclusion: Mexican Americans, Assimilation, and Race*. New York: Russell Sage Foundation.

U.S. Census Bureau. (2006). *U.S. Hispanic Population: 2006.* Retrieved from http://www.census.gov/population/socdemo/hispanic/cps2006/CPS_Powerpoint_2006.pdf

U.S. Census Bureau. (2008). *Hispanic Heritage Month 2008: Sept. 15-Oct.15.* Retrieved from http://www.census.gov/Press-Release/www/releases/archive/cb08ff-15.pdf

Valdes, I. (1991, October). Understanding Hispanic Consumers. *American Demographics*, S1–S4.

Valdes, I. M., & Sadane, M. H. (1995). *Hispanic Market Handbook.* Detroit, MI: Gale Research Inc.

Valencia, H. (1989). Hispanic Values and Subcultural Research. *Journal of the Academy of Marketing Science, 17*(winter), 23-28.

Webster, C. (1992). The Effects of Hispanic Subcultural Identification on Information Search Behavior. *Journal of Advertising Research, 32*(September/October), 54–62.

Yankelovich, S. (1981). *Spanish USA: A Study of the Hispanic Market in the United States*. New York: White, Inc.

Yinger, J. (1994). *Milton, Ethnicity: Source of Strength? Source of Conflict?*New York, NY: State University of New York Press.

This work was previously published in International Journal of E-Business Research Volume 6, Issue 2, edited by In Lee, pp. 45-58, copyright 2010 by IGI Publishing (an imprint of IGI Global).

Chapter 4
A Trade Value Perspective on Ecommerce Research: An Integration of Transaction Value and Transaction Cost Theories

Ruth C. King
University of North Carolina at Greensboro, USA

Ravi Sen
Texas A&M University, USA

Fergle D'Aubeterre
Flint Transfield Services Ltd (FT-SERVICES), Canada

Vikram Sethi
Wright State University, USA

ABSTRACT

The impact of web-based electronic commerce on the process of disintermediation and re-intermediation has been extensively studied. Two major limitations of the existing work are the focus on a single economic measure (i.e., transaction cost minimization) and the examination of channel-structure decisions from only a single perspective (the seller's). This paper introduces transactional value theory in the context of channel-structure research and integrates it with transaction cost theory to generate a trade value framework. The trade value framework considers channel-structure decisions from the perspectives of both buyers and sellers and is used to analyze the impact of web-based e-commerce on intermediated channel-structures. The proposed framework suggests that intermediaries function best in a channel-structure if they can reduce trade-inhibiting factors and improve trade-enhancing factors. Intermediaries may also prosper if they deliver extraordinary value on one side of the trade value framework to the point that inhibiting factors on the other end of the trade can be overlooked. Intermediaries maximize the value of the trade for both the buyers and the sellers by trading through an intermediated channel-structure as opposed to trading directly.

DOI: 10.4018/978-1-61350-462-8.ch004

INTRODUCTION

Electronic commerce (EC) is a significant and pervasive phenomenon that is impacting not only organizations and societies (Numberger & Rennhak, 2005) but also the structure of markets and the economics of distribution channel-structures (Bakos et al., 2005; Benjamin & Wigand, 1995; Eng, 2008; Tsay & Agrawal, 2004). More specifically, the impact of EC on distribution channel-structures is of particular interest to intermediaries that operate in traditionally intermediated offline channels and sellers who are seeking an optimal distribution channel-structure in a changing economic landscape. For example, the Internet provides opportunities for web-based EC that may (a) result in the elimination of existing offline intermediaries, a process commonly called *disintermediation*, (b) provide new opportunities for existing intermediaries, labeled *re-intermediation,* and/or (c) provide opportunities for new forms of intermediaries, called *intermediation*. Multiple arguments exist in the literature that predicts each of these outcomes; this has tempered the predictive ability of any individual theory. This paper focuses on two principal reasons for these incompatible predictions: (1) the limited applicability of transaction cost theory (Williamson, 1991) in analyzing the impact of web-based EC on traditionally offline channel-structures and (2) the inherent shortcomings of transaction cost theory in discounting the effect of channel-structure on product demand.

In addition to using different definitions for intermediaries, the extant literature that examines the impact of EC is predominantly based on transaction cost (TC) theory and argues that information technology (IT) enables the elimination of factors responsible for increasing transaction costs for sellers and buyers (e.g., Benjamin & Wigand, 1995; Gehrig, 1993; Sarkar et al., 1995). However, most of the analysis presented in these studies apply TC theory from the seller's perspective and are predominantly focused on the reduction of transaction costs through reduced search costs. Other factors, such as maintaining long-term supplier loyalty, that may contribute towards increasing transaction costs are generally ignored. The scope of transaction costs to the buyer also needs to be extended to encompass more than just search costs to make the analysis more comprehensive. Therefore, we posit that *intermediaries are present in a trading structure to improve the trade-enhancing factors and reduce the trade-inhibiting factors for both buyers and sellers, and they will continue to exist only if they do not decrease the trade value that the buyers and the sellers could have had by trading directly*. We propose that the optimal trading structure will result from the simultaneous consideration of a set of relevant transaction costs and transactional values for both buyers and sellers.

Transaction cost theory (Williamson, 1975) has its own inherent limitations that hinder a comprehensive analysis of the issue. A major limitation of TC theory is its underlying assumption that organizational forms (i.e., the channel-structure) do not have a significant impact on the demand for a product (Zajac & Olsen, 1993). According to TC theory, sellers choose a channel-structure that minimizes transaction costs and therefore maximizes their profits. However, if the demand for a product is dependent on the channel-structure, choosing a channel-structure to minimize transaction costs will necessarily result in maximum profits.

To address these concerns, this paper introduces the *trade value (TV) framework* to complement transaction cost theory, which enables a more holistic analysis of the impact of EC on channel-structures and develops a theoretical understanding of effective digital intermediation as suggested by Kauffman and Walden (2001). An integrative framework is proposed by building upon existing research in (1) transaction cost theory, (2) transactional value theory, and (3) literature concerning distribution channels and consumer behavior. This integrative approach applies a multi-paradigm lens (Lewis & Grimes, 1999) to analyze complex phenomenon, resulting in a more comprehensive understanding of trade value in e-commerce intermediaries.

This paper contributes to the current channel-structure research in three significant ways. First, we examine the current channel-structure research from multiple research disciplines to create a more comprehensive understanding of the accumulated research tradition. These multi-disciplinary perspectives are valuable in understanding the multi-dimensional view of e-commerce as suggested by Dwivedi et al. (2008). Second, we introduce transactional value theory as an alternative but complementary lens to examine intermediaries; this provides a more holistic understanding of channel-structure research. Third, the development of a new trade value framework is valuable in guiding future research into channel-structure strategy.

EXISTING LITERATURE ON CHANNEL-STRUCTURES

To investigate the nature of channel-structure transformation, we have analyzed the relevant literature from the disciplines of (1) industrial organization, (2) marketing, and (3) information systems.

The Industrial Organization (IO) literature suggests that some activities in the trading process (notably searching) do not add value to the trade for either buyer or seller; these activities are referred to as *trade frictions* (Rubinstein & Wolinksky, 1987). Intermediaries in the trading (or channel) structure reduce these trade frictions (Brynjolfsson & Smith, 2000; Gehrig, 1993) and extract a surplus in return for shortening the period that buyers sellers and have to wait to complete a transaction (Rubinstein & Wolinksky, 1987). In cases where both a direct and an intermediated channel-structure co-exist, the supply and demand in the direct channel-structure (e.g., web-based channel) and the intermediated channel-structure (e.g., traditional offline channels) will be '*interdependent*' (Fingleton, 1997). An increase in the ask price from the direct channel-structure (i.e., the price charged to the customers) brings more buyers to the intermediary and fewer buyers to the direct seller. With fewer buyers trading directly, the profits for the seller from direct trade are reduced, causing some sellers to move to the intermediated channel (i.e. sell through the intermediary), even if the intermediary's bid price (i.e., the price paid to the seller) is unchanged. However, if the buyers and sellers can find ways to reduce trade frictions via direct trade, then intermediaries may have to lower their ask price further, resulting in lower margins. It is possible their margins may be reduced to such an extent that they do not find it financially viable to participate in the trading process.

In the context of web-based EC, it has often been stated that the Internet helps to reduce trade frictions; therefore, Internet-based retailers are able to charge lower prices than their counterparts (Brynjolfsson & Smith, 2000). Moreover, academics and practitioners generally agree that the Internet reduces the search cost between buyers and sellers (Chu et al., 2005; Grover & Teng, 2001). Applying the *interdependence* argument from above (Fingleton, 1997), one can postulate that to compete with a web-based direct channel-structure (with lower trade frictions), traditional intermediaries will have to lower their selling price, which in turn reduces their profit margins. If they must continue to lower their ask price, they will inevitably be squeezed out of the trading structure since an insufficient profit margin would exist to cover their operational costs. In summary, the IO literature suggests that if the Internet lowers search costs, then traditional offline intermediated channel-structures will be replaced by online direct channel-structures.

The marketing literature provides the most comprehensive description of the intermediary's role in a channel-structure. According to Sarkar et al. (1995), intermediaries are "organizations that support exchanges between producers and consumers, increasing the efficiency of the exchange process by aggregating transactions to create economies of scale and scope." In addition, intermediaries assist companies in obtaining access to new markets without those companies committing significant resources (Easingwood & Coelho, 2003) and online intermediaries (i.e., "infomediaries") reduce information asymmetries between buyers and sellers by providing unbiased information about product characteristics and prices (Chu et al., 2005). Services that intermediaries render can be classified into (1) matching services (listing, matching, and discovery) and (2) value-added services (workflow coordination, transaction management, and account management) (Bhargava & Choudhary, 2004). In short, the marketing literature suggests that intermediated channels exist because they (1) are economically more *efficient* than direct channel-structures, (2) expand market coverage, (3) provide routinization and support of transactions, (4) facilitate the search process and reduce asymmetric information, and (5) smooth the flow of goods and services by categorizing them, accumulating them, and bundling them (Bakos et al., 2005). Marketing researchers would argue that although intermediaries can be eliminated, these functions themselves are integral to any trade. When the institutions providing these functions are eliminated (disintermediation), then the responsibility for the functions provided by the intermediary is shifted either forward (toward the buyers) or backward (toward the producers). Thus, web-based e-commerce will lead to disintermediation only if Internet-based applications help sellers and buyers perform the aforementioned intermediary functions.

Despite the promise of Internet technology, not all channel participants have the capability to fully utilize and benefit from it. For example, sellers who have traditionally operated through offline intermediated channels may lack the expertise required for an online channel-structure and may find it difficult to provide a smooth flow of goods and services in a direct online channel (Bain & Company, 2002). In addition, investments in offline intermediated channel-structures make it economically prohibitive for many sellers to completely switch to online channel-structures (either direct or intermediated). Therefore, the marketing literature would suggest that the Internet may lead to the automation of some trading processes but not to complete disintermediation (Sarkar et al., 1995).

The information systems literature in EC research suggests that Internet technology may lead to (1) complete disintermediation, (2) more intermediation, and/or (3) re-intermediation. On one hand, Internet-based e-marketplaces leverage IT to directly match buyers and sellers with increased effectiveness and lower transaction costs, which leads to more efficient and "friction-free" markets. The role of the intermediary may be reduced or even eliminated leading to "disintermediation" as a result (Benjamin & Wigand, 1995; Malone et al., 1989). However, based on empirical studies conducted across several industries, electronic hierarchies seem to be more common than would be expected from the Malone el al (1987) electronic market hypothesis (Bakos & Brynjolfsson, 1993; Grover & Teng, 2001; Wigand et al., 2005). Moreover, disintermediation might not be the only possible outcome since EC has facilitated new roles for intermediaries, such as supply and demand aggregation, trust provision, facilitation, and matching (Grover & Teng, 2001; Zwass, 2003). Complete disintermediation is unlikely because many buyers continue to prefer familiar offline intermediated channel-structures. Bergen, Kauffman, and Lee (2005) explain that a new phenomenon is emerging where traditional firms are first disintermediated by Internet firms and later forced into a process of re-intermediation to stay competitive in the channel-structure. This outcome is termed as *re-intermediation* (Chircu & Kauffman, 2000; Giaglis et al., 2002), wherein the traditional firms develop EC capabilities and start competing in electronic markets as well. The keys to successful re-intermediation for an intermediary are whether it is able to leverage its assets in the traditional market and reconfigure them to be effective in an EC-enabled marketplace (Chircu & Kauffman, 2000). Traditional intermediaries are re-emerging because they still add value to the trading process (Bergen et al., 2005). Therefore, it is evident that the EC literature provides support for all possible outcomes; i.e. intermediation, disintermediation, and re-intermediation. In short, the phenomena of intermediation and/or disintermediation are complex and dependent on various factors like market characteristics and competition intensity. These complexities make predicting which market structure will survive in a particular channel-structure an uneasy task (Kauffman & Walden, 2001; Smits & Janssen, 2008). Appendix A presents major studies and research work from various disciplines and conclusions drawn about the impact of EC on channel-structures.

TRANSACTION COST AND TRANSACTION VALUE PERSPECTIVES

The main tenets of transaction cost (TC) theory focus on 'transactions and the costs that attends completing transactions by one institutional mode rather than another' (Williamson, 1975). The basic argument of transaction cost theory is that transactions will be executed in a manner that minimizes costs. TC theory identifies three forms of transaction governance (market, hybrid, and hierarchy) with each form supported and controlled by a different type of contract law and coordination system. Transactions are characterized by asset specificity, uncertainty, and frequency. TC theory predicts that transactions characterized by low asset specificity will be executed in *market* forms of governance while transactions with high asset specificity will be executed within a *hierarchical structure* (Williamson, 1991).

TC theory is considered one of the leading theories in management and organizational studies; however, controversies about its empirical validity exist (David & Han, 2004). For instance, Dyer (1997) found that although Japanese automakers were more specialized (which implies a high level of asset specificity), they incurred lower transaction costs than their U.S. counterparts. Interestingly, these findings suggest that transaction costs do not increase with investments in specific assets – a finding that runs contrary to the predictions of TC theory. David and Han (2004) conducted a systematic assessment of the empirical evidence for TC theory. Their findings suggest that the predictive power of TC theory is only 47 percent. Most TC theory research has focused on asset specificity and uncertainty while neglecting other important TC constructs. More recently, Carter and Hodgson (2006) analyze 27 empirical articles regarding TC theory, 12 of which are related to vertical integration and 15 to hybrid relationships. No single article regarding vertical integration was consistent with Williamson's TC theory, and only partial consistency with Williamson's TC theory was found across other topics such as hybrid and hierarchical relationships.

Despite the controversies surrounding TC theory, the debate in electronic commerce, industrial organization and marketing literature on distribution channels has nevertheless been profoundly influenced by transaction cost theory. From a TC theory perspective, a seller who chooses to trade through some form of intermediary does so because the intermediated channel-structure is the most efficient mode of organizing that particular economic transaction. However, it has been recognized that another approach to explaining economic organization and governance is given by a resource-based (RB) view, which focuses on the intelligent management of a firm's rare, imperfectly imitable, and non-substitutable resources in the pursuit of sustainable competitive advantage (Barney, 1991). From this perspective, intermediaries provide buyers and sellers with resources that they are unable to purchase through market transactions or develop internally in a timely and cost-effective manner.

Transaction Value Theory provides another lens to approach these issues. Transaction value theory considers how to structure exchange relations in order to maximize transaction value. It argues that organizations transact as they do because they anticipate the generation of increased value for all transacting players (Zajac & Olsen, 1993). Extending this idea to the channel-structures context, buyers and sellers trade through an intermediary because the value generated through an intermediated channel-structure is more than that generated through a direct channel-structure. The advantage of transaction-value (TV) and resource-based (RB) theory over transaction cost (TC) theory is best illustrated by the following example:

A firm's expected profits from a transaction can be expressed as a simple linear function of its expected revenues from the transaction and the cost of that transaction.

Expected Profits (Seller) = Expected Revenues – Seller Transaction Costs

The TC approach implicitly assumes that demand is independent of the channel-structure used, in other words a modification in the channel-structure will not affect *expected revenues*. Under this assumption, a buyer's purchasing behaviors are independent of or have no impact on the channel-structure used. Therefore, sellers have to concentrate only on reducing their own *transaction costs* in order to maximize profit. Let us assume that a seller decides to attempt to lower transaction costs by switching to a different channel-structure (e.g., switching from an offline intermediated channel-structure to an online direct channel-structure). According to the TC theory, this new channel-structure should result in increased profits for the seller. However, in this scenario the reaction of buyers to the new channel-structure is not factored into the equation. If the new channel does not maintain or improve the value that the traditional intermediated channel has offered to buyers, they may not be willing to trade through it (Sen & King, 2003). Therefore, the new direct channel-structure may actually lead to a loss in revenues even though the seller's transaction costs have been reduced. Figure 1 shows that buyers and sellers need to take into account not only the transaction costs associated with a channel-structure but also their expected revenues before deciding on whether to switch or diversify.

Transactional value (TV) theory allows for a more comprehensive analysis since, unlike TC theory, it takes into account the reaction of buyers and the resulting impact on sellers' net profits. According to the RB and TV approaches, inter-

Figure 1. Possible scenarios for a buyer/seller who is thinking about changing its current channel structure

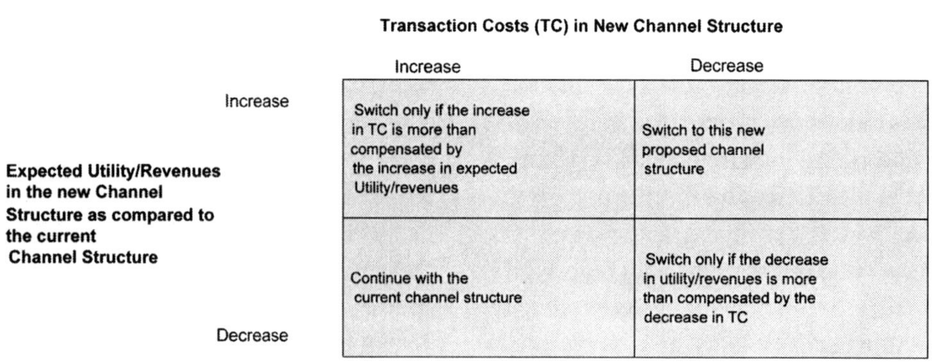

mediaries will be present in a channel-structure as long as they *reduce transaction costs* for the sellers and *discourage buyers* from switching to a direct channel. Thus, sellers will continue to trade through an intermediated channel-structure as long as the reduction in transaction costs provided by the alternative (direct channel) is insufficient to compensate for the resulting loss in revenues brought about by the new channel-structure.

TV theory was originally proposed in the context of inter-organizational collaborations such as joint R&D by two organizations or the development of communication standards among various industrial players (Zajac & Olsen, 1993). This paper extends this framework to include transactions between organizations and individuals since in both instances the trade objectives of participating parties are similar: value maximization from the transaction. Using the core elements of TC and TV theory, we propose the *trade value framework*, which can be used to comprehensively analyze the impact of the Internet on offline intermediated channel-structures and channel-structures supporting business-to-business (B2B), business-to-consumer (B2C), or consumer-to-consumer (C2C) transactions.

TRADE INHIBITORS AND TRADE ENHANCERS

Research has identified three main functions that markets perform: matching buyers and sellers, facilitating transactions, and providing the institutional infrastructure for business. In other words, markets must exist so that trade can successfully take place. However, the *value* of a trade (for both the buyers and the sellers) is affected by several factors, and these factors can be broadly classified as (a) those that decrease this value and therefore inhibit the trading process (*trade inhibitors*) and (b) those that increase the trade value and therefore encourage the trading process (*trade enhancers*).

It is critical to highlight two significant issues pertinent to our classification of these factors. First, a careful analysis of trade-inhibiting and trade-enhancing factors will show that a reduction in trade-inhibiting factors within a channel does not necessarily mean that trade-enhancing factors are improved. The levels of trade-enhancing and trade-inhibiting factors within a channel are not necessarily correlated. For example, both catalogue-based direct channels and web-based direct channels could have relatively low levels of trade friction (i.e., low search cost). However, their trade-enhancing factors differ: catalogue-based direct channels offer location and product transparency while web-based direct channels offer price transparency as well. The second issue relevant to our classification of trade-inhibiting and trade-enhancing factors is our deliberate consideration of the selling prices of products and services as part of the channel-characteristic rather than an individual characteristic controlled by supply and demand. However, we do take price into account when we are discussing the value that a seller and a buyer extract from the trade. The following section elaborates *the trade value framework* and explains various components of this framework.

Trade Inhibitors

Trade inhibitors discussed in the literature include *search costs* and the costs for the *transaction support* that is required to smooth the trading process. The magnitude of trade inhibitors is unique for each channel-structure. For any channel-structure, expected value is reduced by the amount of transaction costs required to overcome these *trade inhibitors*. Therefore, buyers and sellers are discouraged from using channel-structures that exhibit high trade-inhibiting factors. To identify and classify trade-inhibiting factors, we build upon the literature on channel-structures. However, unlike the existing literature, we consider trade inhibitors from the perspective of both buyers and sellers. Based on our research from the marketing and IS literature, we have identified the following as the main trade inhibiting factors: *location, high search costs, information asymmetry, high perceived risk, and inferior transaction support.*

Location

As identified by Krugman (1993), the spatial distribution of the demand has a direct effect on the trade. Physical location can be a major trade inhibitor that may prevent buyers and sellers from discovering each other or may make the discovery process so expensive as to render the trade undesirable. For instance, industries characterized by a large number of buyers and sellers require intermediaries such as retail stores to bring location transparency to the trade. In such industries an intermediated channel-structure is more likely than a direct channel-structure because it reduces the costs involved in bringing together buyers and sellers who are geographically separated and thereby encourages them to participate in the trade. The physical location of buyers and sellers no longer acts as an inhibitor to trade due to the location transparency provided by the Internet. Therefore, web-based direct channel-structures threaten to replace traditional intermediated channel-structures (Benjamin & Wigand, 1995) that did nothing more than eliminate location-related trade inhibitors.

High Search Costs

Costly information search associated with any trade is another factor that may discourage buyers and sellers from doing business (Gehrig, 1993). It has been established that high search costs for buyers can increase advertising costs for sellers (Kauffman & Walden, 2001). Therefore, channels that minimize search costs for buyers create more value for sellers. In this context, when electronic markets are compared with physical markets, the former are more efficient in reducing search costs (Brynjolfsson & Smith, 2000). If an intermediated channel-structure reduces search costs for buyers and sellers by consolidating buyers (e.g., bargain clubs), sellers (e.g., real estate agents), or both (e.g., stock exchange), then intermediated channel-structures will dominate over direct. On the other hand, as the Internet makes it possible for direct channel-structures to incur lower search costs, it is likely that buyers and sellers will switch to a web-based direct channel (Benjamin & Wigand, 1995; Gupta et al., 2004). However, this will occur only if both buyers and sellers are capable of leveraging Internet technology to reduce their search costs. For instance, even though online searching techniques (e.g., search engines, shopping agents) have undergone tremendous technological and usability improvements, not all potential e-commerce consumers are aware of them, let alone know how to benefit from them. Research has emphasized the need for increased development in search engine functionality and terminology to alleviate difficulty for average online consumers (Spink & Jansen, 2008).

Information Asymmetry

An information asymmetry (or information gap) between buyers and sellers can also reduce the amount of trading that takes place. Intermediaries can reduce information asymmetry by aggregating demand and/or product information (Bailey & Bakos, 1997). For buyers, this information asymmetry is reduced by information brokers such as trade magazines that provide product and seller-related information. For sellers, market research firms alleviate information asymmetry by providing insights on consumer behavior and information about target markets. The amount of information available on the Internet has changed the problem of information asymmetry from insufficient information to a glut of information, especially for buyers. This glut of too much unreliable information has become a new inhibitor for trade participants. As a result, *information brokerage sites*, which provide buyers with summaries of relevant information, have emerged (Zhou et al., 2007). In addition, recent research into infomediary and intelligent agents has shown that their use can increase consumer welfare (Wang & Benbasat, 2007).

High Perceived Risk

The concept of trust has always existed in business but its form has evolved from product trust (Doney & Cannon, 1997) and trust in service (Jarvenpaa et al., 2000) to trust in relationships in the networked economy (Doney & Cannon, 1997). In offline intermediated channel-structures, the intermediaries undertook the responsibility to engender trust (Wang & Benbasat, 2007), thereby reducing the risks associated with the trade (Sarkar et al., 1995). The Internet is an anonymous, impersonal medium that lacks many established trust mechanisms, such as physical storefronts, thereby increasing risks to all parties. Intermediaries must become agents of trust by protecting buyers and sellers from each others' opportunistic behaviors (Bailey & Bakos, 1997; Giaglis et al., 2002).

Sellers have to invest in risk management (e.g. providing a secure payment system) and buyers have to minimize their risk by using special mechanisms (e.g. using credit cards with online fraud protection). Furthermore, buyers must make an extra effort to verify the reliability of online sellers in order to develop a sense of security when dealing with them. This uncertainty increases the cost of minimizing risk associated with online channel-structures and may discourage some buyers from participating in online trading.

Inferior Transaction Support

Sellers are not always qualified to provide the services needed for successful completion of a trade (payment processing, shipping, after-sales service, installation, etc.). A lack of economical transaction support services in a channel-structure discourages buyers and sellers from trading through that channel. However, intermediaries can reduce the costs associated with processing and coordination (Bailey & Bakos, 1997) as they do in traditional intermediated channel-structures. The Internet can enable more efficient handling of some transaction support services (for example, order tracking, online payments, CRM) but cannot entirely replace established intermediaries.

Trade Enhancers

Trade-inhibiting factors primarily increase transaction costs associated with trading via a particular channel-structure. However, as suggested by transaction value theory (Zajac & Olsen, 1993), factors that enhance the value of trading through a particular channel-structure are equally significant. Consistent with the literature on consumer behavior and EC, this paper identifies trade enhancers as (1) *convenience*, (2) *recreational value*, (3) *familiarity*, (4) *available legal support system*, and (5) *social value*.

Convenience

Buyers will always prefer channels that they find convenient. Therefore, intermediaries that enhance trade convenience encourage more participation. For buyers, a more convenient channel improves the enjoyment of the shopping experience (Childers et al., 2001; Dillon & Reif, 2006; Lichtenstein & Williamson, 2007). However, convenience may vary according to the individual buyer or seller. For example, offline channels may be more convenient for some buyers due to limited Internet access. On the other hand, technology-savvy shoppers may find new online shopping applications such as computerized sales assistants convenient since it can help focus shoppers towards specific products by customizing the shopper's online product exposure (Shergill et al., 2008). Furthermore, time-starved traditional shoppers may find shopping online extremely convenient due to 24 X 7 availability and accessibility (Xu et al., 2008).

Recreational Value

The recreational value associated with a particular channel-structure refers to how enjoyable or entertaining trading through this channel-structure is for buyers. Some buyers, called *recreational shoppers* in the consumer behavior literature (Belenger & Kargaonkar, 1980), view shopping as an enjoyable activity and therefore want to spend time in offline retail outlets. For these buyers, offline channel-structures carry recreational value. On the other hand, buyers who like to browse the Internet may prefer trading through an online channel-structure, which provides them *recreational* value (Wulf et al., 2006)

Familiarity

Familiarity with the channel also plays an important role in determining expected value from the trade for both buyers and sellers. Sellers that traditionally sold through offline retailers and later attempted to sell through web-based direct channels often did not succeed because of their lack of familarity with direct trade (Williams Group, 2005). Similarly, buyers also prefer familiar channels because of the lower expected risk. Familiarity with the technology of a channel-structure is also an important determinant in choosing it. For example, buyers unfamiliar with computer technology are not likely to become online shoppers. On the other hand, previous successful online shopping experiences have been identified as a key driver for buyers adopting the Internet as a channel (Bhatnagar et al., 2000; Perea y Monsuwe et al., 2004). Analogously, sellers that are unfamiliar with sophisticated Internet-based applications often have problems creating high-quality online trading platforms and providing competent levels of customer support.

Available Legal Support System

Legal regulations may inadvertently enhance trade in a particular channel-structure. For example, strict credit card regulations for buyers in China strengthen offline channel-structures for many consumer products. In the US, the sales tax exemption granted to online trade may enhance the trade value for buyers. Another example is automobile distribution. Some states in the US restrict car manufacturers from selling directly to end consumers if they have offline dealers in that state. This enhances the trade of automobiles through offline intermediated channel-structures and protects these channels from direct online competition. With respect to the legal and regulatory aspects of e-commerce, electronic marketplaces guarantee the transparency of transactions by providing clear rules for trading partners, such as authentication and contract enforcement mechanisms (Giaglis et al., 2002); as a result, e-marketplaces demonstrate less legal friction than their offline counterparts.

Social Value

A long-term relationship among trade participants through a particular channel-structure enhances the trading through that channel-structure. For instance, if a long-term relationship helps suppliers to understand their customers' needs, create better tailored products or deliver higher-quality services, this can often lead to higher customer satisfaction and revenues. Alternatively, trade participants may feel socially obliged to trade through a particular channel-structure. For example, *personalized shoppers*, as defined by Stone (1954), are those who enjoy developing a close relationship with store personnel and tend to shop close to home, and *ethical shoppers* (Tsay & Agrawal, 2004) feel a moral obligation to support local stores. In addition, existing literature recognizes that subjective norms affect the on-line shopper's behavior (Chen & Dhillon, 2003). For example, eBay.com's emphasis on users reviewing

Figure 2. Trade value framework to study channel structure

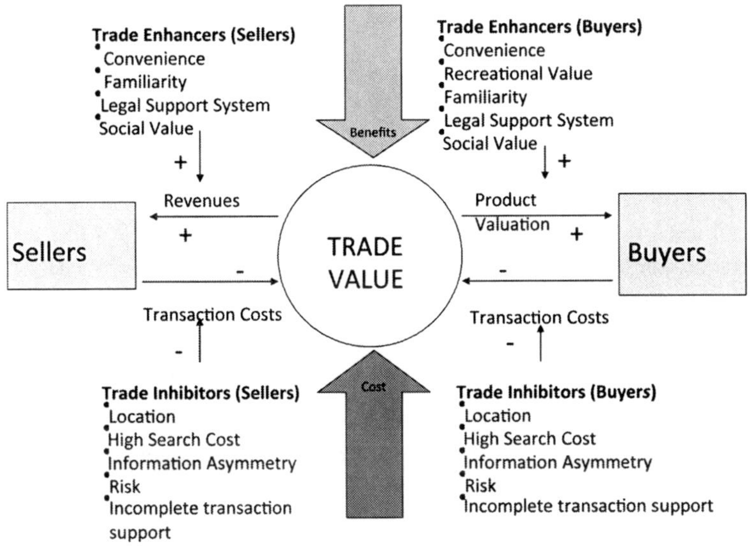

each other generates a community that is highly favorable for encouraging on-line auction trade.

TRADE VALUE FRAMEWORK

The marketing and industrial organization literature suggests that the function of intermediaries is to reduce the economic impact of trade inhibitors whereas the literature on EC suggests that the purpose of intermediaries is to reduce the economic impact of some trade inhibitors while increasing the economic impact of trade enhancers. Our *trade value framework* (Figure 2) proposes that the channel-structure that minimizes the negative impact of trade-inhibiting factors while maximizing the value of trade-enhancing factors for the both buyers and sellers will become the dominant channel. For any product or service, only web-based channels, either direct or intermediated, that are able to reduce trade-inhibiting factors and increase trade-enhancing factors will replace their traditional offline counterparts. A mathematical illustration using the trade value framework can be found in Appendix B.

In summary, this paper proposes that the purpose of intermediaries in a trading structure is *to improve the trade-enhancing factors and reduce the trade-inhibiting factors for both buyers and sellers* and that those *intermediaries can endure only if they do not decrease the trade value that buyers and sellers could have had by trading directly.* This understanding of intermediaries can be easily applied in future studies on channel-structures. Based on the discussion above, we propose the following:

Proposition 1: *The decision of buyers and sellers to diversify from their existing channel-structure is influenced not only by the relative transaction costs associated with these channels, but also by the trade values associated with those channels.*

This is particularly evident in smaller, less well-known sellers. These sellers may establish their own online presences and simultaneously promote their products on another e-marketplace, such as Amazon, buy.com and eBay, even with increased transaction costs. Since these sellers lack immediate name recognition, online buyers do not have the familiarity and trust established by some well-known website vendors. By diversifying their selling channels to include other well-established channels, sellers can create an immediate sense of security, trust and legitimacy to buyers. Large, well-known sellers, such as Target and Borders, also adopt similar strategies to provide multiple selling channels in addition to their brick-and-mortar and online shops. Because e-marketplaces are acting as product aggregators and market portals, large well-known sellers can provide more value to potential buyers in terms of greater convenience and lower search cost. It is expected that a significant number of US and European firms will shift their online trading to e-marketplaces (Yu, 2007). The fact that online buyers have grown exponentially from 25 million in 1996 to nearly 153 million in 2007, according to the US Census Bureau, demonstrates that a growing number of consumers find shopping online an attractive channel-structure in comparison to existing offline choices. In fact, studies have found that consumers are increasingly using multi-channel strategies for purchasing activities (Telzrow et al., 2007).

Proposition 2: *A direct channel-structure will eliminate an intermediated channel-structure only when the net trade value provided by the intermediated channel-structure is significantly less than the one provided by the direct channel-structure.*

The airline industry has long been dependent on intermediaries to handle the reservation and distribution of tickets. However, many airlines have since been aggressively promoting and enticing customers to book travel tickets on the airlines' own websites by giving discounts or other rewards. These strategies clearly demonstrate that it is much more advantageous for airlines to sell tickets directly to their customers and to bypass the intermediated channel-structure since the value these carriers gain from owning the consumers' traveling patterns and behaviors is significantly more valuable than the costs incurred for moving consumers into the direct channel. Recent online micro-financing websites are another example of emerging direct channel-structures in the money lending/borrowing industry. For example, Prosper.com and Kiva.org can provide a market for people who are interested in lending money to people directly and for those who need to borrow money without using traditional money lending/borrowing channels such as banks or pawnshops. Furthermore, e-banking presents new opportunities for banks in how they develop, deliver, and market products over the Internet. In fact, the Internet gives rise to a completely new channel (Wu, 2006). Here, we have seen banks adopt diverse channel strategies ranging from singular direct channel e-banking strategies to a mixture of e-banking with traditional offline direct channel-structures to reach customers.

Proposition 3: *A direct or intermediated channel-structure will survive only if it provides extraordinary value on one side of the trade value framework to the point that inhibiting factors on the other side of the trade can be overlooked.*

This proposition suggests that both direct and intermediaries will survive if the value they provide to buyers and sellers overshadow the disadvantages associated with their trade inhibitors. Traditional intermediaries such as real estate or travel agents were projected to disappear with the emergence of direct, online channels (Giaglis et al., 2002; Sakar et al., 1995). Nonetheless, real estate and travel agents who can provide superior value to specific sectors of consumers, such as first time home buyers, or unique services, such as complicated trip arrangements, continue to exist even though such services are found online by buyers themselves with much lower transaction costs.

Even though bypassing a middleman seems to be the solution to lowering transaction costs and streamlining processes, there are cases where the middleman adds real value to the consumer. For instance, consider the Vineyardswine.com website, which acts as an intermediary between consumers interested in eclectic wines and many small geographically dispersed wineries by providing match-making functions and valuable expert suggestions (Steinfield, 2003).

In summary, web-based EC will initially result in competing and co-existing trading structures. The emergence of a dominant channel-structure (direct or intermediated) will be influenced by (a) the trade value offered by the competing channel-structures, as determined by their trade-inhibiting and trade-enhancing factors, and (b) the relative bargaining power of buyers and sellers as determined by competitive environmental forces. These factors will determine the composition of trading structures; for example, an industry might be characterized by a few dominant trading structures, including web-based EC, or multiple trading structures that co-exist in equilibrium.

CONCLUSION

This paper addresses some major concerns with the existing literature on the impact of the Internet on offline intermediated channel-structures and responds to the call for a better theoretical framework in studying the impact of the Internet on intermediation research. These concerns are centered on the lack of a common definition of intermediaries, the limited applicability of transaction cost theory, and the inherent shortcomings of transaction cost theory. The paper contributes a classification of the inhibiting and enhancing factors inherent in any channel-structure and the impact of these factors on the trade value associated with the channel-structure. The paper proposes a comprehensive *trade value framework* that combines transaction cost-based research from industrial organization, marketing and information systems (particularly electronic commerce) research with the conceptual underpinnings of transaction value theory from the strategic management literature. The proposed framework not only addresses the inherent limitations of transaction cost theory in the study of intermediation and disintermediation, but also extends the application of transaction value theory to study the impact of e-commerce on intermediated trading structures. We use the framework to analyze the emergence of a dominant channel-structure and move forward to developing a new theory to explain the phenomenon of digital intermediation, an area in need of further investigation (Kauffman & Walden, 2001).

Theoretical Implications

The paper reviews and synthesizes the literature on channel-structures from multiple disciplines and proposes a standardized definition of intermediaries. It also provides a critical evaluation of transaction cost theory that has limited the discussion on channel-structures to analyzing sellers' decisions based on a single economic variable: cost. The *trade value framework* proposed by this paper brings a more complete viewpoint to analyze the impact of the Internet on channel-structures. Building upon transaction cost theory, transaction value theory, and consumer behavior research, the trade value framework takes into account both the sellers' and buyers' perspectives across multiple trade-enhancing or -inhibiting factors. Applying the trade value framework allows us to derive several propositions that are empirically testable.

Managerial Implications

The trade value framework provides a comprehensive toolkit to assist sellers in analyzing their existing channel-structures and evaluating alternative channel-structures. Using the proposed framework, managers can arrive at a better understanding about the impact of the Internet on their existing channel-structures. Intermediaries can use the framework to analyze the value they bring to the channel-structure and develop strategies to improve their value propositions. Furthermore, manufacturers and buyers can evaluate the benefits of using different channel-structures based on the multiple aspects of trade value.

Future Research

Using the *trade value framework*, we argue that disintermediation in a traditionally intermediated channel-structure will occur only if a direct channel (e.g. web-based direct channel-structure) is able to reduce the trade-inhibiting factors present in the traditional channel and provide more trade-enhancing factors, thereby encouraging more trade through the new structure. Otherwise, re-intermediation and new forms of intermediation, such as e-marketplaces, are more likely to occur. Although the *trade value framework* is conceptual, the propositions suggested by the framework can be empirically validated. To do so, we need to develop measurement scales for various trade-inhibiting and -enhancing factors. These scales can then be used to measure the magnitude of these factors and explain the dominance of a particular channel-structure in an industry. In addition, further empirical research is needed to determine the weights of various factors among buyers and sellers.

REFERENCES

Bailey, J., & Bakos, J. (1997). An exploratory study of the emerging role of electronic intermediaries. *International Journal of Electronic Commerce*, *1*, 7–20.

Bain & Company. (2002). *Catalog companies set multi-channel standard*. Fighting Fire with Water -From Channel Conflict to Confluence.

Bakos, J., & Brynjolfsson, E. (1993). From vendors to partners: Information technology and incomplete contracts in buyer–seller relationships. *Journal of Organizational Computing*, *3*, 301–328. doi:10.1080/10919399309540206

Bakos, J., Lucas, H., Oh, W., Simon, G., Viswanathan, S., & Weber, B. (2005). The impact of e-commerce on competition in the retail brokerage industry. *Information Systems Research*, *16*, 352–371. doi:10.1287/isre.1050.0064

Barney, J. (1991). Firm resources and sustained competitive advantage. *Journal of Management*, *17*, 99–120. doi:10.1177/014920639101700108

Belenger, D., & Kargaonkar, P. (1980). Profiling the recreational shopper. *Journal of Retailing*, *56*, 77–92.

Benjamin, R., & Wigand, R. (1995). Electronic markets and virtual value chains on the information superhighway. *Sloan Management Review*, *36*, 62–72.

Bergen, M., Kauffman, R., & Lee, D. (2005). Beyond the hype of frictionless markets: Evidence of heterogeneity in price and rigidity on the internet. *Journal of Management Information Systems*, *22*, 57–89.

Bhargava, H., & Choudhary, V. (2004). Economics of an information intermediary with aggregation benefits. *Information Systems Research*, *15*, 22–36. doi:10.1287/isre.1040.0014

Bhatnagar, A., Misra, S., & Rao, H. (2000). On risk, convenience and internet shopping behavior. *Communications of the ACM*, *44*, 98–104. doi:10.1145/353360.353371

Brynjolfsson, E., & Smith, M. (2000). Frictionless commerce: A comparison of internet and conventional retailers. *Management Science*, *46*, 563–585. doi:10.1287/mnsc.46.4.563.12061

Carter, R., & Hodgson, G. (2006). The impact of empirical tests of transaction cost economics on the debate on the nature of the firm. *Strategic Management Journal, 27*, 461–476. doi:10.1002/smj.531

Chen, S., & Dhillon, G. (2003). Interpreting dimensions of consumer trust in e-commerce. *Information Technology and Management, 4*, 303–318. doi:10.1023/A:1022962631249

Childers, T., Carr, C., Peck, J., & Carson, S. (2001). Hedonic and utilitarian motivations for online retail shopping behavior. *Journal of Retailing, 77*, 511–535. doi:10.1016/S0022-4359(01)00056-2

Chircu, A., & Kauffman, R. (2000). Re-intermediation strategies in business-to-business electronic commerce. *International Journal of Electronic Commerce, 4*, 7–42.

Chu, W., Choi, B., & Son, M. (2005). The role of on-line retailer brand and infomediary reputation in increasing consumer purchase intention. *International Journal of Electronic Commerce, 9*, 115–127.

David, R., & Han, S. (2004). A systematic assessment of the empirical support for transaction cost economics. *Strategic Management Journal, 25*, 39–58. doi:10.1002/smj.359

Dillon, T., & Reif, H. (2006). Identifying purchase perceptions that promote frequent e-commerce buying. *International Journal of Electronic Marketing and Retailing, 1*, 48–66. doi:10.1504/IJEMR.2006.010095

Doney, P., & Cannon, J. (1997). An examination of the nature of trust in buyer seller relationships. *Journal of Marketing, 61*, 35–51. doi:10.2307/1251829

Dwivedi, Y., Kiang, M., Williams, M., & Lal, B. (2008). Profiling research published in the Journal of Electronic Commerce Research. *Journal of Electronic Commerce Research, 9*, 77–91.

Dyer, J. (1997). Effective collaboration: How firms minimize transaction costs and maximize transaction value. *Strategic Management Journal, 18*, 535–556. doi:10.1002/(SICI)1097-0266(199708)18:7<535::AID-SMJ885>3.0.CO;2-Z

Easingwood, C., & Coelho, F. (2003). Single versus multiple channel strategies: Typologies and drivers. *Service Industries Journal, 23*, 31–46. doi:10.1080/02642060412331300872

Eng, T. (2008). The effects of the internet on network structures and business performance. *Electronic Markets, 18*, 91–101. doi:10.1080/10196780701797706

Fingleton, J. (1997). Competition between intermediated and direct trade and the timing of disintermediation. *Oxford Economic Papers, 49*, 543–556.

Gehrig, T. (1993). Intermediation in search markets. *Journal of Economics & Management Strategy, 2*, 97–120. doi:10.1111/j.1430-9134.1993.00097.x

Giaglis, G., Klein, S., & O'Keefe, R. (2002). The role of intermediaries in electronic marketplaces: Developing a contingency model. *Information Systems Journal, 12*, 231–246. doi:10.1046/j.1365-2575.2002.00123.x

Grover, V., & Teng, J. (2001). E-commerce and the information market. *Communications of the ACM, 44*, 79–86. doi:10.1145/367211.367272

Gupta, A., Su, B., & Walter, Z. (2004). An empirical study of consumer switching from traditional to electronic channels: A purchase-decision process perspective. *International Journal of Electronic Commerce, 8*, 131–161.

Jarvenpaa, S. L., Tractinsky, N., & Vitale, M. (2000). Consumer trust in internet stores. *Information Technology and Management, 1*, 45–71. doi:10.1023/A:1019104520776

J.C. Williams Group. (2001). *Multi-channel retail report 2001.* Retrieved March 2005, from http://www.shop.org

Kauffman, R., & Walden, E. (2001). Economics and electronic commerce: survey and directions for research. *International Journal of Electronic Commerce, 5*, 5–116.

Krugman, P. (1993). First nature, second nature, and metropolitan location. *Journal of Regional Science, 33*, 129–144. doi:10.1111/j.1467-9787.1993.tb00217.x

Lewis, M., & Grimes, A. (1999). Metatriangulation: Building theory from multiple paradigms. *Academy of Management Review, 24*, 672–690. doi:10.2307/259348

Lichtenstein, S., & Williamson, K. (2006). Understanding consumer adoption of internet banking: An interpretive study in the Australian banking context. *Journal of Electronic Commerce Research, 7*, 50–66.

Malone, T. W., Yates, J., & Benjamin, R. I. (1989). The logic of electronic markets. *Harvard Business Review, 30*, 484–497.

Ng, I. (2007). Establishing a service channel: A transaction cost analysis of a channel contract between a cruise line and a tour operator. *Journal of Services Marketing, 2*, 4–14. doi:10.1108/08876040710726257

Numberger, S., & Rennhak, C. (2005). The future of B2C e-commerce. *Electronic Markets, 15*, 269–282. doi:10.1080/10196780500209077

Perea y Monsuwe, T., Dellaert, B., & Ko, D. (2004). What drives consumers to shop online? A literature review. *International Journal of Service Industry Management, 15*, 102–121. doi:10.1108/09564230410523358

Rubinstein, A., & Wolinksky, A. (1987). Middlemen. *Quarterly Journal of Economics,* 581- 593.

Sarkar, M. B., Butler, B., & Steinfield, C. (1995). Intermediaries and cybermediaries: A continuing role for mediating players in the electronic marketplace. *Journal of Computer-Mediated Communication, 1*, 111–133.

Sen, R., & King, R. (2003). Revisit the debate on intermediation, disintermediation and re-intermediation due to e-commerce. *Electronic Markets, 13*, 51–60. doi:10.1080/1019678032000067181

Shergill, G. S., Sarrafzadeh, A., Diegel, O., & Shekar, A. (2008). Computerized sales assistants: The application of computer technology to measure consumer interest – A conceptual framework. *Journal of Electronic Commerce Research, 9*, 176–191.

Smits, M., & Janssen, R. (2008). Impact of electronic auctions on health care markets. *Electronic Markets, 18*, 19–29. doi:10.1080/10196780701797607

Spink, A., & Jansen, B. (2008). Trends in searching for commerce related information on web search engines. *Journal of Electronic Commerce Research, 9*, 154–161.

Steinfield, C. (2003). *New directions in research on e-commerce.* West Lafayette, IN: Purdue University Press.

Stone, G. P. (1996). City shoppers and urban identification: Observation on the social psychology of the city life. *American Journal of Sociology, 60*, 36–55. doi:10.1086/221483

Teltzrow, M., Meyer, B., & Lenz, H. (2007). Multi-channel consumer perceptions. *Journal of Electronic Commerce Research, 8*, 18–31.

Tsay, A., & Agrawal, N. (2004). Channel conflict and coordination in the e-commerce age. *Production and Operations Management, 13*, 93–110.

Wang, W., & Benbasat, I. (2007). Recommendation agents for electronic commerce: Effects of explanation facilities on trusting beliefs. *Journal of Management Information Systems, 23*, 217–246. doi:10.2753/MIS0742-1222230410

Wigand, R. T., Steinfield, C. W., & Markus, M. L. (2005). Information technology standards choices and industry structure outcomes: The case of the U.S. home mortgage industry. *Journal of Management Information Systems, 22*, 165–191.

Williamson, O. E. (1991). Comparative economic organization: The analysis of discrete structural alternatives. *Administrative Science Quarterly, 36*, 269–296. doi:10.2307/2393356

Williamson, O. E. (1992). *Markets and hierarchies*. New York: Free Press.

Wu, J. (2006). Core capabilities for exploring electronic banking. *Journal of Electronic Commerce Research, 7*, 111–122.

Wulf, K. D., Schillewaert, N., Muylle, S., & Rangarajan, D. (2006). The role of pleasure in web site success. *Information & Management, 43*, 434–446. doi:10.1016/j.im.2005.10.005

Xu, M., Ferrand, B. M., & Roberts, M. (2008). The last mile of e-commerce: Unattended delivery from the consumers and retailers' perspectives. *International Journal of Electronic Marketing and Retailing, 2*, 20–38. doi:10.1504/IJEMR.2008.016815

Yu, C. (2007). What drives enterprises to trading via B2B e-marketplaces? *Journal of Electronic Commerce Research, 8*, 84–100.

Zajac, E. J., & Olsen, C. P. (1993). From transaction cost to transaction value analysis: Implications for the study of inter-organizational strategies. *Journal of Management Studies, 30*, 131–145. doi:10.1111/j.1467-6486.1993.tb00298.x

Zhou, M., Dresber, M., & Windle, R. (2007). Online reputation systems: Design and strategic practices. *Decision Support Systems, 44*, 785–797. doi:10.1016/j.dss.2007.10.001

APPENDIX A: MAJOR RESEARCH ON THE IMPACT OF INTERNET ON CHANNEL-STRUCTURE

Existing Research on The Impact of Internet on Channel-structure				
Research Article	**Theoretical Foundation**	**Intermediary Functions Studied**	**Outcome**	**Research Method**
Gehrig (1993)	Transaction Cost Theory; Industrial Organization	Reduce trade friction, i.e. (search cost)	Supports Disintermediation	Analytical modeling
Rubinstein and Wolinksky, 1987	Transaction Cost Theory; Industrial Organization	Reduce trade friction, i.e. (search cost)	Supports Disintermediation	Analytical modeling
Fingleton (1997)	Industrial Organization	Reduce trade friction, i.e. (search cost)	Supports Disintermediation	Analytical modeling
Stern et al. (1996)	Transaction Cost Theory; Industrial Organization	Improve process efficiency, reduce search cost	Supports Intermediation,. re-intermediation	Analytical with case studies
Sarkar et al. (1995)	Transaction Cost Theory	Aggregation and Transaction support	Supports Disintermediation, Re-intermediation, Cyber-mediation	Conceptual, no empirical support
Benjamin et al. (1995)	Transaction Cost Theory	Location,, reduce search cost	Supports Disintermediation	Case study
Chircu and Kauffman (1999)	No explicit theory. Use of existing works on electronic markets.	Aggregation and transaction support and operational support	Supports Re-intermediation	Qualitative Case study
Malone et al. (1989	Transaction Cost Theory	Reduce search costs	Supports Disintermediation	Conceptual, no empirical support
Bakos (1998)	Transaction Cost Theory	Reduce sSearch costs	Supports Disintermediation, Cyber-mediation	Analytical, no empirical support
Sen and King (2003)	Industrial Organization	Aggregation	Disintermediation is not an absolute possibility	Analytical modeling
Bunduchi (2005)	Transaction Cost Theory and Trust	Type of relationship (competitive vs. collaborative)	Mixed	Qualitative case study
Ng (2007)	Transaction Cost	Aggregation of cruise line services	Mixed	Qualitative case study

Appendix B: Mathematical Illustrations of Trade Value Framework

This paper defines *trade value* as the surplus obtained by buyers and sellers after all the costs associated with trade-inhibiting factors have been deducted from the total gains accrued from the trade. As illustrated in *Figure II*, the transaction cost for the seller when trading through Channel-Structure *i* is assumed to be proportional to the trade-inhibiting factors associated with Channel-Structure *i*:

$$Transaction\_Cost(Seller) \alpha Trade\_Inhibitors(Seller) \quad (1)$$

In addition, the expected revenues are inversely proportional to the magnitude of trade-inhibiting factors for Channel-Structure *i* for *buyers* and directly proportional to the trade enhancing factors for Channel-Structure *i* for buyers:

$$Expected\_\mathrm{Re}venues \alpha \frac{1}{Trade\_Inhibitors(Buyers)} \quad (2)$$

$$Expected\_\mathrm{Re}venues \alpha Trade\_Enhancers(Buyers) \quad (3)$$

Now the trade value for *sellers* V_i^S is defined as the surplus from the trade through Channel-structure *i:*

$$V_i^S = Expected\_\mathrm{Re}venue - Transaction\_Cost(Seller) \quad (4)$$

Substituting from equations (1) – (3) into equation (4) gives the trade value for *the seller* trading through Channel-Structure *i:*

$$V_i^S \alpha \frac{Trade\_Enhancers(Buyers)}{Trade\_Inhibitors(Buyers)} - Trade\_Inhibitors(Seller) \quad (5)$$

Similarly, with respect to any Channel-Structure *i*, where a buyer pays the Price P, the following relationships hold true: $Transaction\_Cost(Buyers) \alpha Trade\_Inhibitors(Buyers)$ (6)

In addition, the expected utility (the value obtained by *the buyer* from the trade) is inversely proportional to the magnitude of trade-inhibiting factors associated with Channel-Structure *i* and directly proportional to the trade- enhancing factors associated with Channel-Structure *i*:

$$Expected\_Utility \alpha \frac{1}{Trade\_Inhibitors(Buyers)} \quad (7)$$

$$Expected\_Utility \alpha Trade\_Enhancers(Buyers) \quad (8)$$

Now the trade value for *the buyer* V_i^b is proportional to the surplus from the trade through Channel-Structure *i:* $V_i^b = Expected\_Utility - Transaction\_Cost(Buyer)$ (9)

Substituting from equations (6) – (8) into equation (9) gives the Trade Value for *the buyer* trading through this channel-structure:

$$V_i^b \alpha \frac{Trade\_Enhancers(Buyers)}{Trade\_Inhibitors(Buyers)} - Trade\_Inhibitors(Buyers) \quad (10)$$

The trade value approach suggests that *a seller* would prefer to choose a Channel-Structure *m* such that $V_m^S = Max(V_1^S, \ldots\ldots\ldots, V_n^S)$ (11) and *the buyer* would prefer to trade through Channel-Structure *k* such that $V_k^b = Max(V_1^b, \ldots\ldots\ldots, V_n^b)$ (12). When *m=k,* then both the seller and the buyer agree upon a common channel-structure, and this channel-structure becomes dominant. However, when $m \neq k$, the outcome depends upon the relative bargaining power of the buyer and the seller.

When the buyer and the seller have almost equal bargaining power, they will share the trade surplus equally, agreeing on the channel-structure that maximizes the *total surplus from the trade:*

$$\text{Max}[(V_1^S, V_1^b), (V_2^S, V_2^b), \ldots, (V_n^S, V_n^b)] \quad (13)$$

For each transaction, the share of the trade surplus for the buyer and seller will be:

$$\text{Max}[\frac{(V_1^S, V_1^b), (V_2^S, V_2^b), \ldots, (V_n^S, V_n^b)}{2}] \quad (14)$$

In this channel-structure, if the magnitude of trade inhibitors is high for any participant (see Equations 5 and 10), then there is an opportunity for an intermediary to reduce or eliminate these trade-inhibiting factors, thereby increasing the trade value generated by this new channel-structure. Let us call this intermediated channel-structure *e*. If V_e^b is the value acquired by the buyer, V_e^S is the value acquired by the seller, and V_e^i is the value acquired by the intermediary when trading through *e*, then this intermediated channel-structure will survive only if V_e^b and V_e^S are more than the total surplus offered by other channel-structures. This means the value acquired by a participant through the intermediated channel-structure must be greater than the value acquired by trading through some other channel-structure. If these conditions are not satisfied, then there is an incentive to switch to a better channel-structure.

When the buyer and the seller have unequal bargaining power, two possible scenarios could exist. First, if the buyer has more bargaining power, then the dominant channel-structure is Channel-Structure *k*, because: $V_k^b = Max(V_1^b, \ldots, V_n^b)$. Second, if the seller has more bargaining power, then the dominant channel-structure is Channel-Structure *m* because: $V_m^S = Max(V_1^S, \ldots, V_n^S)$. In this context, there is still an opportunity for intermediation as long as the intermediary's participation increases the trade value. The surplus value should be distributed in such a manner that the buyer and the seller get at least as much value as they had in the direct trade.

This work was previously published in International Journal of E-Business Research Volume 6, Issue 2, edited by In Lee, pp. 59-77, copyright 2010 by IGI Publishing (an imprint of IGI Global).

Chapter 5
Effects of e-CRM Value Perceptions on Website Loyalty:
An Empirical Investigation from Customer Perspective

Wen-Jang Jih
Middle Tennessee State University, USA

Su-Fang Lee
Overseas Chinese Institute of Technology, Taiwan

ABSTRACT

Web-enabled customer relationship management, or e-CRM, holds the promise to simultaneously achieve the conflicting goals of strategic flexibility and operational efficiency in developing and executing innovative e-commerce marketing strategies. A new multi-disciplinary field involving such diverse areas as marketing, information systems, and consumer behavior, e-CRM seeks to strengthen customer relationship via a portfolio of trust-building practices with the support of powerful information technologies. While most literature in e-CRM has focused on how firms formulate and implement e-CRM initiatives, there is a dearth of literature on viewing the overall quality of an e-commerce firm's e-CRM practices from the consumer point of view. This research proposes such a consumer-oriented concept, e-CRM value, based on existing e-CRM research literature and examines the effect of e-CRM value on website loyalty. It was found that customers' perception of e-CRM value has a positive causal effect on their website loyalty. The implications of this finding for the theory-building of e-CRM and marketing practices of e-commerce firms are discussed.

BACKGROUND

The Internet technology has essentially become an integral component of information infrastructure for business operations and managerial decision making in all business functions. Marketing activities, especially, are increasingly incorporating the Web in the strategic initiatives and relying on the website as the primary platform of interaction with existing and prospective customers that are otherwise difficult to reach for technical or financial reasons. In some information and knowledge-intensive industries (information technology, insurance, and banking, for example),

DOI: 10.4018/978-1-61350-462-8.ch005

the website is becoming an ever more important source of impact upon customers' purchasing decision. Leveraging the power of information and knowledge in the knowledge economy, both companies and customers are using the website as a strategic tool for obtaining maximum value in the e-commerce market.

In light of the strategic value of the Internet as a crucial enabling technology, companies are increasingly placing their emphasis on managing customer relationship (CRM) through effective website design and management. Technological advances in such areas as search engines, blogs, Really Simple Syndication (RSS 2.0), social networking services, highly distributed databases, multimedia, personalization, localization, and computer-telephone integration have provided a rich set of system capabilities and are prompting for implementation of innovative and highly interactive mechanisms to serve prospective as well as existing customers (Chen et al., 2007). Given the growing popularity of Web-enabled CRM (or e-CRM), it is essential for business researchers to develop a systematic understanding of important aspects of this business initiative.

Due to the multidisciplinary nature of the field, a variety of issues surrounding e-CRM have been investigated from different points of view. In particular, research in information systems has identified the importance of evaluation technology applications from the user perspective (Gefen & Straub, 2000). Tung (2007) studied the e-CRM in the context of technology diffusion. Marketing research has examined the role of Internet technologies for generation, distribution, and application of market information (Chen & Chen, 2004; Min et al., 2002). Consumer behavior studies also confirm the value of examining e-CRM issues from customer /user point of view (Lee et al., 2006; Scullin et al., 2004). These studies suggest that, in order to obtain a systematic understanding of the highly composite nature of e-CRM, more research must be conducted along a variety of distinctive dimensions and at deeper levels. This study is a response to that calling. Specifically, this study is conducted to evaluate the website users' perceived value of e-commerce firms' CRM activities. It also examines the effect of e-CRM value on website loyalty.

These two research issues are addressed by answering two research questions:

1. What are the significant components of website users' perceived value of e-commerce firms' CRM activities, or e-CRM value?
2. Does website users' perceived e-CRM value significantly affect their loyalty to the websites?

Answering these questions would shed some light on the value perception of e-CRM value from the e-commerce customer's point of view, and thus contribute to systematic accumulation of knowledge and insight regarding this important business concept.

In the rest of the paper, we describe the concept of CRM and e-CRM. We also address measurement issues of e-CRM as proposed in the existing literature. The concept of customer loyalty is described in the same section. The third section of the paper presents the data collection and data analysis methods employed in this research. We then discuss the result of data analysis. The paper is concluded by the summary of the research and discussion of implications for both e-commerce practices and academic research.

From CRM to e-CRM

As a manifestation of customer and market orientation, CRM is a concept that influences strategic, tactical, and operational levels of business and organizational management. Although the importance of customer orientation has never been ignored in the past decades, it is becoming an ever more critical issue in today's knowledge-driven market. In seeking viable ways to enhance competitive advantage in a business environment characterized by fierce competition and other environmental challenges, firms are not only looking to customer as a major source of market information and knowledge, they are also linking customers with vital aspects of their product value chains. As a result, many important business processes such as new product development and customer service in these customer-centric firms are now even more driven by customer-oriented thinking than ever before.

The concept of CRM can be traced back to a traditional wisdom for good business practice, relationship management (Parvatiyar & Sheth, 2001). Defined as an integrated effort to identify, maintain, and build up a human network with individual customers, relationship management activities seek to strengthen the network for the mutual benefit of both sides through interaction, individualized and value-added contacts over a long period of time (Shani & Chalasani, 1992). The vastly improved information flow enabled by recent advancements of computing and networking technologies has significantly elevated the business value of a systematic management of customer relationship (Parvatiyar & Sheth, 2001).

Multiple definitions of CRM exist in the literature, but no universally accepted one has been agreed upon (Ngai, 2005). This is partly due to the young stage of the field, and partly a reflection of the multidisciplinary nature of the endeavor. For example, defining CRM from customer communications point of view, Swift (2001) views CRM as an "enterprise approach to understanding and influencing customer behavior through meaningful communications in order to improve customer acquisition, customer retention, customer loyalty, and customer profitability (p.12)." Emphasizing the strategic role in resource integration, Kincaid (2003) considers CRM as "the strategic use of information, processes, technology, and people to manage the customer's relationship with your company (marketing, sales, services, and support) across the whole customer life cycle (p. 41)." Kennedy (2006) views CRM as a business philosophy that is "about identifying a company's best customers and maximizing the value from them by satisfying and retaining them (p. 58)." In information management, CRM is an amalgamation of information technology application and relationship marketing. With relationship marketing serving as the strategic foundation for the design of system architecture and functional capabilities, a CRM may incorporate a variety of information technologies to automate and integrate such enterprise-wide processes as sales, customer service, and product configuration (Scullin et al., 2004; Turban et al., 2004). The rapid advancement and universal adoption of Internet technologies have made information and knowledge exchange technologies, especially Internet-based technologies (e-mail, world wide web, chat room, e-forums etc.), a core component of CRM systems (Ad Hamid, 2005). With Internet programs serving primarily as user interface and back office applications providing database processing capabilities, today's CRM systems are increasingly being implemented using Internet technologies, called e-CRM systems, and are making customer relationship management as easy and convenient as never before (Jih et al., 2006)

A relatively recent development, e-CRM has extended in both its scope of coverage and the enabling technologies. In its early stage, automatic tools separately support such marketing functions as identifying potential customer, keeping tract of customer contacts, customizing product configurations for unique customer requirements, and providing other value-added customer services. Currently converging information technologies prompt use of an integrated approach to managing customer relationship as an enterprise-wide effort. The e-CRM for Enterprise CRM (Peppard, 2000) and the e-CRM for electronic CRM (Kennedy, 2006) are mutually reinforcing each other.

e-CRM Value

E-CRM implementations are expected to produce significant value for companies and customers. From business point of view, the value propositions of e-CRM implementations at the operational level include such benefits as reduction of the cost of communicating with customers, reduction of administrative and operational costs, streamlining workflows, improving sales through better segmentation, and improving overall customer interactions (Ad Hamid, 2005). Major strategic benefits are also discussed in the literature. Scullin et al. (2004), for example, specifically emphasized three areas associated with marketing strategy: strategic alliance, customer strategy, and product differentiation. Both strategic and operational benefits are realized via viewing customers as an essential component of major business processes. Customer interactions are embedded in the decision making process, with communication channels designed to encourage free flow of customer opinions and company responses.

From the e-marketer's point of view, quality interaction allows for collection of meaningful customer information that is required to design customized or even personalized products or services. This personal marketing approach is characterized by "the capability of the companies to present their goods and services customized to fit the distinct personal interests and needs of the customer" (Ross, 2005, p. 40). Better personalization generally leads to better customer satisfaction, which may in turn result in better customer loyalty (Anderson & Srinivasan, 2003). The concept of customer value as an essential deliverable is widely discussed in CRM literature (Day, 1994; Slater, 1997). Personalized interactivity of e-CRM provides even more welcomed features in strengthening companies' capability to reach and serve their customers.

Several researchers have proposed specific aspects of customer value in the e-commerce environment. Szymanski and Hise (2000) suggested that e-satisfaction is a function of shopping convenience, product and service offerings, and financial security. Extending the traditional concept of market orientation, Min et al. (2002) coined the term "Internet-mediated Market Orientation" to attribute the benefit of e-commerce to marketers' capability to generate market information, distribute market information, and act on the collected market information. The notion is validated by Lee et al. (2006) in an empirical study, using Taiwanese consumer data, to determine the positive effect of the three aspects of internet-mediated market orientation on customer website satisfaction.

Srinivasan et al. (2002) conducted a study to investigate the factors affecting customer loyalty in an online business-to-consumer context. The model identified eight factors influencing customer loyalty: customization, contact interaction, cultivation, care, community, choice, convenience, character. The result of regression analysis showed that all eight factors significantly influenced customer loyalty. The study suggested the importance of these eight antecedents in forming customers' perceived value of e-commerce firms' CRM activities. In a separate study, Szymanski and Hise (2000) found that e-satisfaction was significantly affected by shopping convenience, product information provision, website design, and financial security. The website design in the latter model may be reconciled with the Character factor in the model developed by Srinivasan et al. In addition, financial security has been found to be an important factor in the e-CRM studies that investigate more security-sensitive industries such as banking (Jih et al., 2004). An integration of the previous research findings and suggestions prompted us to define the e-CRM value from the customer perspective in terms of these nine components: customization, contact interactivity, cultivation, care, community, choice, convenience, security, and market response.

Website Loyalty

Customer loyalty is often associated with repeat purchasing. However, as pointed out by Jacoby and Kyner (1973), although loyal customers tend to purchase repetitively, those who purchase repetitively do not necessarily do so out of loyalty, such as what might happen in the scenario of lacking an alternative offering. Customer loyalty usually is strengthened as a result of satisfaction with the product or service received from the seller. This is applicable in the physical economy and even more so in the Internet-enabled, online commerce. In light of the importance of the customer loyalty in the company's marketing strategy, a number of researchers have examined the nature of as well as factors affecting customer loyalty. Early views of brand loyalty almost exclusively emphasized the customer behavior of repeat purchase over a period of time (Brown, 1952; Cunningham, 1956; Lipstein, 1959; Kuehn, 1962; Day, 1969; Newman & Werbel, 1973; Reynolds & Darden, 1974; Monroe & Guiltinan, 1975). More recent research, however, contended that customer loyalty and repeat purchase do not equal (Dick & Basu, 1994). This latter view suggests that, when examining the impact of customer loyalty on the company's marketing performance, a distinction ought to be made between behavioral and attitudinal aspects of the concept (Chaudhuri & Holbrook, 2001; Jacoby & Kyner, 1973; Lim & Razzaque, 1997; Dick & Basu, 1994). Our research investigated the effect of e-CRM value as perceived by online shoppers on their loyalty to their chosen e-commerce website. Further, he customer-perceived e-CRM value is viewed as a composite variable with multiple features in this research. No distinction, however, is made between different types of loyalty. The effect of e-CRM value as perceived by the customer on the overall customer loyalty is investigated.

Based on the literature review, we proposed to represent customer-perceived e-CRM value in terms of these nine components of website features: customization, contact interactivity, cultivation, care, community, choice, convenience, security, and market response. The literature on CRM and e-CRM reviewed above also provides the rationale for the hypothesis that customer-perceived e-CRM value positively affects their loyalty to their chosen websites. The following section describes the research method we adopted to test these two research hypotheses:

H1: e-CRM value can be adequately measured in terms of customization, contact interactivity, cultivation, care, community, choice, convenience, security, and market response.
H2: e-CRM value positively affects e-loyalty

The research framework and the hypotheses are graphically depicted in Figure 1.

RESEARCH METHOD

Research Variables

The composite variable in the first hypothesis, e-loyalty, is conceptually defined as "the customer's favorable attitude toward an electronic business resulting in repeat buying behavior" (Anderson & Srinivasan, 2003, p. 125). It is operationally defined as customer-perceived importance of multiple features of a website. The hypothesized construct of e-CRM value consisting of nine components is validated by a variety of reliability and validity tests. The hypothesized causal effect of e-CRM value on e-loyalty is validated by the structural equation modeling technique. A number of index values are obtained to evaluate the fitness of the theoretical models with regards to both hypotheses. The definitions of the first seven factors in the measurement of the e-CRM value as presented in Srinivasan et al. (2002) are adopted in this research. The definitions of these seven factors and the remaining two are described as follows.

Figure 1. Conceptual framework

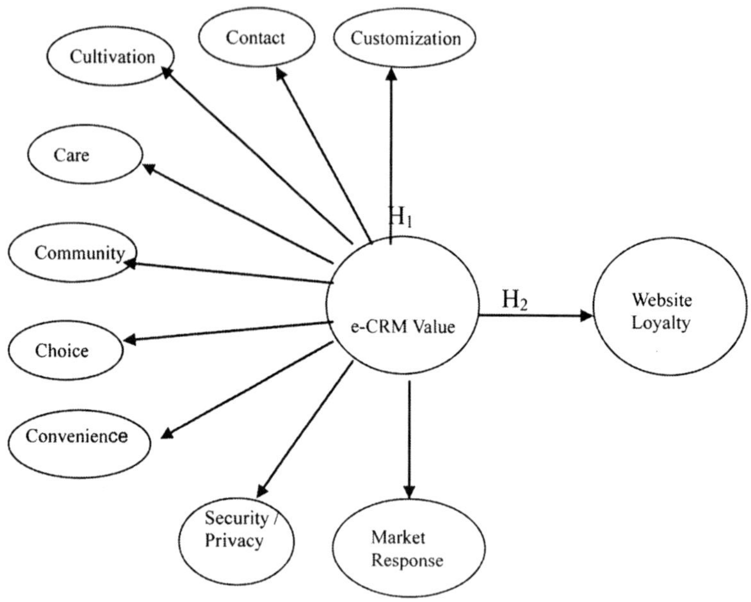

In the Srinivasan et al.'s e-loyalty model, customization was defined as the customer's perception of an e-tailer's website feature to tailor products, services, and user interface to individual customers. Contact interactivity was operationally defined as "the availability and effectiveness of customer support tools on a website, and the degree to which two-way communication with customers is facilitated" (p. 42). Cultivation was defined as the effort of the e-tailer in providing desired information and cross-selling offers to customers. Care referred to "the extent to which a customer is kept informed about the availability of preferred products and status of orders, and the level of efforts expended to minimize disruptions in providing desired services" (p.43). Community was measured in terms of "the extent to which customers are provided with the opportunity and ability to share opinions among themselves through comment links, buying circles, and chat rooms sponsored by the e-retailer" (p. 43). The number of product or service alternatives made available either on the website or through external links characterizes the choice factor of an e-tailer's website. Security was measured by the customer's trust in the e-tailer's protection of customer privacy and security mechanism. Finally, market response was measured in this research by three question items: e-tailer's responsiveness to the change of the customer's product or service requirement, e-tailer's responsiveness to market change, and the e-tailer's responsiveness to customers' opinion and suggestion. Each of these nine independent variable factors and the dependent variable, website-loyalty, is represented by three question items, resulting in a total of 30 questions in the survey questionnaire. This is in contrast with Srinivasan et al.'s research in which five questions were used for each factor. Our choice of three questions was the result of our analysis of the questions' reliability measures (i.e., Cronbach's α values).

Data Collection

The research data were collected in Taiwan during Summer 2007. Ranked the 3rd in Brown University's Global e-Government rankings and the 17th in the Economist Intelligence Unit's e-readiness rankings as of 3rd quarter 2007, Taiwan has aggressively invested in all major aspects of e-commerce, including infrastructures construction, merchant and customer education, and applications development (www.find.org.tw). E-commerce is now a major thrust in Taiwan's economic development, especially in the sector of small businesses (Jih, 2004). While the percentage of very small businesses in the total number of businesses has declined from 66.27% twenty years ago to the current 42.94%, the sales volume of e-commerce has increased 36.4% since 1999. The growth is likely to continue. Among the population of 23 million, there are currently 4.58 million wired broadband subscribers and 1.1 million wireless Internet access subscribers in this island state. Supported mostly by the 3rd-generation wireless networks and personal handyphone systems (PHS), Taiwanese shoppers are well equipped and conveniently linked with the online merchandisers worldwide. These ten million Taiwanese users of Internet-enabled services are potential participants in online merchandisers' e-CRM programs.

The questionnaire was e-mailed to volunteering online shoppers with an incentive offering of a small amount of donation being made to a charity on behalf of the effective respondents. Since the volunteers have indicated in advance their willingness to participate in the study, most participants completed and returned the questionnaire. A total of 564 volunteers returned their questionnaires. However, 174 returned questionnaires appeared to be filled out mindlessly and were removed. The remaining 390 were useful responses and used as the effective sample in our subsequent data analysis. The response rate was about 69%. The sample consists of 36% of males and 64% of females. The majority (60.3%) were in the age group from 20 to 29 years old. The second largest age group (24.6%) was from 30 to 39. There were 12.6% of the respondents that were 40 or above, and 2.6% aging 19 or below.

Data Analysis

In determining the adequacy of e-CRM value measurement, the reliability of each of the nine aspects of e-CRM value and that of the website loyalty was assessed using Cronbach's α value. First-order confirmatory factor analysis was then employed to evaluate the discriminant validity and convergent validity (two essential aspects of construct validity) of the measurement instrument. Finally, second-order confirmatory factor analysis was used to establish the adequacy of the e-CRM value measurement. Overall, our analysis revealed that the instrument we used to measure e-CRM value passed reliability test, validity test, and also satisfied the adequacy criteria.

The reliability test of the measurement was done via Cronbach's α value. As shown in Table 1, the Cronbach's α value of each of the nine aspects of e-CRM value was 0.714, 0.702, 0.761, 0.843, 0.833, 0.829, p.849, 0.876, and 0.802, respectively. According to Cuieford (1965), in measuring a variable such as what's investigated in this research, a Cronbach's α value that is greater than 0.7 indicates high reliability and a measurement scale with Cronbach's α value less than 0.35 represents unacceptable reliability. A Cronbach's α value between 0.35 and 0.7 has fair but acceptable reliability. Since all aspects of our e-CRM measurement have Cronbach's α values greater than 0.7, a high reliability of the measurement instrument is evident. In addition, the Cronbach's α value of website loyalty measurement, 0.917, also shows high reliability.

The convergent validity expresses the degree to which measures of construct that are related to each other in theory are observed to be related to each other. In this study, the convergent validities of e-CRM value measurement were indicated by the squared multiple correlations (SMC), construct composite reliabilities, and the average variance extracted values. The fact that all SMC values are greater than 0.5, all construct composite reliability values are greater than 0.7, and average

Table 1. Reliability and convergent validity of research constructs

Constructs	Items	Standardized Regression Weights	Cronbach's α	Squared Multiple Correlations	Average Variance Extracted	Composite Reliabilities
Customization	1	0.724	0.714	0.524	0.517	0.762
	2	0.713		0.509		
	3	0.719		0.517		
Contact / Interaction	4	0.722	0.702	0.521	0.513	0.759
	5	0.716		0.513		
	6	0.710		0.504		
Cultivation	7	0.720	0.761	0.518	0.547	0.784
	8	0.728		0.530		
	9	0.770		0.593		
Care	10	0.920	0.843	0.547	0.649	0.864
	11	0.865		0.749		
	12	0.806		0.650		
Community	13	0.778	0.833	0.605	0.660	0.852
	14	0.925		0.856		
	15	0.721		0.520		
Choice	16	0.789	0.829	0.623	0.631	0.836
	17	0.877		0.769		
	18	0.707		0.500		
Convenience	19	0.756	0.849	0.571	0.656	0.814
	20	0.863		0.744		
	21	0.807		0.652		
Security	22	0.832	0.878	0.692	0.716	0.883
	23	0.938		0.879		
	24	0.760		0.577		
Market Responsiveness	25	0.784	0.802	0.614	0.598	0.817
	26	0.807		0.652		
	27	0.730		0.527		
Website Loyalty	28	0.804	0.917	0.646	0.548	0.916
	29	0.947		0.897		
	30	0.900		0.810		

variances extracted are all greater than 0.5, suggests acceptable convergent validity of the e-CRM value constructs measurement (Table 1). Another important aspect of construct validity, discriminant validity, indicates the extent to which measures of constructs that are not related to each other in theory are observed to not be related to each other. Fornell and Larcker (1981) contended that acceptable discriminant validity can be demonstrated by the average variance extracted of a construct exceeding the squared correlation between this construct and each of other constructs.

Table 2. Discriminant validity evaluation of research constructs

Constructs	Customiza-tion	Contact / Inter-action	Cultiva-tion	Care	Com-munity	Choice	Conve-nience	Secu-rity	Market Responsive-ness	Web-site Loy-alty
Customiza-tion	0.517									
Contact / Interaction	0.354	0.513								
Cultivation	0.403	0.329	0.547							
Care	0.345	0.324	0.475	0.649						
Community	0.186	0.262	0.207	0.250	0.660					
Choice	0.297	0.319	0.247	0.244	0.244	0.631				
Convenience	0.319	0.286	0.262	0.259	0.211	0.465	0.656			
Security	0.294	0.178	0.300	0.272	0.141	0.256	0.348	0.716		
Market Re-sponsiveness	0.362	0.342	0.404	0.494	0.266	0.275	0.291	0.339	0.598	
Website Loyalty	0.325	0.278	0.362	0.331	0.166	0.270	0.317	0.287	0.304	0.548

Note: Measures on the diagonal are extracted variance percentages. The rest are squared correlation coefficients.

Table 2 presents these average variance extracted values and squared correlation coefficients. The average variance extracted of customization, for example, is 0.517, with its squared correlations with other constructs being 0.354 (Contact Interactivity), 0.403 (Cultivation), 0.345 (Care), 0.186 (Community), 0.297 (Choice), 0.319 (Convenience), 0.294 (Security), 0.362 (Market Response), and 0.325 (Website Loyalty), respectively. Table 2, therefore, provides evidence of acceptable discriminant validity of our measurement. With both convergent validity and discriminant validity being acceptable, the construct validity of our measurement is, therefore, deemed acceptable.

The adequacy of the e-CRM value measurement was further evaluated through a second-order confirmatory factory analysis. A set of fitness indices produced by the analysis provides the basis of the evaluation, including goodness-of-fit (GFI), adjusted goodness-of-fit (AGFI), normed fit index (NFI), comparative fitness index (CFI), and root mean square error of approximation (RMSEA). According to the analyses of Bollen (1989), Hair et al. (1998), and Bentler (1995), a valid theoretical model is characterized by the values of GFI, AGFI, and NFI greater than 0.9, GFI value greater than 0.95, and RMSEA less than 0.08. The values of these indices produced by the second-order confirmatory factor analysis were GFI = 0.937, AGFI = 0.900, NFI = 0.931, CFI = 0.941, and RMSEA = 0.077, all passing the evaluation criteria stated above. Furthermore, Table 3 also shows that the t-values of all standard structure coefficients are significant. The validity of the e-CRM value measurement model adopted by this research is confirmed. Hypothesis 1 is accepted.

Hypothesis 2 was tested using structural equation modeling. All model fit indices, as shown in Table 4, pass the evaluation criteria: GFI = 0.923 (> 0.9), AGFI = 0.910 (> 0.9), NFI = 0.962 (> 0.9), CFI = 0.951 (> 0.9) and RMSEA = 0.065 (, 0.08). The standardized structure coefficient value is 0.754 (t-value = 11.055, significant at 0.01 level). These results lead to acceptance of

Table 3. Second-order confirmatory factor analysis of e-CRM value

Construct	Question Item	Confirmatory Factor Analysis			
		First-Order		Second-Order	
		Standardized Coefficient	t value	Standardized Coefficient	t value
Customization	1. The advertisement and promotion from this website fit me perfectly.	0.913	---a	0.653	---
	2. I can find products and services that satisfy my personal requirements from this website.			0.699	13.01
	3. This website makes me feel like a VIP.			0.647	13.33
Interactivity	4. The website allows me to browse products from different points of view.	0.880	11.02	0.662	10.68
	5. The search feature of the website makes it very easy to find the products I look for.			0.650	10.53
	6. The website is very responsive and effective in answering my questions.			0.660	---
Cultivation	7. The website's various campaigns really appeal to me.	0.897	12.68	0.649	12.04
	8. I feel this website acts aggressively to cultivate customer relationship.			0.747	13.86
	9. This website sends me information to save my effort and time in information search.			0.751	---
Care	10. This website is sensitive in responding to my problems.	0.854	13.29	0.765	16.22
	11. This website is hospitable and friendly to its visitors.			0.830	17.86
	12. I trust this website does its very best to serve customers.			0.820	---
Community	13. This website provides a forum in support of customer sharing of shopping experience.	0.656	9.80	0.778	13.98
	14. This website encourages me to express my own opinions and comments.			0.895	14.94
	15. The customers of this website actively participate in the discussions on its forum.			0.709	---
Choice	16. This website facilitates one-stop shopping due to its wide variety of product/ services.	0.774	11.15	0.790	14.37
	17. I am satisfied with the choices that are available on this website.			0.866	15.304
	18. I don't feel constrained by the choices available on this website.			0.721	---
Convenience	19. I easily completed my purchasing the first time I logged onto the website	0.789	12.33	0.789	16.39
	20. It's very convenient to shop on this website.			0.828	17.24
	21. Shopping on this website saves me lots of time.			0.811	---
Security	22. I trust this website in protecting user privacy.	0.733	11.37	0.865	17.72
	23. I don't worry about this website's security mechanisms.			0.903	18.33
	24. I feel comfortable to provide my private information to this website.			0.764	---
Market Response	25. This website is serious about changes of customer requirements and their demand s for products.	0.900	12.78	0.768	14.58
	26. This website is very responsive to competition's actions in the marketplace.			0.753	14.30
	27. This website quickly responds to customer comments and suggestions.			0.751	---

---a: Loading fixed

Table 4. Validation of research model

Causal Effect Path	Expected Result	Standardized Coefficient	t-value
e-CRM Value -> Website Loyalty	+	0.754	11.055*
Fitness Indices	Criteria		Fitness Assessment
GFI	> 0.9	0.923	Good
AGFI	> 0.9	0.910	Good
NFI	> 0.9	0.962	Good
CFI	> 0.95	0.951	Good
RMSEA	< 0.08	0.065	Good

*p< 0.01

the hypothesis, i.e., E-CRM value positively affects website loyalty.

CONCLUSION AND SUGGESTIONS

Innovative implementation of customer-centric CRM program is crucial to an e-commerce company's strategy. Measuring customers' perception of how well the company's e-CRM practice goes, however, often is challenging due to the complex and dynamic nature of the interactions between the website and the customer. This research explores the concept of e-CRM value as a measure of consumer-perceived value of CRM practices conducted by e-commerce firms by integrating and extending the works found in the existing literature. Specifically, the eight-factor antecedents model of Srinivasan et al. (2002) was slightly modified and expanded to come up with a nine-component framework for e-CRM value. A measurement instrument is developed to operationalize the proposed framework. Using both first- and second-order confirmatory factor analyses, the conceptual framework and the measurement instrument are empirically validated with the primary data collected from Taiwanese online shoppers. It was found that the framework formed by the proposed nine dimensions – customization, contact interactivity, cultivation, care, community, choice, convenience, security, and market response – does adequately measure customer-perceived value delivered by the e-commerce company through its e-CRM implementation. Using structural equations modeling analysis, the study also reveals the positive causal effect of e-CRM value on customers' website loyalty.

The research findings have important implications for practicing managers of e-commerce firms. Most of the factors that determine customers' perceived e-CRM value are closely related to the information and knowledge flow between product/service providers and customers. Simply providing basic product or service information to facilitate consumers' purchasing decision making is not enough for development of long-term customer loyalty. Today's online customers demand in-depth information and even proprietary knowledge in their purchasing decision-making process. E-commerce firms must act innovatively and aggressively to engage customers in two-way and open communications, using Web 2.0 tools such as blogs or social networking services to strengthen mutual trust.

For e-commerce researchers, this study tests an efficient measurement instrument to assess e-CRM value perception, with each factor only consisting of three question items. The external validity of the instrument is subject to validation in different contextual settings, especially different cultural environments. Another promising research direction involves how various categories of information technologies, such as Web 2.0 tools, can be used to enhance e-CRM value perception. To be fruitful, research addressing these issues must consider industrial, organizational, and individual (website visitor) characteristics in the research design. Websites selling auto parts, for example, may be perceived differently from the websites that sell entertainment items. Consumers may have different expectations for and thus exhibit different attitudes toward large websites and much smaller websites. As another example, regular online shoppers may perceive differently from ad hoc online shoppers of the same website. Including data of finer granularity in the analysis would shed light on how the effect of e-CRM value on consumers' website loyalty is mediated by institutional and individual characteristics.

REFERENCES

Ad Hamid, N. R. (2005). E-CRM: Are we there yet? *Journal of American Academy of Business*, *6*(1), 51–57.

Anderson, R. E., & Srinivasan, S. (2003). E-satisfaction and e-loyalty: A contingency framework. *Psychology and Marketing*, *20*(3), 123–138. doi:10.1002/mar.10063

Bentler, P. M., & Hu, L. T. (1995). Evaluating model fit. In Hoyle, R. H. (Ed.), *Structural Equation Modeling* (pp. 76–99). Thousand Oaks, CA: Sage Publications.

Bollen, K. A. (1989). *Structural Equations with Latent Variables*. New York: Wiley.

Brown, G. H. (1952, June 9). Brand loyalty – fact or fiction? *Advertising Age*, *23*, 53–55.

Chaudhuri, A., & Holbrook, M. B. (2001). The chain of effects from brand trust and brand affect to brand performance: the role of brand loyalty. *Journal of Marketing*, *65*(2), 81–93. doi:10.1509/jmkg.65.2.81.18255

Chen, H. M., Chen, Q., & Kazman, R. (2007). The affective and cognitive impacts of perceived touch on online customers' intention to return in the Web-based eCRM environment. *Journal of Electronic Commerce in Organizations*, *5*(1), 69–91.

Chen, Q., & Chen, H. M. (2004). Exploring the success factors of eCRM strategies in practice. *Journal of Database Marketing & Customer Strategy Management*, *11*(4), 333–343. doi:10.1057/palgrave.dbm.3240232

Cuieford, J. P. (1965). *Fundamental Statistics in Psychology and Education* (4th ed.). New York: McGraw Hill.

Cunningham, R. M. (1956). Brand-Loyalty-What, Where, How Much. *Harvard Business Review*, *34*(1), 116–128.

Day, G. S. (1994). The capabilities of market-driven organizations. *Journal of Marketing*, *58*(4), 37–52. doi:10.2307/1251915

Dick, A. S., & Basu, K. (1994). Customer loyalty: Toward an integrated conceptual framework. *Journal of the Academy of Marketing Science*, *22*, 99–113. doi:10.1177/0092070394222001

Fornell, C., & Larcker, D. F. (1981). Evaluating Structural Equations Models with Unobservable Variables and Measurement Error. *JMR, Journal of Marketing Research*, *18*, 39–50. doi:10.2307/3151312

Gefen, D., & Straub, D. (2000). The relative importance of perceived ease of use in IS adoption: A study of e-commerce adoption. *Journal of the Association for Information Systems*, 1–8. Retrieved from http://jais.isworld.org?articles/1-8/article.htm.

Hair, J. F., Anderson, R. E., Tatam, R. L., & Black, W. C. (1998). *Multivariate Data Analysis* (5th ed.). Upper Saddle River, NJ: Prentice-Hall.

Jacoby, J., & Kyner, D. B. (1973). Brand loyalty vs. repeat purchasing behavior. *JMR, Journal of Marketing Research, 10*(1), 1–9. doi:10.2307/3149402

Jih, W. J., Lee, S. F., & Fang, S. R. (2006). Investigating the Impact of Customer Relationship Management Practices of E-Commerce on Online Customers' Website Satisfaction: A Model-Building Approach. *International Journal of E-Business Research, 2*(4), 61–77.

Jih, W. J., Wong, S., & Chang, C. B. (2004). Effect of Perceived Risk on Adoption of Internet Banking: An Empirical Investigation. *International Journal of Electronic Business Research, 1*(1), 91–105.

Kennedy, A. (2006). Electronic customer relationship management (eCRM): Opportunities and challenges in a digital world. *Irish Marketing Review, 18*(1-2), 58–68.

Kincad, J. W. (2003). *Customer Relationship Management: Getting It Right!*Upper Saddle River, NJ: Prentice-Hall.

Kuehn, A. (1962). Consumer brand choice as a learning process. *Journal of Advertising Research, 2*, 10–17.

Lee, S. F., Jih, W. J., & Fang, S. R. (2006). Investigating the impact of customer relationship management practices of e-commerce on online customers' website satisfaction: A model-building approach. *International Journal of E-Business Research, 2*(4), 61–77.

Lim, K. S., & Razzaque, M. A. (1997). Brand loyalty and situational effects: An interactionist perspective. *Journal of International Consumer Marketing, 9*(4), 95–115. doi:10.1300/J046v09n04_06

Lipstein, B. (1959). The dynamics of brand loyalty and brand switching. In *Proceedings of the Fifth Annual Conference of the Advertising Research Foundation* (pp. 101-108). New York: Advertising Research Foundation.

Min, S., Song, S., & Keebler, J. S. (2002). An Internet-mediated market orientation (IMO): Building a theory. *Journal of Marketing Theory and Practice, 10*(2), 1–11.

Monroe, K. B., & Guiltinan, J. P. (1975). A path-analytic exploration of retail patronage influences. *The Journal of Consumer Research, 2*, 19–28. doi:10.1086/208612

Newman, J. W., & Werbel, R. A. (1973). Multivariate analysis of brand loyalty for major household appliances. *JMR, Journal of Marketing Research, 10*(4), 404–409. doi:10.2307/3149388

Ngai, E. W. T. (2005). Customer relationship management research (1992-2002): An academic literature and classification. *Marketing Intelligence & Planning, 23*(6-7), 582–605. doi:10.1108/02634500510624147

Parvatiyar, A., & Sheth, J. N. (2001). Customer relationship management: Emerging practice, process, and discipline. *Journal of Economic & Social Research, 3*(2), 1–34.

Peppard, J. (2000). Customer Relationship Management (CRM) in Financial Services. *European Management Journal, 18*(3), 312–326. doi:10.1016/S0263-2373(00)00013-X

Reynolds, F. D., Darden, W. R., & Martin, W. (1974). Developing an image of the store-loyal customer. *Journal of Retailing, 50*(4), 73–84.

Ross, D. (2005). E-CRM from a study chain management perspective. *Information Systems Management, 22*(1), 37–44. doi:10.1201/1078/44912.22.1.20051201/85737.5

Scullin, S. S., Fjermestad, & Romano, N. C., Jr. (2004). E-relationship maketing: changes in traditional marketing as an outcome of electronic customer relationship management. *Journal of Enterprise Information Systems*, *17*(6), 410–415.

Shani, D., & Chalasani, S. (1992). Exploiting niches using relationship marketing. *Journal of Consumer Marketing*, *9*(3), 33–42. doi:10.1108/07363769210035215

Slater, S. F. (1997). Developing a customer value-based theory of the firm. *Journal of the Academy of Marketing Science*, *25*(2), 162–167. doi:10.1007/BF02894352

Srinivasan, S. S., Anderson, R., & Ponnavolu, K. (2002). Customer loyalty in E-commerce: An exploration of its antecedents and consequences. *Journal of Retailing*, *78*(1), 41–50. doi:10.1016/S0022-4359(01)00065-3

Swift, R. S. (2001). *Accelerating Customer Relationships Using CRM and Relationship Technologies*. Upper Saddle River, NJ: Prentice-Hall.

Szymanski, D. M., & Hise, R. T. (2000). E-satisfaction: An initial examination. *Journal of Retailing*, *76*(3), 309–322. doi:10.1016/S0022-4359(00)00035-X

Tung, F. C. (2007). Using the e-CRM information system in the hi-tech industry: Predicting salesperson intentions. *Journal of American Academy of Business*, *11*(2), 131–136.

Turban, E., King, D., Lee, J., & Viehland, D. (2004). *Electronic Commerce: A Managerial Perspective*. Upper Saddle River, NJ: Prentice-Hall.

This work was previously published in International Journal of E-Business Research Volume 6, Issue 3, edited by In Lee, pp. 1-14, copyright 2010 by IGI Publishing (an imprint of IGI Global).

Chapter 6
An Exploratory Look at Attributes of Internet Use and Adoption by Franchisees

Kelley O'Reilly
Utah State University, USA

Zsolt Ugray
Utah State University, USA

ABSTRACT

This case study explores the attitudes and perceptions of franchise owner-operators in regard to their acceptance of advanced Internet innovation and technologies. Because these franchisees serve a dual role as both the decision maker and the end user of new technology and innovation, they provide a dichotomy of perspectives that yield insights into many aspects of business leadership, customer service, and operational proficiency. Findings suggest five key attributes of Internet use and adoption by franchisees. The data in this exploratory case study also reveals three areas of disparity regarding franchisee behavior worthy of consideration by practitioners and academics: (1) The inward focus of franchisees, (2) project costs are considered superficially, and (3) the micro SME as change agent. This research is significant and accretive by capturing the voice of franchisees in regard to Internet acceptance and by providing a strategic look at how the franchise micro SME is significantly different than non-franchised and larger SMEs.

INTRODUCTION

The world in which small and medium sized enterprises (SMEs) compete has been forever changed by the Internet (McGowan, Durkin, Allen, Dougan, & Nixon, 2001; Simmons, Armstrong, & Durkin, 2008; Terziovski, 2003; Yang, Lin, Liu, Chao, & Chen, 2008). While the benefits of the Internet for SMEs are well supported in the literature, Internet use and adoption in SMEs has been relatively slow (Al-Qirim, 2007; Brown, Lockett, & Schubert, 2005; Fillis & Wagner, 2005; Hashim, 2009; Levy, Powell, & Worrall, 2005); particularly for advanced Internet inno-

DOI: 10.4018/978-1-61350-462-8.ch006

vations that go beyond the mere use of websites and mass emails and utilize more sophisticated two-way interaction and analytical data processing (Bengtsson, Boter, & Vanyushyn, 2007; Hashim, 2009; Poon & Swatman, 1999).

While much of the SME literature detailing technology adoption and innovation reports nuances and certain discrepancies as to the key factors that either drive or inhibit Internet adoption and use by SMEs, certain agreements do exist. Factors commonly cited as drivers and/or inhibitors of Internet use and adoption are typically considered from three perspectives: internal, external, and organizational. Within each perspective, common drivers and/or inhibitors for Internet use can be found in the literature. Internal factors typically include perceived usefulness, marketing benefits, existence of a project champion, and project cost (Al-Qirim, 2007; Beckinsale, Levy, & Powell, 2006; Bengtsson et al., 2007; McGowan & Durkin, 2002; Mehrtens, Cragg, & Mills, 2001; Poon & Swatman, 1997, 1999; Simmons et al., 2008; Wymer & Regan, 2005). External factors commonly listed are competitive pressure, customer demand, and vendor support (Al-Qirim, 2007; Beckinsale et al., 2006; Levy & Powell, 2003; Mehrtens et al., 2001). Finally, organizational factors usually include organizational size, technical experience, and organizational readiness (Bengtsson et al., 2007; Burke, 2005; Cragg & King, 1993; Karakaya & Khalil, 2004; Mehrtens et al., 2001).

A variety of approaches, methodologies, and contexts have been used to identify, investigate, and verify technology adoption and use and the drivers and/or inhibitors of Internet use by SMEs through empirical work (Wymer & Regan, 2005). From the aspect of technology adoption and use, several prevailing frameworks and theoretical models are commonly used to frame this research such as the Theory of Reasoned Action (Ajzen & Fishbein, 1980), the Technology Acceptance Model (Davis, 1989; McFarland & Hamilton, 2006), the Adoption, Innovation, and Diffusion Theory (Rogers, 1995), Social Cognitive Theory (Bandura, 1986), and the Unified Theory of Acceptance and Use of Technology (UTAUT) (Venkatesh, Morris, Davis, & Davis, 2003). However, "no single model or theory dominates" (Wymer & Regan, 2005, p. 439).

From an operational perspective, the drivers and/or inhibitors of Internet use by SMEs have been approached in the literature through the development of conceptual frameworks and models (Levy & Powell, 2003; Mehrtens et al., 2001; Simmons et al., 2008). However, these models, conceptualizations, and frameworks have never been applied to franchises. Adding to the complexity, the literature spans several functional areas such as marketing, information systems (IS), management, organizational behavior, and communications to name but a few. These drivers and/or inhibitors of Internet use and adoption by SMEs have commonly been explored in the context of larger SMEs versus the "micro" SME, which is typically defined as an enterprise with fewer than 10 employees. Considering in contrast that "medium" SMEs can have 100-499 employees, it is not surprising that "the variability between firms within the SME sector is such that the distinction between very small firms and large small firms is too important to ignore" (Brown et al., 2005, p. 77).

The purpose of this study is to learn more about a relatively sparsely researched group, micro SMEs. Our focus is the technological Internet innovation that is being conducted within service franchise SMEs at the 'front line': the entrepreneurial Internet use and innovation driven by the owner-operator of a franchised SME. This research is centered on an e-marketing pilot project utilizing advanced Internet innovation to solicit customer feedback and provide new means for two-way interactions between franchisee and customer. The participant data also reveals insight into other Internet uses for two-way customer-company interactions such as blogging and instant messaging.

We explore the innovation in a context previously not studied: micro SMEs with a service franchise network as the backdrop. Franchise businesses are ideal for this type of exploration because they are commonly run by an owner-operator. By focusing our research on the franchise owner-operator, we obtain the perspective of the SME president (franchise owner) who also provides the knowledge and perspective that comes from being closely connected to, and interacting with, customers on a daily basis. This owner-operator perspective is important to capture in order to answer our fundamental research question: what are the key attributes which affect a franchisee's use and adoption of advanced Internet innovation? In particular, we focus on more advanced uses of the Internet and innovation that is conceptualized, funded, and driven by the franchise owner independent of franchise headquarters.

In organizing the narrative for this study, we first discuss the relevant literature in this area. Next, we highlight the research methods and analysis used, and then we present our findings on the attributes related to Internet use and adoption by franchisees. Finally, we offer insight on the implications of these findings, and conclude with a discussion for future research.

REVIEW OF LITERATURE

Factors commonly cited as drivers and/or inhibitors of Internet use and adoption are typically considered from three perspectives: internal, external, and organizational. Within each perspective, common drivers and/or inhibitors for Internet use are frequently found in the literature. Additionally, these perspectives have commonly been explored in the context of larger SMEs versus the "micro" SME with little regard to the business entity type (e.g., franchise vs. non-franchised). For these reasons we have structured our review of the literature so that each of these perspectives, as well as the similarities and differences between SME owners and SME franchise owners, can be discussed in turn.

Internal Drivers and/or Inhibitors of Internet Use and Adoption

Internal factors listed in the literature typically include perceived usefulness, marketing benefits, existence of a project champion, and project cost. The perceived usefulness/benefits of the Internet and subsequent innovation of Internet technologies is well supported as a key driver of use and adoption by SMEs (Levy & Powell, 2003; Mehrtens et al., 2001; Poon & Swatman, 1997). Many very small firms tend to be 'customer led' and therefore view the usefulness of the Internet mainly as an additional marketing tool (Beckinsale et al., 2006; Fillis & Wagner, 2005; Simmons et al., 2008). In their work, Beckinsale et al. (2006) report that "overall, perceived benefit is the main driver for Internet adoption, with some evidence of external pressure particularly for SMEs with close customer relationships" (p. 363). It can be argued that the existence of close customer relationships relates to the marketing orientation of the firm and therefore, perceived benefits and perceived usefulness can be viewed synonymously with marketing benefits in the case of many SMEs. As means of example, "the marketing function has been identified as a key driver of website adoption, with websites viewed as essentially marketing tools" (Simmons et al., 2008, p. 353).

CEO or other senior management support and involvement is another key driver of Internet use and adoption (Bengtsson et al., 2007; Poon & Swatman, 1999). SMEs that are attracted to Internet commerce tend to be more entrepreneurial, risk takers, innovative, and creative. Poon and Swatman (1999) found that the most 'hands-on' and involved managers were from micro-sized firms. Likewise, Beckinsale et al. (2006) reported that "as with other Internet and communication technologies (ICT) decisions in small firms, the owner is key" (p. 361).

Within a small enterprise, the entrepreneur is often recognized as having "limited access to information and a limited capacity, in any formal sense, for processing what information is available" (McGowan & Durkin, 2002, p. 363). This often results in marketing decision-making that is informal and ad hoc in character. So while the entrepreneur of the small firm considers the potential marketing benefits for determining Internet use and adoption, it may be within the context of limited or constrained information that the entrepreneurs' decision-making regarding the vision and value perceptions of adoption occurs. This finding may help to explain why cost does not consistently surface as an inhibitor of Internet use and adoption within SMEs. Al-Qirim (2007) found that the majority of firms that were considered 'high adopters' rejected the notion that electronic commerce is expensive, which may be more indicative of the firm's perception of email and Web page use rather than advanced Internet use and adoption - a notion that adds support to McGowan and Durkin's (2002) portrayal of entrepreneurs having limited access to information and limited resources for processing that information.

External Drivers and/or Inhibitors of Internet Use and Adoption

External factors typically considered in the literature include competitive advantage, competitor pressure, and customer demand as key external drivers of adoption and use. Findings related to competitive advantage that can be gained through Internet use and adoption for SMEs seems contradictory. Most literature in this area fails to consider the existing alternate forms of business entities and as a result many findings seem confusing when viewed through the lens of franchise operations. For instance, in their work, Levy et al. (2005) suggest that "those SMEs remaining in their existing markets are the least likely to invest, primarily due to the Internet not being seen as necessary for growth" (p. 18). Yet, for a franchise business, market territories are commonly defined by contract, and market penetration is a typical strategy to achieve growth within a franchisee's prescribed market area. The Internet is certainly a viable means by which to drive this growth.

Studies considering competitive pressure as a factor for innovation often argue that due to resource or capital constraints, very small firms react to pressure rather than proactively seeking out innovative solutions that might deliver competitive advantage (Bengtsson et al., 2007). Other studies report little empirical evidence that links competitor pressure as a driver of Internet use and adoption. Beckinsale et al. (2006, p. 369) state that "competitor pressure has little influence on Internet adoption by SMEs, while customer pressure is central." This may be more indicative of the marketing orientation of the firm, with those firms closely connected to customers more influenced by customer demand, and those firms with more distanced connections to customers perceiving competitors to be more influential.

Organizational Drivers and/or Inhibitors of Internet Use and Adoption

In the literature, the typical organizational factors affecting Internet use and adoption are the following: organizational size, technical experience, and organizational readiness. Organizational size has been commonly cited as a key predictor of Internet use and adoption, with larger firms more likely than smaller firms to adopt and use Internet technologies (Al-Qirim, 2007; Bengtsson et al., 2007; Burke, 2005). Burke (2005) reported that size was a better predictor of Internet use and adoption than CEO characteristics or industry factors. This finding may indicate that larger firms are more innovative than smaller firms. As an alternative explanation, it is conceivable that larger SMEs possess greater resources, can afford higher investment, or have more experience with Internet technologies, therefore giving larger firms an advantage in implementing newer technologies and innovations.

Supporting the notion that other factors may be present in addition to organizational size, Bengtsson et al. (2007) reported that "firm size is a multi-dimensional factor that reflects a set of complicated inter-relationships" (p. 38) and suggest that in addition to size, the presence of a project champion, top management commitment, and entrepreneurial support are key differentiators for adopters versus non-adopters in very small firms. It has been argued that organizational factors may not be able to adequately capture the adoption processes and rather, factors related to individual preferences and perceptions of individuals may be necessary for a more complete explanation, especially in the case of micro SMEs (Bengtsson et al., 2007).

Technical experience and organizational readiness are also commonly identified predictors of Internet adoption and use. Not surprisingly, firms with higher IT capabilities, i.e. firms that employ personnel who understand the technology, tend to adopt the technologies more rapidly (Karakaya & Khalil, 2004; Mehrtens et al., 2001). Very small firms may look externally to outsource their skill gaps, perhaps more often than larger firms who are more likely to employ full-time trained IT personnel. While discussions regarding the outsourcing of Internet IT initiatives are fairly common in the literature, it is rare that the impact and influence of the outsourcing partner is considered as either a driver or inhibitor of adoption and use (Beckinsale et al., 2006; O'Reilly & Paper, 2009). However, it has been found that the support from technology vendors is often viewed negatively by SMEs (Al-Qirim, 2007), and in those situations, SMEs rely more heavily on their own and locally available knowledge (Beckinsale et al., 2006).

Similarities between Typical SME Owner and Franchised SME Owner

There are similarities between advanced Internet innovation and implementation within SMEs and franchise networks: in both cases, these innovations tend to be 'top-down' in nature and initiated by partners, suppliers, or other parties (such as the franchisor) related to SMEs and not by the SMEs themselves (Beck & Weitzel, 2005; Sharma, Wickramasinghe, & Singh, 2005). This phenomenon has been explained as the result of SMEs' lack of one or more of the following: technology sophistication; IT expertise; or belief in the strategic value of e-business as a tool for profitability or competitive advantage (Beck & Weitzel, 2005; Terziovski, 2003). Within franchised systems, large-scale IT innovation tends to occur by franchisor-driven IT initiatives with expectations (and demands in some cases) of franchisee implementation 'pushed down' from the corporate franchisor. Under this scenario, a franchisee's role is similar to an employee within an organization, whereby the use of technology is mandated into the work processes, procedures, or workflow of the franchisee or employee.

Differences between Typical SME Owner and Franchised SME Owner

While the similarities of SMEs and franchise operations are easy to grasp, it is also important to remember the differences as well. SMEs and franchises are both small business entities, ultimately driven by self-interest. However, in the case of franchisees, large parts of the operations are governed by a franchise contract between franchisor and franchisee. So while studies of SMEs are focused on top management as the decision maker for new technologies, it is important to recognize that innovation within SMEs follows a 'top-down' strategy with the business owner or top management governing the innovation, not the end user of the technology. On the other hand, for franchised SMEs, the ultimate acceptance of new technologies often rests on the franchisee owner-operator as both the decision maker *and* the end user of new technology and innovation.

STUDY PURPOSE AND RESEARCH QUESTION

Because of the dual role of the franchisee (decision maker and end user), there is a great deal of innovation taking place at the 'front line,' where franchisees themselves are implementing IT innovation. This innovation may be driven by internal, external, or organizational factors of the franchisee, or due to a simple unmet need that is not being addressed by the franchisor. This type of franchisee-driven innovation could certainly be considered more 'bottom-up' in nature and for the most part, tends to be overlooked by academic studies. It is with this gap in mind that we embarked on an exploratory case study focused on a US-based service franchise and its franchisees. This study is motivated by the need to understand the perceptions and attitudes of franchisees in regard to their acceptance of advanced Internet innovation and technologies as both decision-makers and end-users.

STUDY DESIGN AND METHODOLOGY

Selection of Research Method, Company, and Franchisees

The case company is a van-based service franchise with a long history of franchising and over 30 years in the service business. With thousands of franchisees across the US as well as internationally, this company is recognized as one of the top franchise networks in the world based on number of franchised units. In addition, this company's 'loose' franchise requirements for hardware, software, and technical specifications for system compliance creates rich opportunities to explore the nature of 'bottom-up' Internet use and innovation by franchisees. It should be noted that prior to the research study, the lead author served as the Vice President of Sales and Marketing at the company for five years; however, all data collection occurred after the lead author had left the employ of the company. In an effort to provide additional information regarding the current structure of the franchisor's business and its subsequent requirements for franchisee's businesses, Table 1 is provided.

To investigate the perceptions and attitudes of franchisees in regard to Internet use and adoption as well as innovation, we chose the case study methodology (Yin, 2003). We believe this qualitative approach, grounded in data, is appropriate due to its origins in interpretive research and its mission to "achieve interpretive research goals and insights concerning meanings" (Gephart, 2004, p. 458). The value of information that will be gleaned rests on our orientation as qualitative researchers and the ontological beliefs that reality is "socially constructed, complex, and ever changing" (Glesne, 2006, p. 6). What is real becomes relative to the specific phenomenon and context being studied (Denzin & Lincoln, 2005). This epistemology relies on assumptions of coming to know and understand this 'reality' through in-depth subjectivist explorations of participants and/or settings and it values a perspective that seeks to situate the data into a socio-political-cultural milieu with the intent to understand, contextualize, and interpret (Glesne, 2006).

We focus on the experiences of the franchisees as they contemplate the design, development, and implementation of new advanced Internet innovation related to electronic customer relationship management (eCRM). Specifically, the innovation in question relates to an electronic feedback mechanism for franchisees to secure feedback and ratings from customers post-service. At the time of data collection, this innovation was in the early pilot stages for all observed franchisees with continued investment expected based on results from the pilot program. By studying this initial pilot innovation initiated by franchisees (as well as other innovation they are designing) we study the phenomenon within its real context.

Table 1. Case company IT requirements and marketing expectations

IT Requirements and Marketing Expectations:	Case Company Internal Marketing (marketing to franchisees)	Case Company External Marketing (franchisees' marketing to consumers)
IT REQUIREMENTS Information Technology and Systems:	Primarily phone and fax based ordering systems. Electronic catalog with basic shopping cart functionality for franchisees to order supplies. No full integration at headquarters between inventory, sales, shipping, compliance, and A/R.	Franchisee's required to maintain database of customers electronically. No universal system requirements for hardware or supporting Point-of-Sale software at franchisee's location. Roughly 20% of franchisees without basic Internet service.
MARKETING EXPECTATIONS Core Service Offering:	Patents and proprietary system elements in place for franchisee competitive advantage in marketplace.	Automotive service
Personalized Services for Customers:	Individualized prospective franchisee sales and follow-up appointments, calls and "Discovery Day" meetings at the corporate office.	One-on-one phone appointment booking and service satisfaction follow-up call to each customer. Individual customer thank-you and reminder cards by mail.
Additional Services	Addition of other auxiliary services franchisee can offer via existing franchise territory without incremental fees.	Addition of other repair and maintenance services such as mobile windshield repair and detailing.
Pricing strategies to encourage customer loyalty	"Pioneer" discounts for existing franchisees when they purchase additional franchised territory(s).	Discounts for pre-booking next scheduled maintenance appointment at time of service.
Marketing to employees so that they, in turn, will perform well for customers	"Spiff" paid to internal associates at corporate headquarters whenever a friend or family member purchases a franchise.	"Spiff" paid to service technicians whenever their "clients" recommend the company's service to friends or family.

*For purposes of anonymity, service description is fictional, while marketing activities are factual.

The selection of franchisees was based on purposeful sampling to select franchisees that are "cases within the case" (Denzin & Lincoln, 2005, p. 451). Purposeful sampling is a viable method for this inquiry because qualitative analysis embraces unfolding discovery, iterative refinement of the research focus, and flexibility in generating concepts and ideas that are substantiated by the data (Glaser & Strauss, 1967). We made our selection based on the following criteria: franchisees are currently involved with the eCRM pilot program, each has experience with online marketing activities, and each is considered to have some technical sophistication by the franchisor. During the selection process we strived to achieve diversity in age, revenue size, experience, and geographical location within our target micro SME franchise category. Over a dozen franchisees were recommended to the researchers by the case company, and each was invited to participate via an email communication. Those franchisees who were interested in participating were called by the lead author to discuss the proposed study and to schedule an interview. All franchisees who agreed to participate were interviewed and through these interviews, the researchers reached saturation of insights (Denzin & Lincoln, 2005).

We observed six franchisee participants (cases). Each participant is the franchise owner, with each owner representing their own organization. The use of six cases is within the range from four to 10 cases as recommended by Eisenhardt (1989). The franchisee participants in this study include two franchisees that have been in the franchise system for over 20 years, three with roughly ten years of experience and one franchisee in his first full year of operation. Two of the franchisee participants run $1 million franchise operations, three have businesses with roughly $500K in annual revenues, and the newest franchisee's operation will do roughly $150K in its first year. The franchisee participants operate their franchises in Florida, New Mexico, New Jersey, New York, Texas, and Washington. In future references, franchisees are referred to as 'a franchisee' with a number (e.g., franchisee [01]) designating which of the six franchisee participants the data is related.

Data Collection

Data were collected through initial interviews with each participant. Interviews were semi-structured with the following interview questions asked of all participants:

1. Describe your understanding of CRM "Customer Relationship Management?
2. Does your understanding change if the word "electronic" is added?
3. What is the primary objective of having an eCRM system for your franchise?
4. What is the benefit a customer would receive through the eCRM system?
5. How would you define eCRM success?
6. Would a fully functioning eCRM system replace or augment your CRM activities?
7. Is there a competitor that you think is really "doing it right" in regard to eCRM?

In addition to the semi-structured open-ended questions listed above, the franchisee participants were asked to share their experiences regarding any other Internet innovation they were testing or currently using. In all cases, these interviews were conducted over the phone. Consistent with Glesne (2006), all interviews were approximately one hour in duration, conducted one-on-one, tape-recorded, and transcribed. In addition, to add to the trustworthiness of the data, member-checking interviews were conducted with each participant (Glesne, 2006) to ensure the accuracy of the interview transcript. In preparation for the member-checking interview, the individual's interview transcription notes were emailed to each as a Microsoft Word document. Participants were asked to review the transcript thoroughly and encouraged to correct the document for any errors and to also include any additional thoughts by adding them to the document. Once the participant transcript review process was completed, an email exchange took place with each participant to finalize the member-checking interview process. In all cases, the transcript was verified by the participant and in two of the six cases, material changes were made to the interview transcription file by the participant. The member-checking cycle occurred within 2-3 weeks of the initial interview and took place via email in all instances. In these ways, data were triangulated (Glesne, 2006).

Data Analysis and Methods

The unit of analysis for our data was the organization or franchise. In addition to the interview and member-checking interview, all interview transcriptions and online communications of the participants were analyzed and coded, in accordance with traditional qualitative methods (Glesne, 2006). For purposes of this study, our coding techniques and paradigms relied on Strauss' systematic schema for coding qualitative data (Strauss & Corbin, 1990). We adopted this coding schema because its systematic nature makes it easy for interested readers to see the codes that emerged from the data. Throughout the coding process, we recognize that the codes are based on our judgment and interpretation, and "limitations are consistent with the always partial state of knowing in social research" (Glesne, 2006, p. 169).

Limitations

While the franchisee attitudes, beliefs, and acceptance of Internet use and adoption illuminated by this exploratory case study are grounded in the data, this research is limited by the general nature of case-based research. The data provides a rich and valuable source of information that can be trusted for the case company in particular. A clear limitation of this exploratory study is that this case is focused on six specific franchisees within a single organization. By the nature of the methodology applied, this study is not meant to infer generalizability to all SMEs or franchisees. It is intended to shed light on the issues surrounding the acceptance of Internet use and initiatives for the franchisee participants in this case study.

KEY FINDINGS

The data revealed five factors that emerge as attributes of Internet use and adoption by franchisees; three of which are related to internal drivers, and one factor each aligning with external or organizational drivers. These factors are: (1) (Internal) degree of useful value the technology will deliver; (2) (Internal) franchisee's comfort with change; (3) (Internal) bottom-line incentive-based motivation; (4) (External) perceived need for innovation, and; (5) (Organizational) access to ad hoc experts. Each of these factors will be discussed below.

Degree of Useful Value the Technology Will Deliver

In line with findings from previous research, the data revealed a forward looking view as the key means by which franchisees determined the value of Internet use and adoption. Franchisees consider the value of the Internet and new innovation more in regard to what they *might* receive, rather than what they *have* received. Poon and Swatman (1999) reported this as "perceived benefits" from Internet use in contrast to "actual benefits." From the data we found that franchisees are continuously evaluating the value of new Internet innovation by the degree to which it will improve their business, enhance their relationships with customers, or fill a current operational gap. For instance as said by one franchisee [05]:

It [e-business] would certainly help in the booking process to eliminate phone time which is important especially in the busy times when you are spending 5-7 minutes on the phone with each customer and during the busy season, and as much as you would like to, you really need to spend 1-2 minutes.

Another franchisee [01] said, "Today we do reminder phone calls and even if someone says on the follow-up phone call, 'Yes, I would like you to call me in 6 months' - there's nothing in place that says don't send this customer a postcard reminder." The idea of adopting the Internet to help drive efficiency, enhance productivity, or improve relationships with customers is well supported throughout the interviews with franchisees [03]:

I think really it's about saving time...I think the more electronic we become in our age, the more, almost invasive personal [one-on-one] interactions become and I've seen a progression of people [customers] who say, 'Don't call me, don't follow-up with me, I think you do great job, but don't talk to me... I'll call you.' And that mentality is increasing; so as a customer, it would minimize the interruptions in their day if we were more electronic...

Some franchisees view the value of the Internet by the degree to which it will deliver something not currently present in their operation: in this example, a means to communicate electronically with customers. Another franchisee [02] discussed other useful benefits of his new online customer survey tool:

Another huge pro for me is that we are maintaining our quality since we are finding out what our customers think about us - it is an easy way to see trends, and you know, I found one trend pretty easy. I found a trend that I had one technician that is not cleaning up after himself on the job. It was great to find. I don't think I would have found out any other way. Although we have our survey cards, it is not compiled in one easy place to see, so I can see trends easier and head off problems.

These statements suggest that the Internet's ability to fill an operational gap is important for franchisee acceptance. This logic by franchisees is intuitive since the type of Internet innovation explored in this study is 'bottom-up' in nature and funded by franchisees. We would expect this rational action by business owners since if new value was not expected, it is unlikely that innovation would be funded or supported. However, this innovation seems to be occurring without consideration for integrating the information and data collected with existing point-of-sale systems to ensure adequate knowledge management. The short-term views of the franchisees suggest either an overly optimistic view of the value of this information silo, or a simplistic view of the process for integrating their Internet innovation with back shop systems and seem to highlight one area where the micro SME owners may differ from larger SMEs.

Franchisees' Comfort with Change

For all franchisee participants, openness to change, a willingness to innovate, and a vision for future Internet initiatives is apparent. For example [04]:

I don't really have a firm grasp on the whole e-thing. But, it would be nice if things were all integrated. You know, all the way around from our relationship tools whether it is our paper stuff, the things we do in our office, to our e-marketing, to having GPS on our trucks and have it all work together. I think that would be a very cool system - I think that would be very powerful...but I have no idea how to do it...

Similar to earlier findings (McGowan et al., 2001), the small business owners (franchisees) in this study exhibited the characteristics inherent in entrepreneurs: an innovative nature, continuously opens to new opportunities, and comfort with change. These franchisee participants were actively seeking new opportunities for Internet innovation independent of the franchisor. One franchisee [02] excitedly described a new texting program he was testing with his technicians:

Did I tell you about XXX.com? It is a free service that you can sign up for and enter your cell phone and your contacts...you call this 800 number and it is the XXX operator and she says, 'Who would you like to contact?' and you say, 'Jane Doe'... and then you say your message and it records it and then sends it as a text message to Jane's cell phone and email. We are testing this right now to keep in contact with our technicians when they are on the road. They contact us when they are on each job and give us a status. We are only testing with one technician right now - it is not a perfect system because a tech might not speak totally clearly or there is background noise, but you can actually go to your email and just listen to the message...

The data suggests that franchisees feel free to experiment within their respective franchised businesses. This may be due to the open culture within this franchise system or may be the inherent nature of these small business owners. Regardless, the idea of rapid innovation and low cost experimentation was common for the franchisees in this study. This is a new phenomenon we found in our set of micro SME owners that has not yet been observed by previous studies in the literature.

Similar to their openness to innovate and experience change, franchisees were also comfortable with the notion of change creating some operational 'pain.' One franchisee [03] proclaimed: "I just think that anytime you go and make a major transformation, you are going to have things fall through the cracks - we still do today. We've had our share..." With this acceptance of a learning curve for change is the implicit value franchisees seem to place on the idea of change itself. In every participant interview, franchisees reported "trying different things to make something happen" and were very animated and positive in discussing new innovation and experimentation within their respective businesses. As franchisee [01] put it:

Paul manages the blog... he just adds as the season changes or as his mood strikes him. He goes in and adds to the blog, so that is a new thing for us and [has] only been up for two months... we are just kind of getting our groove there.

For many, the notion of 'being ahead' of the franchisor and not waiting for a franchisor solution to be delivered to franchisees carried pride for franchisees. "If we had not developed our Web like we've done the past two years, I am not so sure we would be here today," declared one franchisee [03]. In fact, for innovation initiated by the franchisees themselves, complexity seemed to be valued or imply 'bragging rights' for the franchisee [02] involved:

It is complex. There are a lot of emails and logins and passwords to remember, and I think I've got eight tabs across my Internet explorer, with eight home page ads and you know, I think in the future it will become even more challenging [laughing].

A common finding from all franchisees in this study is that they expect new Internet technology and innovation to have problems. For instance, "When you are making a major change like that, things fall through the cracks and they still do and they still will in the future," expressed a franchisee [01]. The expectation that new technology should be easy to use is simply not supported in the data. In fact, many considered the learning curve to be a given with any new innovation. As expressed by one franchisee [06], "It is just a lot of time, a lot; a lot of time... time is the biggest obstacle." Another franchisee [03] seemed almost proud of some of his start-up problems with a new Website function in his operation:

Some things are going to fall through the cracks. Like I said, we've got four different things coming from the email, but if we've got a problem, like we had one time with the server and when people were clicking to get information or a call back, something happened and everything was being diverted and nobody knows where it even went and it went to never-never land... and that was in the very beginning when we were trying to get everything set up, so that caused us some grief...

It is interesting to reflect on the nature of change and innovation being discussed here. Specifically, it is a type of change that the franchisee controls, initiates, and approves. For this 'bottom-up' innovation, ease of use is simply not a factor. These findings seem to be in line with previous research from Poon and Swatman (1999) supporting the notion that SMEs that are attracted to Internet commerce tend to be more entrepreneurial, risk takers, innovative, and creative.

Bottom-Line Incentive-Based Motivation

Like many 'customer-led' entrepreneurs, the franchisee participants in this study view Internet use and adoption more from the perspective of what activities and innovation they believe would ultimately drive bottom line improvements rather than what out-of-pocket costs might be. This is interesting to consider since cost has emerged in larger SMEs as the key inhibitor of Internet use and adoption (Cragg & King, 1993; Wymer & Regan, 2005). However, in line with Al-Qirim (2007), the improvements mentioned by franchisees in this exploratory study were described as costing little, reducing expenses, or improving the bottom line by performing more effectively than other marketing tools currently used.

How Much does it Cost? How Much Can We Save?

For two of the franchisees, the risk of Internet innovation was low due to their perception of the minimal costs associated with implementation. For all but one franchisee participant, the cost equation was more about 'soft costs' like time or labor than true out-of-pocket expenditures. One franchisee [05] expressed this as, "It is such a low cost way of advertising, as long as we are getting a good return, like 3:1 which isn't hard when it costs almost nothing except for a few minutes of time..." Many directly related the success of Internet marketing to the fact that it is inexpensive. For instance, in the words of one franchise [04], "I would say there is no question that it [e-marketing] is a success because your cost is minimal - it is mainly your time to create a template or ad so it doesn't take much to get a good return." In addition to low costs and low risk on innovating, Internet innovation was a way to save expenses. One franchisee [03] explained, "Since we've done that [email customer surveys], and since the guys are getting better about mentioning it, we are getting a lot more responses...and it saves us about $400 a month in postage." Franchisees seemed to focus on very superficial costs for implementation that fail to evaluate vendor support, training, system maintenance, or customer risks due to radical innovation. However, in contrast to fully understanding the cost side of the equation, they all reported very detailed numbers regarding the revenue enhancements the Internet innovation delivered as compared with their traditional offline methods for communicating with customers.

Effectiveness

When asked about the effectiveness of his new online e-marketing activities, one franchisee [03] responded, "It is our most valuable channel for customer acquisition." For another, it was clear that a bottom-line focus and detailed tracking all weigh into the consideration for e-marketing (online) versus traditional (offline) marketing results and evaluation [06]:

We did receive out of the 100 emails we sent out; three of the jobs we got came from people that we didn't send the email to - meaning that the email was forwarded to them by the person we sent it to.

From results of the data analysis, it appears that these franchisees were reaping some benefits from their customer-centric Internet initiatives and all seemed comfortable in the notion that the benefits outweighed the costs.

Perceived Need for Innovation

Many franchisees expressed a 'need' for Internet innovation to ensure their service level could meet the changing expectations of their customers. This sentiment was explained by one franchisee [06] in these words: "The way customers want to be managed is different…not just the way they want to respond to advertising, but the way they want to be managed by a service provider is changing…" Another franchisee [04] supported this same idea:

I think that as a small business person, we have to gain the knowledge of how it works and what the Internet does. I am just amazed everyday how many emails Coco [pseudonym] gets to request an estimate or schedule an appointment…I don't know, I really see this as the way to go in the future.

For one of the larger franchisee [01] operators, surprise and delight accompanied her new entry into the online environment, "We probably get 5, 6, or 7 calls a day [from our Website]. It is amazing to me, now that we've started paying attention to our World Wide Web presence and all the tools that are available, it's amazing." Even in cases where the Internet model didn't make life 'easier,' a franchisee [06] still believed that it was what customers were demanding:

It just seems like everybody just wants to communicate by email… it just seems like throughout the day and throughout the night, people are just waiting to send an email and request service. It makes it difficult to communicate via email because you have to try to understand what they are asking and then you have to try to reply through typing… you want to try to get your benefits across like you can over the phone so it can be difficult…

Regardless of the catalyst for a franchisee's acceptance of Internet use and adoption, the majority referred to their efforts to innovate as a means to achieve competitive advantage by attracting more customers to their business or to satisfy existing customers. As aptly put by one franchisee [04], "If they book online and are already accustomed to doing stuff online, why not keep them that way?" Franchisees in this study appeared motivated and energized by the need to innovate and their progressive orientation to change via the Internet was a surprising emergent factor in the data. In fact, this indicated one more area of departure from traditional SME owners and these franchise owners – the participants in this study at least did not report a *need* to react to competitor pressure, but rather a *desire* to proactively meet the needs of their customers.

Access to Ad Hoc Experts

Access to ad hoc 'experts' for guidance and support emerged as a strong theme for franchisees and seems to provide a base of confidence for these business owners. Unlike previous research (Karakaya & Khalil, 2004; McGowan et al., 2001; Mehrtens et al., 2001) which suggests that a lack of technical expertise within a SME may limit an owner's sense of the Web providing any strategic edge for the business, the franchisee data in this study suggests that even without technical expertise, the value of the Web and its e-enabling characteristics is valued universally by study participants.

Perhaps this can be explained by the 'network' of ad hoc experts that were present for each franchisee's specific business. Some expertise was acquired by franchisees through outside Web companies, some through relatives, friends, and employees. When asked how she handled her new online marketing reminder cards, a franchisee [05] explained, "My Mom is retired and she is really the one who is handling this and is very handy with this kind of stuff." For other franchisees, similar sentiments referencing fathers, brothers, cousins, and wives as the subject matter experts for all things technical emerged. Statements such as "My dad is retired from IBM"… "My sister works for Microsoft"… "My cousin sells ERP systems"… were used as explanation for how they were implementing e-business innovation even without a personal understanding of the underlying technology. In several instances, franchisees referred to another franchisee that was helping them with online marketing [01], "We've been really active with him [franchisee] and we've seen our [Web] presence and actually people booking off of him [his online tool], and that is really amazing to me." Interestingly, in all but one instance, franchisees referred to the same web-company they hired to help them with e-business innovation.

This outside web-company was common to five of the six franchisee participants and was reportedly working on similar projects for each franchisee and billing each separately. Franchisee [04] stated: "[franchisee] is doing our electronic newsletter piece but we rely on [web-company] for all of our e-marketing activities." Another franchisee [03], referring to the same web-company said:

I know the other things that they [web-company] can do, and they've been very willing to help us even if they haven't tried something before, they have been very willing to put forth the effort and to try and make things work for us and develop whatever we need.

Of note is that many of the projects this outside web-company was handling were support functions provided by the franchisor at no cost to the franchisee. This is a key point to consider since it seems to strongly support the notion that when innovation is initiated by the franchisee, franchisees will take whatever action necessary to implement, suggesting low resistance to new Internet initiatives. However, even in cases where the franchisor was providing free programs and assistance, franchisees were choosing to pay for an outside expert. This may suggest a lack of confidence in the company managers' handling the Internet support programs, or simply a desire on the part of these franchisees to keep the franchisor 'at arm's length.'

Considering this, one might argue that it is this network of ad hoc experts that provide the confidence that enables franchisee owners to move forward with Internet innovation. Without these ad hoc subject-matter-experts, the franchisees' attitudes toward Internet use and adoption may vary; particularly if it creates more reliance on the franchisor as the innovation driver. A competitive and independent spirit seemed present in the decisions of these franchisees [03]:

I bet we are farther along than the majority of [other franchisees in system] in that same category. So, from where I am sitting, for us, some of the things that we've got [web-company] working on right now...is to find a way to capture at least 80% of the emails that our current customer database has right now, and we want to do it in you know, in less than a year's time.

This comment, representative of most franchisees in this study, clearly articulated the key drivers and attributes in place for these franchisees to accept new Internet initiatives: a perceived need and openness to change based on customer demand, a bottom-line focus, and heavy reliance on outside 'experts' for guidance and confidence.

FRANCHISEE CONTEXT SPECIFIC BEHAVIOR AND IMPLICATIONS FOR FRANCHISORS

The data in this exploratory case study reveal three areas of franchisee context specific behavior worthy of consideration by practitioners and academics alike: (1) The inward focus of franchisees, (2) project costs are considered superficially, and (3) the micro SME as change agent. In addition we provide two recommendations for franchisors who desire to drive advanced Internet use and adoption by franchisees. These disparities and recommendations will be discussed in turn below.

The Inward Focus of Franchisees

Unlike many studies reporting the drivers or inhibitors of adoption use by SMEs, the franchisees in the case study are much more inwardly focused. In previous work external and/or organizational drivers such as organizational size, readiness, competitive advantage and marketplace issues have been reported to be a stronger driver leading to SME adoption of the Internet than internal drivers (Burke, 2005; Fillis & Wagner, 2005; Karakaya & Khalil, 2004; Mehrtens et al., 2001). In our study, three of the five drivers which emerged from the data were internal (degree of useful value the technology will deliver; franchisee's comfort with change, and bottom-line incentive-based motivation). One possible explanation to this phenomenon is the franchise business structure. The ownership structure between franchisor (parent company) and franchisee (micro business owner) is both symbiotic and regulated by contract. The franchisor is the "owner" of the brand and licenses the brand for use by franchisees. The franchisor, in most cases, is responsible for product innovation, new R&D, marketing and brand communication strategies, and training and operational support. Consequently, franchisees tend to be more focused on "micro" issues related directly to their local operations. This fundamental difference between franchise SMEs and non-franchise SMEs adds a level of complexity to the phenomenon of SME Internet use and adoption that has yet to be fully explored in research.

Project Cost Considered Superficially

An additional confounding finding is the superficial consideration these franchisees give to the cost of Internet innovation and experimentation which is in direct conflict with previous research findings by Wymer and Regan (2005). Cost has been reported to be the most important barrier for ICT adoption (Marasini, Ions, & Ahmad, 2007), yet cost as a barrier or inhibitor never surfaced in our data. There may be two explanations for this finding. First, this may be related to the low-cost/low-risk types of innovation and experimentation these franchisees were individually involved with. Additionally, the pilot program for the eCRM tool to interact with customers was being provided by the vendor at no charge with the ultimate fee being based on a 'pay on performance model,' where franchisees would only pay a fee for leads or jobs they get but would not pay any software, licensing, or use fees to the external vendor. For these reasons, the franchisees may have become 'accustomed' to considering cost only as it relates to the expected revenue the innovation will deliver. In either case, the superficial consideration of possible costs revealed by the data points to an additional discrepancy between these franchised SME owners and the non-franchised SME owners reported in previous work.

The Micro SME (Franchisee) as Change Agent

Previous research reports that many SMEs adopt the Internet since they are suppliers to a larger organization that 'drives' or demands electronic commerce and/or advanced Internet use and innovation (Beckinsale et al., 2006). For instance, many large organizations have insisted on trading partners being electronic data interchange (EDI) capable (Mehrtens et al., 2001, p. 169). In the franchise context underlying our investigation it appears from the data that the franchisees, and not the franchisor, have assumed the role of leadership for advanced Internet use and adoption. In fact, the eCRM pilot program was the result of significant franchisee 'pressure' to the franchisor and the vendor was introduced to the case company by one of its franchisees. This suggests that it is the amalgamation of many micro SME franchise owners, with a common goal, united in action, which is the key driver for change within the larger SME, the franchisor. So while Bengtsson et al. (2007) argue that there is a "positive association between the size of the firm and its propensity to adopt advanced Internet for marketing operations" (p. 31), we found evidence of the reverse in this case, with the small SME (franchisee) leading the larger SME (franchisor).

Levy et al. (2005) reported that "Product innovation rather than market penetration drives e-business" (p. 18). However, due to the franchise contract governing service and trade areas, franchisees are driven by the need to increase market share within their respective areas, while the franchisor is focused on product and brand innovation. These basic business entity structural differences seem to negate some findings from previous research and speak to the need to consider these franchised micro SME owners differently than other non-franchise SMEs.

While this case is focused on six specific franchisees within a single organization, their perspectives and beliefs regarding Internet use and adoption provide valuable lessons and insights for other companies contemplating similar IT innovation and 'best practices' for driving acceptance. It is important to reiterate and recognize that for these participants at least, resistance to Internet innovation varies depending upon the 'driver' of change. For Internet initiatives driven internally (by the franchisee), little to no resistance exists.

However, it seems apparent that initiatives driven externally (by the franchisor) are more likely to experience resistance. For this reason the following two recommendations are based on the franchisee insights of this case and are intended to provide a guideline for driving Internet acceptance and innovation by similar companies, in similar situations.

The Customer as External Force for Change

As seen in the data, the perceived need for change by franchisees is motivated primarily by their sense that 'customers demand it.' Understanding that the voice of the customer is the most valued by franchisees, acceptance of franchisor-driven innovation will be difficult without consideration of customer needs. This notion of customers being the most valuable source of assistance in achieving improved business performance is both intuitive and supported by earlier studies (Terziovski, 2003). Knowing this, franchisors will be well-served to conduct formal data-gathering and market research projects to ensure that what they propose is both wanted and expected by customers. If so, the customers, via secondary data (for example focus group, video, or audio recordings), can 'sell' franchisees on the need for change. Without this customer voice, the franchisor is left struggling to secure acceptance by franchisees with only the 'stick' of demand making.

Control as a Motivator or Inhibitor

Those familiar with franchising understand that franchisees are more likely to listen to fellow franchisees than to the franchisor. 'Control' is one factor impacting Internet use and adoption in this study. When franchisees control the type, pace, and path of innovation, resistance is minimized. One tool in the arsenal of franchisors and managers is to enlist the help of franchisees who support the IT innovation and carry influence with other franchisees in the network. Clearly, the franchisees in this study are not content to simply find comfort in the fact that "someone in the business is looking at the possibilities offered by the Web" – a finding unlike previous research (McGowan et al., 2001, p. 133). As small business owners, they desire to 'get in the middle of change' and innovate to 'make something happen.' Pulling influential franchisees into the process will help break down the barriers and resistance that will likely accompany franchisor-driven innovation. The franchisees in this study demonstrate a high degree of interest and desire for experimentation with Internet initiatives, supporting the notion that "Managers who have a positive attitude toward the adoption of e-commerce also perceive e-commerce as adding strategic value to the firm" (Grandon & Pearson, 2004, p. 209). Franchisors will benefit by recognizing the desire for, and acceptance of, franchisee-controlled innovation. By tapping into this support, franchisors can increase the probability of acceptance for their own innovation initiatives.

CONCLUSION AND FUTURE RESEARCH

Franchise systems, and franchisees specifically, provide fertile research ground. Franchisees, as both customers (of the franchisor) and CEOs (of their respective franchise), provide a dichotomy of perspectives that yield insights into many aspects of business leadership, customer service, and operational proficiency. Other than franchised hotel or fast-food format franchises, little research exists on the exploration of franchisee operations. We think there is an opportunity for academics and practitioners to fill this gap by adding to the literature base on franchising, franchisors, and especially, franchisees.

This same opportunity applies to the types of Internet use, adoption, and innovation discussed by the franchisees in this case study. Perhaps because much of the innovation discussed here is 'low or no' out-of-pocket expense, one can argue that this type of Internet innovation is becoming prolific amongst SMEs and franchisees. The new accessibility to previously expensive technologies may level the playing field for all (Sands, 2003). It is probable that the Internet will continue to be a driver in SMEs for improving customer relations and operational efficiency – particularly for those SMEs with close customer relationships (Beckinsale et al., 2006).

We would be remiss if we did not point out one observation during our analysis that raised more questions than it answered. Throughout our data analysis, similarities between our findings and certain aspects of the Technology Acceptance Model (TAM) and its ensuing forms (Davis, 1989; Venkatesh et al., 2003; McFarland & Hamilton, 2006) surfaced. TAM and its extensions have long held as a prevailing framework for understanding user acceptance of new technologies. The two specific variables as key determinants of user acceptance employed throughout the prevailing literature are "*perceived usefulness*" and "*perceived ease of use.*" *Perceived usefulness* is a straightforward concept and is rooted in the notion that if a particular technology is perceived to help a user perform a job better, then that user is more likely to use the system. In contrast, *perceived ease of use* is reliant on the premise that even if a technology is perceived to be useful by the end user, if the technology is perceived to be too difficult to use and the *benefits* of using the technology are not greater than the *effort* of using it, the technology is less likely to be employed.

Within most franchised systems, large-scale IT innovation tends to occur by franchisor-driven IT initiatives with expectations (and demands in some cases) of franchisee implementation 'pushed down' from the corporate franchisor. Under this change scenario, a franchisee could be viewed in the same light as an employee within an organization, whereby the use of technology is mandated into the work processes, procedures, or workflow of the employee or franchisee. For this scenario, TAM fits neatly into the confines of this condition and one would intuitively expect to find *perceived usefulness* of the new technology as well as the *perceived ease of use* to be key factors in a user's (franchisee's) decision to accept the new technology. However, the innovation studied here was driven by franchisees and was 'bottom-up' in nature.

To find a direction for future investigations we looked to the more recent work by Venkatesh et al. (2003) in which a more unified model, the Unified Theory of Acceptance and Use of Technology (UTAUT) was formulated. Building on the survey questions presented there we asked the six participants in our study to respond to the UTAUT survey instrument. While the number of our participants is too low for meaningful statistical analysis, the results mirrored that of our qualitative data and suggest a need to consider an alternative framework for understanding franchisee acceptance of Internet use and adoption. This finding is in line with the general conclusion originating from social cognitive theory that behavior is selective for the individual and a specific environment (Bandura, 1986). McFarland and Hamilton (2006) attempted to address the issue by introducing contextual variables but their model does not capture features of the environment that give a distinctive nature to the context where franchisees operate.

We believe there are many questions uncovered by our study that merit future investigations: for instance, would other franchisees from this same franchise network share the beliefs and attitudes of Internet innovation as expressed by the participants here? Would franchisees from other companies share these same beliefs and attitudes? How does the 'demand making' by the franchisor regarding IT initiatives impact these attitudes and beliefs? And finally, is there a conceptual model to explain the phenomenon of Internet use and adoption by franchisees more directly than the current prevailing models?

The phenomenon of 'bottom-up' innovation and acceptance by SMEs and franchisees is worthy of additional research and exploration. Understanding how the Internet is accepted and innovation is considered by franchisees in other systems (with both tight and loose IT requirements as defined by the franchise contracts) is of interest to either support or challenge the findings of this case and to illuminate additional 'best practices' for ultimately driving acceptance of varying advanced Internet initiatives. Clearly, there is much yet to know.

REFERENCES

Ajzen, I., & Fishbein, M. (1980). *Understanding Attitudes and Predicting Social Behavior*. Englewood Cliffs, NJ: Prentice Hall.

Al-Qirim, N. (2007). A research trilogy into E-Commerce adoption in small businesses in New Zealand. *Electronic Markets*, *17*, 263–285. doi:10.1080/10196780701635872

Bandura, A. (1986). *Social Foundations of Thought and Action: A Social Cognitive Theory*. Englewood Cliffs, NJ: Prentice Hall.

Beck, R., & Weitzel, T. (2005). Some economics of vertical standards: Integrating SMEs in EDI supply chains. *Electronic Markets*, *15*, 313–322. doi:10.1080/10196780500302781

Beckinsale, M., Levy, M., & Powell, P. (2006). Exploring Internet adoption drivers in SMEs. *Electronic Markets*, *16*, 361–370. doi:10.1080/10196780600999841

Bengtsson, M., Boter, H., & Vanyushyn, V. (2007). Integrating the Internet and marketing operations. A study of antecedents in firms of different size. *International Small Business Journal*, *25*, 27–48. doi:10.1177/0266242607071780

Brown, D. H., Lockett, N., & Schubert, P. (2005). Preface to the focus theme section 'SMEs and E-Business. *Electronic Markets*, *15*, 76–78. doi:10.1080/10196780500083720

Burke, K. (2005). The impact of firm size on Internet use in small businesses. *Electronic Markets*, *15*, 79–93. doi:10.1080/10196780500083738

Cragg, P. B., & King, M. (1993, March). Small-firm computing: Motivators and inhibitors. *Management Information Systems Quarterly*, 47–60. doi:10.2307/249509

Davis, F. D. (1989, September). Perceived usefulness, perceived ease of use, and user acceptance of Information technology. *Management Information Systems Quarterly*, 319–340. doi:10.2307/249008

Denzin, N. K., & Lincoln, Y. S. (2005). *The Sage Handbook of Qualitative Research* (3rd ed.). Thousand Oaks, CA: Sage Publications.

Eisenhardt, K. M. (1989). Building theories from case study research. *Academy of Management Review*, *14*, 532–550. doi:10.2307/258557

Fillis, I., & Wagner, B. (2005). E-business development. An exploratory investigation of the small firm. *International Small Business Journal*, *23*, 604–634. doi:10.1177/0266242605057655

Gephart, R. (2004). Qualitative Research and the Academy of Management Journal. *Academy of Management*, *47*, 454–462.

Glaser, B. G., & Strauss, A. L. (1967). *The Discovery of Grounded Theory*. Hawthorne, NY: Aldine De Gruyter.

Glesne, C. (2006). *Becoming qualitative researchers* (3rd ed.). Boston: Pearson Education, Inc.

Grandon, E. E., & Pearson, J. M. (2004). Electronic commerce adoption: An empirical study of small and medium US businesses. *Information & Management*, *42*, 19–216.

Hashim, N. A. (2009). E-commerce and SMEs – The need for caution. *Prometheus*, *27*, 125–140. doi:10.1080/08109020902895268

Karakaya, F., & Khalil, O. (2004). Determinants of Internet adoption in small and medium-sized enterprises. *International Journal of Internet and Enterprise Management*, *2*, 341–365. doi:10.1504/IJIEM.2004.005568

Levy, M., & Powell, P. (2003). Exploring SME Internet adoption: Towards a contingent model. *Electronic Markets*, *13*, 173–181. doi:10.1080/1019678032000067163

Levy, M., Powell, P., & Worrall, L. (2005). Strategic intent and e-Business in SMEs: Enablers and inhibitors. *Information Resources Management Journal*, *18*, 1–20.

Marasini, R., Ions, K., & Ahmad, M. (2008). Assessment of e-business adoption in SMEs. A study of manufacturing industry in the UK North East region. *Journal of Manufacturing Technology Management*, *19*, 627–643. doi:10.1108/17410380810877294

McFarland, D. J., & Hamilton, D. (2006). Adding contextual specificity to the technology acceptance model. *Computers in Human Behavior*, *22*, 427–447. doi:10.1016/j.chb.2004.09.009

McGowan, P., & Durkin, M. G. (2002). Toward an understanding of Internet adoption at the marketing/entrepreneurship interface. *Journal of Marketing Management*, *18*, 361–377. doi:10.1362/0267257022872451

McGowan, P., Durkin, M. G., Allen, L., Dougan, C., & Nixon, S. (2001). Developing competencies in the entrepreneurial small firm for use of the Internet in the management of customer relationships. *Journal of European Industrial Training*, *25*, 126–136. doi:10.1108/EUM0000000005443

Mehrtens, J., Cragg, P. B., & Mills, A. M. (2001). A model of Internet adoption by SMEs. *Information & Management*, *39*, 165–176. doi:10.1016/S0378-7206(01)00086-6

O'Reilly, K., & Paper, D. (2009). The role of vendor in the development of eCRM tools. *Qualitative Market Research -. International Journal (Toronto, Ont.)*, *12*(4), 404–427.

Poon, S., & Swatman, P. M. C. (1997). Small business use of the Internet. Findings from Australian case studies. *International Marketing Review*, *14*, 385–402. doi:10.1108/02651339710184343

Poon, S., & Swatman, P. M. C. (1999). An exploratory study of small business Internet commerce issues. *Information & Management*, *35*, 9–18. doi:10.1016/S0378-7206(98)00079-2

Rogers, E. M. (1995). *Diffusion of Innovations* (4th ed.). New York: The Free Press.

Sands, M. (2003). Integrating the Web and e-mail into a push-pull strategy. *Qualitative Market Research –. International Journal (Toronto, Ont.)*, *6*, 27–37.

Sharma, S. K., Wickramasinghe, N., & Singh, M. (2005). Building systems around reverse value chain: A new system development approach. *International Journal of Management and Enterprise Development*, *2*, 93–105. doi:10.1504/IJMED.2005.006032

Simmons, G., Armstrong, G. A., & Durkin, M. G. (2008). A conceptualization of the determinants of small business website adoption. *International Small Business Journal*, *26*, 351–389. doi:10.1177/0266242608088743

Strauss, A., & Corbin, J. (1990). *Basics of qualitative research: grounded theory procedures and techniques*. Newbury Park, CA: Sage Publications.

Terziovski, M. (2003). The relationship between networking practices and business excellence: A study of small to medium enterprises (SMEs). *Measuring Business Excellence*, *7*, 78–92. doi:10.1108/13683040310478011

Venkatesh, V., Morris, M. G., Davis, G. B., & Davis, F. D. (2003). User acceptance of information technology: Toward a unified view. *Management Information Systems Quarterly*, *27*, 425–478.

Wymer, S. A., & Regan, E. A. (2005). Factors influencing e-commerce adoption and use by small and medium businesses. *Electronic Markets*, *15*, 436–453. doi:10.1080/10196780500303151

Yang, M. H., Lin, W. S., Liu, S. C., Chao, H. Y., & Chen, S. H. (2008). Developing the partner relationship management system for franchised electronic stores. *International Journal of Technology Management*, *43*, 176–193. doi:10.1504/IJTM.2008.019414

Yin, R. (2003). *Case study research: Design and methods*. London: Sage Publications.

This work was previously published in International Journal of E-Business Research Volume 6, Issue 3, edited by In Lee, pp. 15-33, copyright 2010 by IGI Publishing (an imprint of IGI Global).

Chapter 7
Towards an Understanding of User Acceptance to Use Biometrics Authentication Systems in E-Commerce:
Using an Extension of the Technology Acceptance Model

Fahad AL Harby
University of Bradford, UK

Rami Qahwajim
University of Bradford, UK

Mumtaz Kamala
University of Bradford, UK

ABSTRACT

Security is an increasingly important issue for business, and with it, the need for authentication. The use of biometric systems for personal authentication is a response to the rising issue of authentication and security. As with any new technology, user acceptance is often hard to measure, thus, in this work, a study of user acceptance of a biometrics authentication system in e-commerce, such as online banking within the Saudi society, was conducted. The study examined whether Saudis are practically willing to accept this technology by involving 306 participants in a large scale laboratory experiment that actively tested a biometric authentication system in combination of a survey. The Technology Acceptance Model (TAM) was adopted as the theoretical basis to develop the research framework in which the model has proven its efficiency as a good predictor for the study's application.

DOI: 10.4018/978-1-61350-462-8.ch007

Table 1. A comparison of the levels of online banking use in UK and Saudi Arabia

Country	Population[1]	Number of Internet Users	Registered banks[2]	The percentage of online banking users of total banks users	Industry Net Income (£ mil)
UK	60,093,000	31,906,960	382 (2005)	85%	8440 (2004)
SA	22,673,538	6,400,000[3]	16 (2006)	20%	3429 (2006)

INTRODUCTION

Due to the rapid development of e-commerce during the last decade, e-commerce activities such as online banking have become widely accepted in industrialised countries such as UK and partially accepted in less industrialised countries such as Saudi Arabia. The following table shows a comparison of the levels of online banking use in the two countries (Alsajjan, 2008) (Table 1).

Securing online banking log in access is becoming increasingly important. Identity theft, hacking, and viruses are growing threats to online banking users. As more people use the internet, more identity theft cases are being reported. This could harm not only the users but also the reputation of the organisations whose names were used in these illegal acts. Over 8.4 million Americans had their personal information compromised during 2007[4]. For the same year, the loss caused by referred complaints was $239.09 million. In 2007, the latest UK statistics have estimated losses due to online banking fraud at £22.6 m[5].

The use of biometric authentication is a response to the rising issue of security. Biometrics devices have the obvious benefit of not falling prey to many of the well known vulnerabilities of traditional methods (James et al., 2006). While a biometric device uses a unique biological trait to distinguish an individual, it is very difficult and often impossible for the identifier to be lost, stolen, duplicated, or given away (Liu & Silverman, 2001; James et al., 2006). This advantage makes biometric authentication an attractive opportunity for individuals and organisations that wish to adopt a new security technology. The most widely used method of biometric authentication is fingerprint recognition. Fingerprints are well known to be unique for each person, and, for this reason, they are considered a secure method of authentication (Jain, 2004). The choice of fingerprints over other biometric technologies was due to their low cost, ease of use, reliability and high accuracy (Jain, 2004). Choosing fingerprints could also make it easier to overcome cultural issues that may arise if other features such as speech or faces are used.

Online banking is among to which biometrics may be applied (Liu & Silverman, 2001). Several companies look forward to using biometric technologies for authentication as it could reduce losses because of fraud (James et al., 2006). Biometrics technology can help prevent illegal financial transactions and identity theft (Jain et al., 2000; Herman, 2002).

Many benefits can be gained from using a fingerprint biometrics authentication within online banking applications, such as ease of use (since no data input (user ID and password) are required from the user), and reducing data vulnerability. Another benefit is increased security and decreased risk of viruses, as the browser is built in such a way that there is no need to type the Uniform Resource Locator (URL). As a result, phishing would be reduced due the lack of data input by user. Moreover, the benefits of investigating biometric authentication systems in online banking will secure the log in process to the system and removes password vulnerabilities; enhanced convenience such as employees quickly log in using their finger; reduced help desk costs by eliminates calls for password resets.

The acceptance of new technology is often hard to gauge. For this work, a study was conducted of user acceptance of biometrics authentication system in online banking within Saudi society. This study focuses upon Saudi Arabia, which has a diverse immigrant population, a Sharia legal system and a developing economy (Al-Somali et al., 2008), and therefore makes an interesting and unique case study. From the economists' point of view, Saudi Arabia is the powerhouse of the Middle East. It has the leading regional economy, and, even though it is still relatively young, it has achieved a rapid rate of growth. It has a young and rapid growing population, 60% of the population are under the age of 30 years, 49% of the population are female and 51% are male. Based on report by the Centre of Excellence for the security of information at King Saud University[6]; statistics for the information security on the web context of the 2008 showed that Saudi Arabia was ranked ninth globally in the number of users who have been attacked over the Internet. That figures is alarming and reflect the security of information among online banking users in Saudi Arabia as well as government agencies and private, is required a serious stand to raise the level of secure the online banking at the private or personal. This makes Saudi Arabia an attractive potential market for all kinds of online banking applications (Png et al., 2001). Having said that, with 60% of the population under the age of 30 and are more to be expected to take the risk of accepting the new technology (Straughn & Albers-Miller, 2001). From the researcher's point of view, the study findings will support the online banking industry to develop strategic plans in order to introduce more significantly acceptance of this technology.

A robust literature review has revealed only a limited amount of research related to authentication on online banking (Koufaris, 2002; Tamilia et al., 2002; James et al., 2006). In order to shed more light on these perspectives, a study based on Davis' (1986) Technology Acceptance Model (TAM) was performed to examine the acceptance of using biometric authentication system in online banking application.

TAM has received wide acceptance for studying the acceptance of new technologies (Davis, 1989). We extend TAM to determine the acceptability and potential for using biometric technologies, specifically fingerprint devices. There have been many attempts to extend existing models by incorporating additional explanatory variables in order to account for more for the variance. The authors argue that integrating new product attributes and social influences into TAM will offer better prediction of users' acceptance of biometric authentication systems. In fact, most of the previous cross-cultural studies have recommended using TAM, as it does not hold invariants across cultures (Straub et al., 1998; Pavlou & Chai, 2002). It is worth noting that the majority of biometric authentication studies have focused on research surveys (Laux et al., 2007). However, this study is different. We created a laboratory experiment that actively tested a biometric authentication system in combination with a survey which then measured the likeliness of the user's acceptance of this technology.

This paper is organised as follows: the literature review and research motivation are introduced in Section 2, while the research model and the hypothesis development are discussed in Section 3. In Section 4, we discuss the experimental approach followed by data analysis, results and the evaluation of the hypothesised model in Section 5. Section 6 includes discussion. The limitations of this study and possible directions for future work are provided in Section 7, followed by the conclusions.

LITERATURE REVIEW

Biometric Authentication Methods

The process of authentication is the process of verifying and confirming an identity. It is accomplished by using any one or a combination of the following traditional identification techniques: knowledge-based (something you know, e.g., a secret password), object-based (something you have, e.g., a token), and ID-based (something you are, e.g., a biometric) (O'Gorman, 2003; Laurie et al., 2007).

Biometric authentication is considered the automatic identification, or identity verification, of an individual using either a biological feature they possess (physiological characteristic like a fingerprint) or something they do (behavioural characteristic, like a signature) (Wayman & Alyea, 2000). There is increasing interest in using biometrics: the International Biometric Group (IBG) predicts the market for biometrics to rise from \$2.1B in 2006 to \$5.7B in 2010[7], driven by large-scale government programs and dynamic private-sector initiatives. However, general awareness of biometric technologies is low because of their limited applications (Furnell & Evangelatos, 2007).

Woodward et al. (2001) study identified religious objections as one of main socio-cultural concerns about biometrics Different countries have different cultures and religious beliefs which govern business and social practices, For example, women's facial recognition would be prohibiting in some Muslim countries such as Saudi Arabia. Based on a study exploring the feasibility of using biometrics authentication within Saudi culture, AL-Harby, Qahwaji and Kamala (2008) stated that most Saudis would prefer the use of fingerprint authentication methods over other methods such face and voice recognition.

Finally, the perceived usefulness of a biometric in a particular situation plays a major role in user acceptance, as does the perceived ease of use of the biometric (Giesing, 2003; James et al., 2006)

Technology Acceptance Model

The current research extends and adapts the original technology acceptance model to study the user acceptances of biometrics within biometrics system to secure online banking applications. To study this, we need to extend the TAM model to account for the new product attributes and social influences that are not included in the original TAM model. While we focus on biometric authentication technology, new product attributes and social influences are not completely unique to biometrics authentication, but are also valid to other technologies (Mukherjee & Hoyer, 2001; Cheng et al., 2006; Al-Gahtani et al., 2007; Laux et al., 2007).

The TAM model was examined in a variety of technologies, services, products and environments, including IT technology (Davis 1989), the WWW (Atkinson & Kydd, 1997; Moon & Kim, 2001; Chen et al., 2002; Cheng et al., 2006), e-commerce (Chen et al., 2002; Koufaris, 2002), online banking (Pikkarainen et al., 2004), online shopping (Gefen, 2003), software adoption (Szajna, 1994), smart cards (Plouffe et al., 2001), and biometric devices (James et al., 2006).

RESEARCH MODEL AND HYPOTHESIS DEVELOPMENT

The research framework and the relevant hypotheses for user acceptance of biometrics authentication in online banking were mainly developed based on the intrinsic features of the TAM model. We have modified the model to meet the goals of our research by adding other intrinsic factors, new product attributes and social influences. The introduction of new product attributes (biometrics system characteristics) to the model aims to reflect the uncertainty currently associated with the online banking secure authentication (Yogesh & Malhotra, 1999; Paul, 2003; Cheng et al., 2006). Social influence is introduced to the model and is represented as a subjective norm in many theories (Venkatesh et al., 2003). It acts as one of the determinants of behavioural intentions to accept the system. The proposed model is

Figure 1. The proposed model

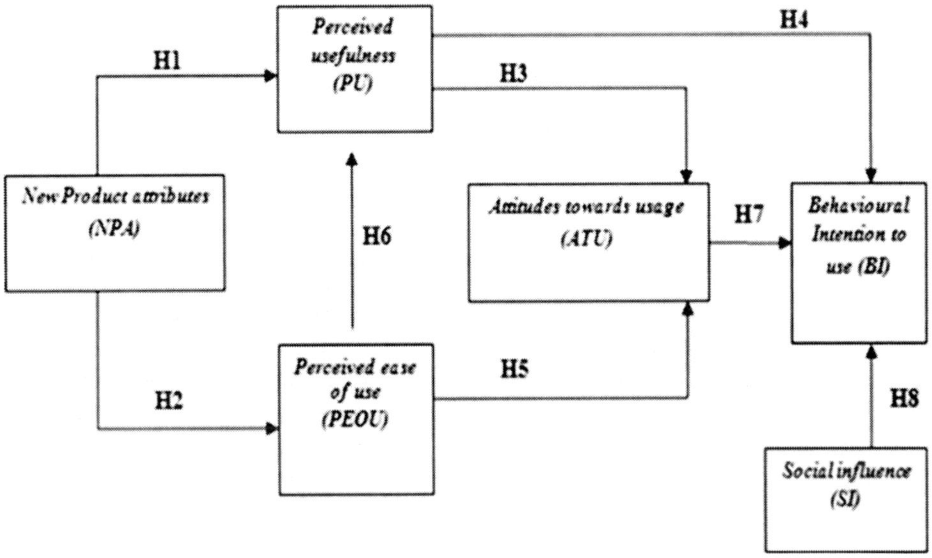

illustrated in Figure 1, and the related hypotheses are discussed below.

The new added variable "new product attributes (biometrics system characteristics)" was adopted from Mukherjee and Hoyer (2001). The purpose of this variable was to measure the extent to which the biometrics system characteristics could provide additional benefits and value to the process. Many high-tech innovations introduce attributes that are novel or totally unfamiliar to a large number of users. For example, recently introduced attributes such as Global Positioning Systems (GPS) in cars or biometric access control in airports are likely to have been novel to many users. Past research suggests that the addition of novel attributes is likely to enhance product acceptance, since consumers interpret these attributes as additional benefits provided by the provider/manufacturer (Mukherjee & Hoyer, 2001).

Companies frequently make an effort to develop and distinguish their products by adding additional features or attributes. These additional attributes may vary in their degree of familiarity to users. Most users may not be familiar with some of these attributes. Hence, it is important to investigate the impact of these novel attributes on users' judgment and choices (Lehmann, 1999; Mukherjee & Hoyer, 2001). Many researchers have revealed that adding familiar attributes to a product generally improves the product acceptance (Meyers-Levy & Tybout, 1989; Nowlis & Simonson, 1996). Nevertheless, other research indicates that adding familiar attributes may not always have a positive effect (Broniarczyk & Gershoff, 1997). However, one study (Mukherjee & Hoyer, 2001) revealed that the positive effect of new attributes holds only in the case of low complexity products/services. For high complexity products, the addition of new attributes can negatively impact the evaluation because of learning costs associated with these attributes. Furthermore, the positive and negative effects of new attributes on product acceptance are affected by the perceived usefulness of the technology. In the same context, it is revealed that the negative effect of new attributes on the evaluation of high complexity products persists even after users are given explicit information about the benefits of the new attributes. (Mukherjee & Hoyer, 2001) concluded that new attributes may contribute to technophobia, or user resistance towards technological innovation.

Davis (1989) defines the perceived usefulness of a system as the degree to which a person believes that using the system would enhance his or her job performance. Attitude refers to a kind of perceptual symbol, which includes the evaluation of an 'attitude object' (e.g., self, others, incidents, activities, events or viewpoints), Intention refers to the willingness of an individual to perform certain behaviours (Zanna & Rempel, 1988). In the early 1960s, there were many academic debates on the influence of perceived usefulness on users' attitudes towards usage and intention to use. Katz (1960) indicated that attitude is the perceptual symbol that an individual constructs internally; hence, perception of usefulness affects attitude. Moreover, when an individual attempts to integrate his/her behaviour and attitudes, he/she will try hard to form an intention to take a certain action; thus perception of usefulness will affect intention. According to the study of Barua and Whinston (Barua & Whinston, 1996, 1996), as soon as an individual uses a function of a system, the function's value (its usefulness) is determined by its cost and productivity. If the cost is lower, or the productivity (i.e., usefulness) is higher, the value of the function will be greater, and, the attitude towards using it will then be more positive. Hence, the intention to use it will be higher. From the above, it can be concluded that perceived usefulness affects user attitudes towards usage and intention to use. The study of Atkinson and Kydd (1997) revealed that the greater the perceived usefulness, the more positive the attitudes of a user towards online information collection. As biometrics authentication system is an application of the WWW and online banking, and online information collection is one kind of online channel function, it can therefore be assumed that the same principle will apply.

Davis (1989) has associated the "perceived ease of use" with the personal perception of simplicity in using a function. Based on Pintrich's study (1989), the greater the perceived ease of use of a product, the more positive the attitude formed towards usage will be. From these studies we can show that perceived ease of use has a positive effect on consumer attitudes towards usage. According to Barua and Whinston (1996), product function with lower usage complexity will be associated with a lower cost, which increases the value of using this function. Thus, there is a positive correlation between perceived ease of use and perceived usefulness. The study of Ajzen and Fishhbein (1980) showed that an individual's behavioural intention will be affected by his/her attitudes, which ultimately determine the individual's actions and responses. Liao and Shao's (Liao et al., 1999, 1999) study showed that an individual's intention to use online banking services is affected by his/her attitudes. It is clear that there is a positive relationship between attitudes towards the usage of online banking and the intention to do so. Both Venkatesh and Davis (1996), and Pavlou (2003) also show that intention affects actual behaviour.

An extension of these studies examined the impact of social influence or subjective norms as predictors of behavioural intention to use. A subjective norm, originally developed from the Theory of Reasoned Action (TRA), predicts general behavioural intentions in strictly voluntary contexts; it assumes that people's perception that others important to them approve or disapprove of certain behaviour has an impact on their actions. Social influence acts as a determinant of behavioural intentions and usage behaviour. Social influence is defined as "the degree to which an individual perceives that important others believe that he or she should use the new system" (Venkatesh et al., 2003, p. 451). In this context, social influence is system/application specific, whereas subjective norm relates to non-system specific behaviour (Al-Gahtani et al., 2007). Yogesh Malhotra and Dennis F. Galletta's 1999 study revealed that social influences play a significant role in determining the

Table 2. The research hypotheses

Hypothesis 1	The new product attributes of the biometric authentication system in online banking have a positive relationship with perceived usefulness.
Hypothesis 2	The new product attributes of the biometric authentication system in online banking have a positive relationship with perceived ease of use.
Hypothesis 3	The perceived usefulness of using the biometric authentication system in online banking will positively influence the attitudes towards using the technology.
Hypothesis 4	The perceived usefulness of using the biometric authentication system in online banking will have a positive impact an individual's behavioral intention to use the technology.
Hypothesis 5	The perceived ease of use of the biometric authentication system in online banking will positively influence the perceived usefulness of using the technology.
Hypothesis 6	The perceived ease of use of the biometric authentication system in online banking will positively influence the perceived usefulness of using the technology
Hypothesis 7	Perceived attitudes towards using the biometric authentication system in online banking will have a positive impact an individual's behavioral intention to use this technology
Hypothesis 8	Social influence will positively influence the individual's behavioral intention to use the biometric authentication system in online banking.

Table 3. The relationships functions

Function 1: Behavioral Intention to use (BI) = f (Perceived usefulness (PU), Attitudes towards usage (ATU), Social influence (SI)).
Function 2: Attitudes towards usage (ATU) = f (Perceived usefulness (PU), Perceived ease of use (PEOU)).
Function 3: Perceived usefulness (PU) = f (Perceived ease of use (PEOU), New Product Attributes (NPA)).
Function 4: Perceived ease of use (PEOU)) = f (New Product Attributes)).

acceptance of new technologies. Biometric devices such as fingerprint scanners can be intimidating to users at the beginning (James et al., 2006). The perception of a biometric fingerprint device will possibly affect the usage behaviour of the biometrics authentication system, in addition to having an impact on the perception of ease of use of the fingerprint device. Based on the empirical findings above, we can formulate the following eight hypotheses (Tables 2 and 3).

The relationships are represented in the following functions (adopted from Lin & Lu, 2000):

EXPERIMENTAL APPROACH

To explore the biometric authentication system in online banking, we designed an experiment, which included the development of a fingerprint log in an authentication system. The experiment also has a survey instrument to measure the likelihood of individual adoption of such technology. The stakeholders in our project were the online banking industry and the customers using these services. About 306 Saudis participated in the experiment, which was held in June-September 2008. Three laptops with a fingerprint reader were made available. The "participant" was asked to verify his or her identity with a fingerprint reader and enroll in the system. After completing the experiment, every participant was asked to complete a 43-question survey to measure the user acceptance of using the biometric authentication system.

The Biometrics Authentication System

Based on the results of our research on the usage of this technology around the world, we found that the United Banker's Bank (UBB)[8] in the USA is

Figure 2. The model of biometrics authentication system

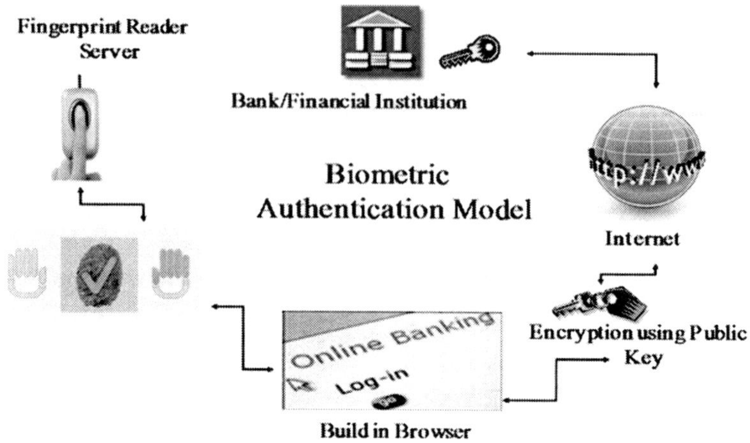

one of many financial organisations that use this technology such as Affinity plus Federal Credit Union, Banco Azteca Mexico, The Bankers Bank, California Commerce Bank, LendingTools.com. As well, Dutch bank ING has implemented fingerprint biometrics in their log in system; hence, we tested the main hypotheses of this research, based on the system used by UBB. The UBB has developed strength in biometrics, with the deployment of fingerprint readers that for some 2,500 employees and UBB clients to securely log into online resources locally and over the Internet, and they do not have to remember multiple passwords - just which finger to press onto the small USB-connected scanner (Figure 2).

The system used in the experiment is quite simple. It consists of a fingerprint sensor, fingerprint authentication software, and a computer. In this case the Digitalpersona U.are.U® 4000 Sensor, which uses a USB connection, was chosen as a sensor, and the Griaule Fingerprint SDK Software from Griaule was chosen as software. Two HP Pavilion dv3000 and Sony Vaio VGN-C2S personal laptops running Windows XP were chosen to run the experiments. The laptops were set up with the following specifications: Intel® Core® 2 Duo processor and 1 GB RAM. This application was developed using C#.NET 2005 under the Windows XP platform. The Griaule fingerprint SDK was chosen to integrate the sensor with this application, and the programming code to integrate the sensor with the system was taken from Wei-Meng Lee (2007). In particular, we built a biometric authentication system whereby users can register and enrol into their accounts with their fingerprints. Once a user is registered, the next time he/she logs in to the system, the user can simply scan his/her fingerprint and the system will grant access. The whole process of scanning and identifying the fingerprint takes about 10 seconds.

Sample Size

The target populations of this study consist of the end users of online banking services of Saudi Arabian financial services. The people selected to take part in a questionnaire are a critical factor for the success of the experiment. The group of participants is called a sample, which is a smaller group that is supposed to represent the larger population. One of the common sampling methods used is random sampling, which is designed to guarantee random and equal representations across the population (Miller, 1991). The equation below is used to determine the minimum sample size[9] (Figure 3):

Figure 3. The sample size

$$\text{Minimum Sample Size}(n) = \frac{t^2 \times p \times (1-p)}{m^2}$$

Where: n= minimum sample size,
t= Confidence level at 95% (standard value of 1.96)
p= Estimated fractional population of subgroup (Almogbil 2005)
m = Margin of error at 5% (standard value of 0.05)
The sample size was calculated based upon these factors:
Population of Saudi Arabia 100%[1] = 22,673,538
Banking customers: 45% est. (Almogbil 2005) = 10,203,029
Online banking users: 20% (SAMA statistics 2008) = 2,040,606
p = 2,040,606/10,203,029= 0.20
Customers minimum Sample Size (n) = 246.

The researcher chose a sample from the population, deciding that 306 participants should be sufficient for the laboratory experiments. To increase the reliability of the experiment, the authors ensured that people with various backgrounds and different experiences and ages took part in this experiment. A survey was developed to investigate the level of user acceptance for biometric authentication systems. The experiment and survey would last approximately 30 minutes. Five respondents were asked to help in pre-testing the experiment tasks and the questionnaire before starting the actual experiment. They commented that the questions had to be presented in a more readable, and less technical, manner. All the questions were re-phrased to take this point into account.

Experimental Procedure

An interactive interface for an online fingerprint login system was developed using PHP. The system would simulate account creation and allow users to login using a fingerprint sensor instead of a user ID and password. A laptop PC was equipped with a Digitalpersona U.are.U® 4000 Sensor. At the beginning of any experiment session, preliminary questions were asked to determine the participants' familiarity with biometrics and their comfort level in performing the experiment. The participants were then asked to complete the following three tasks:

- **Task 1.** Start by creating their own account, which will include your personal data.
- **Task 2.** Place the index fingerprint on the fingerprint sensor to register it for the first time.
- **Task 3.** After the registration process is completed, login to your account by placing the registered fingerprint on the sensor again.

Upon completion of the above tasks, the participants were asked to answer a list of questions about their perspective of the fingerprint system, in terms of its perceived usefulness and perceived ease of use, their attitudes towards its usage, new product attributes (biometrics characteristics), social influence and their behavioural intention to use the system. The reason for carrying out the questionnaire is to examine the hypotheses. The complete questionnaire items, originally produced in English, were translated to Arabic by several bilingual professionals in an iterative process to ensure that both versions converged. Figure 4 shows the items used to estimate the Saudi predictor latent constructs.

The questionnaire starts by asking for demographic data. These questions cover factors such as age, sex, education level, region, familiarity with computers, frequency of use of computers for work purposes, and frequency of use of online banking. Most of the questioner items used the 5-point Likert scale format, ranging from strongly agree to strongly disagree.

DATA ANALYSIS AND RESULTS

Overview of Participant

About 135 (44.1%) of the participants were female and 171 (55.9%) were male. Subject ages ranged from 18 to over 55 years. More than a third of the participants were between 26 and 35 (35.3%). Slightly over half of the participants (168) were

Figure 4. The items used to estimate the Saudi predictor latent constructs

Perceived Usefulness	
PU1	Using the system in my job would enable me to accomplish tasks more quickly.
PU2	Using the system would enhance my effectiveness on the job.
PU3	Using the system would make it easier to do my job.
PU4	I would find the system useful in my job.
PU5	If I use the system, I will spend less time on routine job tasks.
Perceived Ease of Use	
PEOU1	Learning to operate the system is easy for me.
PEOU2	My interaction with the system would be clear and understandable.
PEOU3	I would find the system flexible to interact with.
PEOU4	It would be easy for me to become skilful at using the system.
PEOU5	I would find the system easy to use.
PEOU6	Using the system takes too much time from my normal duties.
PEOU7	Overall, I believe that the system is easy to use.
Attitude toward usage	
ATU1	Using the system is a bad idea.
ATU2	The system makes work more interesting.
ATU3	Working with the system is fun.
ATU4	I like working with the system.
ATU5	This service feature is a great idea.
ATU6	The service feature would be fun to use.
ATU7	This is the best way to improve the quality of service.
ATU8	This service feature is here to stay.
ATU9	Many people will use this service feature.
ATU10	This service feature fills a real need for me.
ATU11	This service feature is a big improvement over existing login methods in online banking.
ATU12	This service feature can provide real value to me.
ATU13	This service feature is just another gimmick.
ATU14	This system fills a need for many people.
ATU15	Many people will believe that this system is worth the cost.
New Product Attributes	
NPA1	It is likely that this new feature will offer advantages to the users.
NPA2	The new feature is likely to perform well.
NPA3	It is likely that the new feature will add value to the online banking industry.
Behavioural Intention to Use	
BI1	Assuming that I have access to this service feature, I intend to use it.
BI2	I will definitely try this service feature.
BI3	It is very likely that I will use this service feature.
BI4	I intend to use the system in the next 12 months.
BI5	I predict I will use the system on a regular basis.
BI6	I plan to use the system in the next 12 months.
Social influence	
SI1	People who influence my behaviour think that I should use the system.
SI2	People who are important to me think that I should use the system.
SI3	In general, my organisation has supported the use of the system.

from the central region of Saudi Arabia. More than half of participants (178) were studying at the undergraduate level, with only 7.8% at the post-graduate level. Figure 5 shows the demographic information for the sample used.

Reliability Analysis of the Instrument

A variety of statistical methods based on the Statistical Package for the Social Sciences (SPSS) was used to analyse the collected data from our experiment. Two separate reliability measures

Figure 5. Demographic information for the sample used

Total Number of Participants	306	
Gender		
Male	171	(55.9%)
Female	135	(44.1%)
Age		
Under 18	26	(8.5%)
18-25	63	(20.6%)
26-35	108	(35.3%)
36-50	93	(30.4%)
Over 50	16	(5.2%)
Education level		
Pre Secondary	32	(10.5%)
Secondary	72	(23.5%)
Undergraduate	178	(58.2%)
Postgraduate	24	(7.8%)
Region		
Central region	168	(54.9%)
West region	58	(19.0%)
East region	27	(8.8%)
North region	15	(4.9%)
South region	38	(12.4%)

were used to illustrate the reliability of data. Veal (2005) defines reliability as the degree to which research findings would be the same if the research were to be repeated with a different sample or at a later date. For our work, we have used the most popular test of inter-item consistency, which is the Cronbach's coefficient alpha (Cronbach, 1951; Nunnally, 1979; Peter, 1979; Sekaran, 2000; Pallant, 2005). The Cronbach's coefficient alpha tests the consistency of participants' responses to all the measurement items. To the degree that the items are independent measures of the similar concept, they will be correlated with one another (Sekaran, 2000; Kripanont, 2007). At first round, the Cronbach's alpha values for PU, PEOU, ATU, NPA, BI and SI were found to be 0.887, 0.758, 0.831, 0.625, 0.885 and 0.760, respectively. Based on (Sekaran, 2000), reliabilities less than 0.6 are considered to be poor, those in the 0.7 range are acceptable, and those over 0.8 are good. It is generally agreed that the minimum acceptable value of Cronbach's alpha is 0.70 (Peter, 1979; Pallant, 2005), but this could be reduced to 0.6 for exploratory research (Robinson et al., 1991a). Reliability tests suggested that screening the data along Churchill's (1979) recommendations would improve reliability levels (Aladwani & Palvia, 2002). We screened the collected data by discarding items that showed very low corrected item-total correlations, i.e., <0.40. After the screening process, 4 items were discarded from our pool of items, namely PEOU6, ATU9, ATU12, and NPA3. The reliability levels came as 0.887, 0.894, 0.914, 0.741, 0.885 and 0.760 for PU, PEOU, ATU, NPA, BI and SI. The Cronbach's alpha coefficients, before and after the deleted items, are illustrated in Figure 6.

In addition to the Cronbach's alpha, a second measure to assess internal consistency is applied. This measure is the item-to-total correlation, which measures the correlation of the item to the total scale and the inter-item correlation (i.e., the correlation among items) (Hair et al., 2006). Figure 7 shows that all item-total correlation values

Figure 6. The Cronbach's alpha coefficients

Measurement Items (Interval Scale)	*Items*	*Cronbach's Alpha*	*Cronbach's Alpha After items deleted*	*Reliability Results*
Perceived usefulness (PU)	5	0.887	0.887	Good
Perceived ease of use (PEOU)	7	0.758	0.894	Good
Attitude toward usage (ATU)	15	0.831	0.914	Good
New Product Attributes (NPA)	3	0.625	0.741	Acceptable
Behavioural Intention to Use (BI)	6	0.885	0.885	Good
Social influence (SI)	3	0.760	0.760	Acceptable

Figure 7. All items correlation

Measurement Items (Interval Scale)	Items	Inter-Item Correlation	Item- total correlation
Perceived usefulness (PU)	5	0.487-0.723	0.625-0.797
Perceived ease of use (PEOU)	7	0.428-0.749	0.581-0.810
Attitude toward usage (ATU)	15	0.245-0.674	0.547-0.753
New Product Attributes (NPA)	3	0.594-0.594	0.594-0.594
Behavioural Intention to Use (BI)	6	0.599-0.594	0.661-0.728
Social influence (SI)	3	0.390-0.747	0.437-0.694

exceed 0.5, except some items in social influence (SI), and all inter-item correlation values exceed 0.3, except a few items in attitude toward usage (ATU), almost all of which exceed 0.30, with only a few below 0.30. It is recommended that the item-to-total correlations should exceed 0.50 and the inter-item correlation should exceed 0.30 (Robinson et al., 1991a). As suggested by Cohen

Figure 8. Correlation coefficients among the variables

		BI	ATU	PU	PEOU	NPA	SI
BI	Pearson Correlation	1	.718**	.594**	.519**	.595**	.350**
	Sig. (2-tailed)		.000	.000	.000	.000	.000
	N	306	306	306	306	306	306
ATU	Pearson Correlation	.718**	1	.674**	.502**	.762**	.404**
	Sig. (2-tailed)	.000		.000	.000	.000	.000
	N	306	306	306	306	306	306
PU	Pearson Correlation	.594**	.674**	1	.493**	.587**	.391**
	Sig. (2-tailed)	.000	.000		.000	.000	.000
	N	306	306	306	306	306	306
PEOU	Pearson Correlation	.519**	.502**	.493**	1	.435**	.386**
	Sig. (2-tailed)	.000	.000	.000		.000	.000
	N	306	306	306	306	306	306
NPA	Pearson Correlation	.595**	.762**	.587**	.435**	1	.419**
	Sig. (2-tailed)	.000	.000	.000	.000		.000
	N	306	306	306	306	306	306
SI	Pearson Correlation	.350**	.404**	.391**	.386**	.419**	1
	Sig. (2-tailed)	.000	.000	.000	.000	.000	
	N	306	306	306	306	306	306

**. Correlation is significant at the 0.01 level (2-tailed).

(1988), correlation (r) = 0.10 to 0.29 implies a small correlation (both positive and negative correlation), (r) = 0.30 to 0.49 implies a medium correlation, and (r) = 0.50 to 1.0 implies a strong correlation. The questionnaire proved to be reliable according to the Cronbach's Alpha coefficient. The bi-variate relationships indicate that all the variables were significantly correlated with each other. This suggests that the correlations are positive between the behavioural intention to use and its independent variables, i.e., attitude toward usage, perceived usefulness, perceived ease of use, new product attributes and social influence, and they are significant at 0.01% level. Figure 8 presents the first-order correlation coefficients among the variables.

In addition, consistent as suggested by Fornell and Larcher (1981), the average variance extracted (AVE) for each measure exceeded 0.50 showed on Figure 9. Figure 10 illustrates the results of testing the discriminate validity of constructs. The entire indicators weight more highly on their own constructs than on other constructs which is supporting the discriminate validity of the constructs.

Analytic Strategy for Assessing the Model

Path analyses (Land, 1969; Li, 1975; Kenny, 1979) were used to analyse the data. Path analysis is a multivariate analytical method of empirically examining sets of relationships in the form of linear causal models (Li, 1975; Duncan, 1986; Lin & Lu, 2000). The initial step in path analysis is to identify a path diagram. The hypothetical causal relationships are indicated by unidirectional arrows linking two variables together. The research model describes the path diagram in this study. According to the research model, New Product Attributes (NPA) are posited to exert simultaneous, direct, and linear influences on both perceived usefulness and perceived ease of using the biometric authentication system in online banking. The user's beliefs further influence his/her attitude towards using the system, as well as the behavioural intention to use the system. The significance of the path coefficient related with each path represents the strength of each linear influence. While the path coefficients can be estimated in several ways (Kenny, 1979), multiple regression analyses have been used by most empirical applications of this methodology. Li (1975) stated that the path coefficient has been revealed to be identical to the standardised regression coefficients. These coefficients are comparable among variables because they are normalised into the same unit (Lin & Lu, 2000).

Figure 9. Average variance extracted (AVE)

Variable constructs	Average variance extracted
Perceived usefulness (PU)	0.51
Perceived ease of use (PEOU)	0.54
Attitude toward usage (ATU)	0.53
New Product Attributes (NPA)	0.52
Behavioural Intention to Use (BI)	0.61
Social influence (SI)	0.59

We conducted path analysis with a series of multiple regression analyses to test our research model. Notice that the variable (BI) has paths to it only from the variables (ATU, PU, and SI). To get the coefficients for those paths we simply conduct a multiple regression to predict behavioural Intention to Use (BI) from attitude toward usage (ATU), perceived usefulness (PU) and social influence (SI). The Beta weights are the path coefficients leading to BI: 0.568 from ATU, 0.189 from PU and.057 from SI. In the model ATU has paths to it from PEOU and PU, therefore we predict ATU from PEOU and PU. Here is the output (Figure 11):

The path coefficients leading to ATU are: 0.564 from PU, and 0.225 from PEOU. The following output shows the path coefficients leading to PU: 0.293 from PEOU, 0.459 from NPA. In the model PU has paths to it from PEOU and NPA. Finally, The Beta weights are the path coefficients leading to PEOU: 0.435 from NPA. Based on the

Figure 10. Discriminant validity

Latent variables	1	2	3	4	5	6
Social influence (SI)	**.769**					
New Product Attributes (NPA)	.290	**.719**				
Perceived ease of use (PEOU)	.245	.250	**.735**			
Perceived usefulness (PU)	.276	.391	.305	**.717**		
Attitude toward usage (ATU)	.228	.394	.241	.375	**.731**	
Behavioural Intention to Use (BI)	.256	.387	.314	.416	.390	**.784**

Figure 11. Path coefficients to BI

Model Summary

Model	R	R Square	Adjusted R Square	Std. Error of the Estimate
1	.735ª	.541	.536	.56439

a. Predictors: (Constant), SI, PU, ATU

Coefficientsª

Model		Unstandardised Coefficients		Standardised Coefficients	t	Sig.
		B	Std. Error	Beta		
1	(Constant)	.084	.124		.675	.500
	ATU	.716	.068	.568	10.505	.000
	PU	.185	.052	.189	3.543	.000
	SI	.054	.041	.057	1.332	.184

a. Dependent Variable: BI

model PEOU has paths to it just from NPA, hence we predict PEOU from NPA (Figure 12).

EVALUATING THE HYPOTHESISED MODEL

In terms of goodness-of-fit indicators, the model accounts for 54% of the variance in the acceptance to use the biometric authentication system. In addition, usefulness and ease of use account for 49% of the variance in attributes towards use the system. New product attributes account for 41% and 18% of the variance in perceived usefulness and perceived ease of use of the biometric system, respectively. Figure 13 presents the path analysis of the model.

The above results can be summarised as follows:

- **Function 1:** In the multiple regression equation predicting the users' behavioural intention, the effect of attributes towards usage is fairly strong, as indicated by the path coefficient of 0.57 ($p < 0.001$). The other path coefficients for the perceived usefulness and social influence of using the system, which are 0.19 and 0.06 ($p < 0.001$) respectively, are less statistically significant. All these three paths account for approximately 54% of the observed variance in the behavioural intention to use the system. Notably, *Hypotheses* 4, 7

Figure 12. Path coefficients leading to PU

Model Summary

Model	R	R Square	Adjusted R Square	Std. Error of the Estimate
1	.643ᵃ	.414	.410	.65155

a. Predictors: (Constant), NPA, PEOU

Coefficientsᵃ

Model		Unstandardised Coefficients		Standardised Coefficients	t	Sig.
		B	Std. Error	Beta		
1	(Constant)	.493	.132		3.739	.000
	PEOU	.339	.057	.293	5.999	.000
	NPA	.494	.053	.459	9.406	.000

and 8 are supported at the 0.05 level of significance.

- **Function 2:** Attitudes towards usage are statistically significantly related to both the perceived usefulness and the perceived ease of use of the biometric authentication system. The effect of perceived usefulness on attitudes is strong, as indicated by the path coefficient of 0.56 ($p < 0.001$). The other path coefficient, perceived ease of use (0.22) is statistically significant at $p < 0.001$. Forty-nine percent (49%) of the observed variance in attitude towards using the system can be explained from these two paths. Accordingly, *Hypothesis* 3 and 4 are confirmed at the 0.05 level of significance.
- **Function 3:** The perceived usefulness of the biometric authentication system is sta-

Figure 13. The path analysis of the model

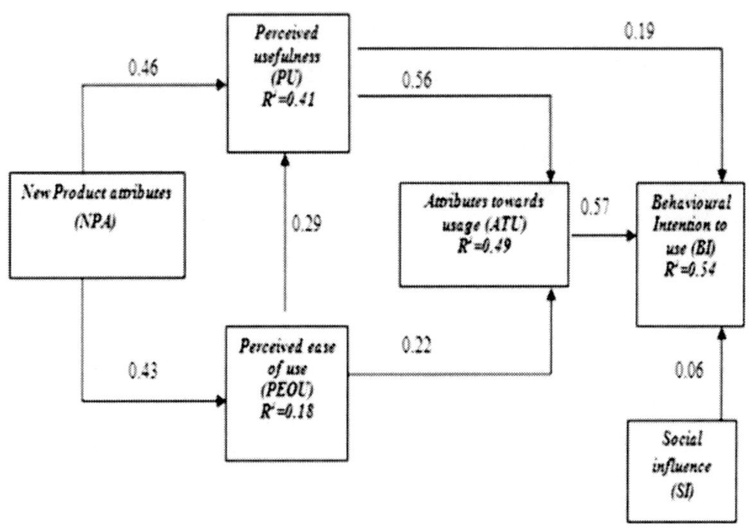

tistically significantly predicted by two paths, from the perceived ease of use and new product attributes. The path coefficients from the above paths are 0.29 ($p < 0.001$) and 0.46 ($p < 0.001$), respectively. These two paths together explain approximately 41% of the observed variance in perceived usefulness. Thus, *Hypotheses* 1 and 6 are supported at the 0.05 level of significance.

- **Function 4:** The perceived ease of use the biometric system can be statistically significantly predicted by the path from new product attributes. The path coefficient from the above is 0.43 ($p < 0.001$). The path accounts for 18% of the observed variance in perceived ease of use for the biometric system. Again, *Hypothesis* 2 is confirmed at the 0.05 level of significance.

DISCUSSION

This study examines the modified TAM model for the biometric authentication system in the online banking log in system by adding the new product attributes and social influence as external variables. The introduction of new product attributes to the model is intended to reflect the uncertainty currently associated with the online banking environment (Yogesh & Malhotra, 1999; Paul, 2003; Cheng et al., 2006). Social influence is introduced to the model and is represented as a subjective norm in many theories (Venkatesh et al., 2003). It acts as one of the determinants of behavioural intentions and usage behaviour. This study has provided some valuable global impact insights into user acceptance of a biometric authentication system from the perspective of the online banking industry. In addition, we have tested the descriptive validity of TAM in the domain of online banking. Based on the results obtained here, several inferences can be made.

The findings of this study show that the acceptance to use the biometric authentication system in the online banking environment can be predicted by using TAM (R^2=0.54).

The perceived usefulness and the perceived ease of use had significant effects on the attitudes towards using the system ($\beta = 0.56$; $\beta = 0.22$), supporting Hypotheses 3 and 5. The implication here is that if the participants believe that the biometric authentication system is useful in a particular situation, then they will be more likely to use the technology. Moreover, participants might be more likely to use the system if they feel that it is easy to use. The perceived ease of use of the system was found to have a significant impact on the perceived usefulness ($\beta = 0.29$), supporting Hypothesis 6. A participant's perception of how easy the system is to use may represent a large determining aspect of its perceived usefulness. This implies that the usage complexity of the biometric system will significantly impact its perceived usefulness.

Based on this study, one finding is the relative strength of the relationship between perceived usefulness and intention of using the biometric system compared to the relationship between ease of use and intention of using the system. In fact, the regression analysis explains that the usefulness-intention relationship remains large, while the ease of use-intention relationship is reduced. Based on previous findings of (Davis, 1989; Gefen & Straub, 1997; Lin & Lu, 2000; Chen et al., 2007) there is a more significant link between usefulness and the intention to use than ease of use, whereas, ease of use exerts a significant effect on usefulness.

Furthermore, the significant correlation between the ease of use-intention and ease of use-attitude relationships may be attributed to the possible causal chains (shown in Figure 4):

1. Ease of use → usefulness → intention
2. Ease of use → usefulness → attitudes

For that reason, perception of ease of use can facilitate in perceiving usefulness as well as developing attitudes towards using the system. Perceived ease of use, hence, exerts an indirect impact on the user's intention to use the biometric system.

Although, the finding from the path analysis shows that the path from social influence to behaviour intention to use is less statically significant than other paths, we can accept that social influence still has direct determinants of intention to use the system. As a result, it can be concluded that if the people, in the user's social environment, use a technologic innovation the user will be more likely to decide to use it.

LIMITATIONS AND DIRECTIONS FOR FUTURE RESEARCH

Although these results can be considered to be statistically significant, the research has some limitations that affect the reliability and validity of the findings. First, the experiment is not representative of the whole of Saudi Arabia. This study was conducted only in the main cities. To generalise the findings, we need to study the acceptance to use the biometric system in other parts of Saudi Arabia, particularly in rural areas (about 20% of whole population)[10]; rural areas might be less awareness and served in telecommunications and IT services than the urban areas. Future comparative studies could target these areas to achieve comparative national results. The research could even be extended to include other countries operating under similar conditions to observe if comparable results are attained and to study the impact of culture, if any, on our findings.

One direction for future research is to use Structured Equation Modelling (SEM) to appraise this model. The methodology employed in the current research examines the relationships in the model and presents key insights about the relationships between the constructs in the model. On the other hand, the model could be further examined in sub-sequent studies by utilising SEM or other techniques to investigate the entire relationships between the constructs of the model simultaneously.

It is important to realise that other factors may also play an important role in the user's acceptance to use the biometric system. Examples of such factors include performance expectancy, effort expectancy and facilitating conditions. These constructs are postulated by the Unified Theory of Acceptance and Use of Technology (UTAUT). UTAUT also posits other moderator variables such as gender, age, experience, and voluntariness of use.

In addition, in terms of cultural differences that affect the individual's acceptance of biometric authentication systems in Saudi society, we need to investigate the theory of cultural dimensions (Hofstede, 1980; Hofstede, 2001). Cultural factors could affect the adoption of new technology in a certain society. Resistance to accept the implementation of a new technology may indicate that some cultural dimensions need to be dealt with. Approaches that take culture into account can be developed to overcome resistance to change and to explore different reactions. However, this issue will be considered later on future publication as it is out of the scope of this paper.

CONCLUSION AND MANAGERIAL IMPLICATIONS

The research model developed in this study is revealed to be a good predictor for the acceptance to use biometric authentication system. The experiment developed for this study could be used as an indicator for the success of the acceptance of using biometric technologies in online banking log in system for both individuals and organisations. Gaining an idea of the acceptance success could be valuable in avoiding unsuccessful efforts to develop a biometric authentication technology, which could prevent major financial losses.

Based on this study, the user acceptance of using biometric technology shows that the technology has the potential to be applied in the online banking industry. The data collected and analysed by this experiment, indicated that participants believed in the benefits and the potential of this technology. The results of the regression analysis supported the model, and all of the eight proposed hypotheses were confirmed.

The finding outline two main areas associated with the acceptance of biometric authentication systems. First; the careful consideration of a specific biometric technology implemented is important, as the new system has characterises that may lead to unwanted acceptance behaviour. In addition, this might affect the perceived ease of use of the biometric system in an unpredictable way, possibly requiring an extensive training on the biometric technology or the use of different approaches for acceptance procedures. Second, both individuals and organisations have to secure their online services; security plays an important role in the acceptance to use of the biometric authentication system. Hence, acceptance procedures that highlight this advantage of the technology possibly will give support to improving the implementation procedure.

The strength of this research lies in its derivation of its own factors from prior theoretical and empirical research. The applied measures utilised in this research were developed through an extensive review of the literature. Moreover, the study variables, which have been used in many previous technology and authentication systems within online banking acceptance researches, were found to have adequate reliability and validity.

The above mentioned results seem to present the following implications:

- Usefulness is still the most important concern in the acceptance to use the system.
- Ease of using the system, nevertheless, has an indirect effect on the development of behavioural intentions to use. That is, the easier the biometric system is to use, the more useful it is perceived to be.
- To encourage the users' acceptance to use the system, the organisation should promote the system's usefulness more than promoting its ease to use.
- Increasing only the ease of using the system would not enhance the users' acceptance to use the system.

ACKNOWLEDGMENT

The authors thank Dr. Said S. Al-Gahtani an associate professor of computer information systems in the Department of Administrative Sciences at King Khalid University, Abha, Saudi Arabia for his valuable comments. Additionally, the authors wish to express their gratitude towards the Saudi government for the financial support received in order to conduct the study's experiment.

REFERENCES

Al-Gahtani, S. (2007). Information technology (IT) in Saudi Arabia: Culture and the acceptance and use of IT. *Information & Management*, *44*, 681–691. doi:10.1016/j.im.2007.09.002

AL-Harby. F., et al. (2008). The feasibility of biometrics authentication in e-commerce: User acceptance. In *Proceedings of the IADIS International Conference WWW/Internet*, Freiburg, Germany.

Al-Somali, S. A. (2008). *An investigation into the acceptance of online banking in Saudi Arabia.* Technovation.

Aladwani, A., & Palvia, P. (2002). Developing and validating an instrument for measuring user-perceived web quality. *Information & Management*, *39*(6), 467–476. doi:10.1016/S0378-7206(01)00113-6

Alsajjan, B. (2008). *Internet banking acceptance model across cultures:The case of England and Saudi Arabia*. Unpublished doctoral dissertation, Brunel Business School, Brunel University, UK.

Atkinson, M., & Kydd, C. (1997). Individual characteristics associated with world wide web use: An empirical study of playfulness and motivation. *The Data Base for Advances in Information Systems*, *28*(2).

Barua, A. C., & Whinston, A. B. (1996). Creating a collaboratory in cyberspace: Theoretical foundation and an implementation. *Journal of Organizational Computing*, *5*(4), 417–442. doi:10.1080/10919399509540261

Broniarczyk, S., & Gershoff, A. (1997). Meaningless differentiation revisited. *Advances in Consumer Research. Association for Consumer Research (U. S.)*, *24*, 223–228.

Chen, C.-D. (2007). Predicting electronic toll collection service adoption: An integration of the technology acceptance model and the theory of planned behavior. *Transportation Research Part C, Emerging Technologies*, *15*(5), 300–311. doi:10.1016/j.trc.2007.04.004

Chen, L. (2002). Enticing online consumers: An extended technology acceptance perspective. *Information & Management*, *39*, 705–719. doi:10.1016/S0378-7206(01)00127-6

Cheng, J. M.-S. (2006). Consumer acceptance of the internet as a channel of distribution in Taiwan—a channel function perspective. *Technovation*, *26*(7), 856–864. doi:10.1016/j.technovation.2005.01.001

Churchill, G. (1979). A paradigm for developing better measures of marketing constructs. *JMR, Journal of Marketing Research*, *16*(1), 64–73. doi:10.2307/3150876

Cronbach, L. J. (1951). Coefficient alpha and the internal structure of tests. *Psychometrika*, 31.

Davis, F. D. (1986). *Technology acceptance model for empirically testing new end-user information systems: Theory and results, Massachusetts Institute of Technology*. Unpublished doctoral dissertation.

Davis, F. D. (1989). Perceived usefulness, perceived ease of use and user acceptance of information technology. *Management Information Systems Quarterly*, *13*(3), 319–340. doi:10.2307/249008

Duncan, O. D. (1986). Path analysis: Sociological example. *American Journal of Sociology*, *72*(1), 1–16. doi:10.1086/224256

Fornell, C., & Larcker, D. (1981). Evaluating structural equation models with unobservable variables and measurement error. *JMR, Journal of Marketing Research*, *18*, 39–50. doi:10.2307/3151312

Furnell, S., & Evangelatos, K. (2007). Public awareness and perceptions of biometrics. *Computer Fraud & Security*, *1*, 8–13. doi:10.1016/S1361-3723(07)70006-4

Gefen, D. (2003). TAM or just plain habit: A look at experienced online shoppers. *Journal of End User Computing*, *15*(3), 1–13.

Gefen, D., & Straub, D. (1997). Gender differences in the perception and use of e-mail: An extension to the Technology Acceptance Model. *Management Information Systems Quarterly*, *21*(4), 389–400. doi:10.2307/249720

Giesing, I. (2003). *User perceptions related to identification through biometrics within electronic business*. Unpublished master's thesis, University of Pretoria, Pretoria, South Africa.

Hair, J. (2006). *Multivariate data analysis*. Upper Saddle River, NJ: Pearson Education, Inc.

Herman, A. (2002). Major bank signs up for digital signature verification technology. *Biometric Technology Today, 10*.

Hofstede, G. (1980). *Culture's consequences: International differences in work-related values*. Beverly Hills, CA: Sage.

Hofstede, G. (2001). *Culture's consequences – Second edition: Comparing values, behaviors. Institutions and organizations across nations*. London: Sage.

Jain, A. (2004). Biometric recognition: How do I know who you are? In *Proceedings of the IEEE 12th Signal Processing and Communications Applications Conference.*

Jain, A. (2000). Biometric identification. *Communications of the ACM, 43*(2), 91–98. doi:10.1145/328236.328110

James, T. (2006). Determining the itention to use biometric devices: An application and extension of the Technology Acceptance Model. *Journal of Organizational and End User Computing, 18*(3), 1–24.

Kenny, D. A. (1979). *Correlation and causality*. New York: Wiley.

Koufaris, M. (2002). Applying the technology acceptance model and flow theory to online consumer behavior. *Information Systems Research, 13*(2), 205–223. doi:10.1287/isre.13.2.205.83

Kripanont, N. (2007). *Examining a Technology Acceptance Model of Internet usage by academics within Thai business schools. School of Information Systems*. Unpublished doctoral dissertation, Victoria University, Melbourne, Australia.

Land, K. C. (1969). *Principles of path analysis*. San Francisco, CA: Jossey-Bass.

Laurie, A. J., et al. (2007). Towards understanding user perceptions of authentication technologies. In *Proceedings of the 2007 ACM workshop on Privacy in electronic society*, Alexandria, VA. New York: ACM.

Laux, D., et al. (2007). *Customer acceptance of biometric technology in the hotel industry.*

Lehmann, D. (1999). Consumer behavior and Y2K. *Journal of Marketing Channels, 63*(5), 14–18.

Li, C. C. (1975). *Path analysis: A primer*. Pacific Grove, CA: The Boxwood Press.

Liao, S. (1999). The adoption of virtual banking: An empirical study. *International Journal of Information Management, 19*, 63–74. doi:10.1016/S0268-4012(98)00047-4

Lin, J. C. C., & Lu, H. P. (2000). Towards an understanding of the behavioural intention to use a web site. *International Journal of Information Management, 20*, 197–208. doi:10.1016/S0268-4012(00)00005-0

Liu, S., & Silverman, M. (2001). A practical guide to biometric security technology. *IT Professional*, 27–32. doi:10.1109/6294.899930

Meyers-Levy, J., & Tybout, A. (1989). Schema congruity as a basis for product evaluation. *The Journal of Consumer Research, 16*(1), 39–54. doi:10.1086/209192

Miller, D. (1991). *Handbook of research design and social measurement*. Newbury Park, CA: Sage Publications.

Moon, J., & Kim, Y. (2001). Extending the TAM for a world-wide-web context. *Information & Management, 38*, 217–230. doi:10.1016/S0378-7206(00)00061-6

Mukherjee, A., & Hoyer, W. (2001). The effect of novel attributes on product evaluation. *The Journal of Consumer Research, 28*(3), 462–472. doi:10.1086/323733

Nowlis, S., & Simonson, I. (1996). The effect of new product features on brand choice. *JMR, Journal of Marketing Research, 32*, 36–46. doi:10.2307/3152011

Nunnally, J. O. (1979). *Psychometric theory*. New York: McGraw-Hill.

O'Gorman, L. (2003). Comparing passwords, tokens, and biometrics for user authentication. *Proceedings of the IEEE, 91*, 2021–2040. doi:10.1109/JPROC.2003.819611

Pallant, J. (2005). *SPSS survival manual. Milton, Keynes*. UK: Open University Press.

Paul, A. P. (2003). *Consumer acceptance of electronic commerce: Integrating trust and risk with the Technology Acceptance Model* (*Vol. 7*, pp. 101–134). Armonk, NY: M. E. Sharpe, Inc.

Pavlou, P. A., & Chai, L. (2002). What drives electronic commerce across cultures? A cross-cultural empirical investigation of the theory of planned behavior. *Journal of Electronic Commerce Research, 3*(4), 240–253.

Peter, J. P. (1979). Reliability: A review of psychometric basics and recent marketing practices. *JMR, Journal of Marketing Research, 16*, 6–7. doi:10.2307/3150868

Pikkarainen, T. (2004). Consumer acceptance of online banking: An extension of the technology acceptance model. *Internet Research, 14*(3), 224–235. doi:10.1108/10662240410542652

Plouffe, C. R. (2001). Research report: Richness versus parsimony in modelling technology adoption decisions-understanding merchant adoption of a smart card-based payment system. *Information Systems Research, 12*(2), 208–222. doi:10.1287/isre.12.2.208.9697

Png, I. (2001). Dimensions of national culture and corporate adoption of IT infrastructure. *IEEE Transactions on Engineering Management*, 48.

Robinson, P. (1991a). *Criteria for scale selection and evaluation. Measures of personality and social psychological attitudes*. San Diego, CA: Academic Press.

Sekaran, U. (2000). *Research methods for business: A skill-building approach*. New York: John Wiley and Sons.

Straub, D. (1998). Testing the technology acceptance model across cultures: A three country study. *Information & Management, 33*(1), 1–11. doi:10.1016/S0378-7206(97)00026-8

Straughn, R., & Albers-Miller, N. (2001). An international investigation of cultural and demographic effects of domestic retail loyalty. *International Marketing Review, 18*(5), 521–541. doi:10.1108/EUM0000000006044

Szajna, B. (1994). Software evaluation and choice: Predictive evaluation of the technology acceptance instrument. *Management Information Systems Quarterly, 18*(3), 319–324. doi:10.2307/249621

Tamilia, R. D. (2002). Conventional channels of distribution and electronic intermediaries: A functional analysis. *Journal of Marketing Channels, 9*(3-4), 27–35. doi:10.1300/J049v09n03_03

Veal, A. J. (2005). *Business research methods: A managerial approach*. Upper Saddle River, NJ: Pearson Education Australia.

Venkatesh, V. (2003). User acceptance of information technology: Toward a unified view. *Management Information Systems Quarterly, 27*(3), 425–478.

Venkatesh, V., & Davis, F. D. (1996). A model of the antecedents of perceived ease of use: Development and test. *Decision Sciences, 27*(3), 181–451. doi:10.1111/j.1540-5915.1996.tb01822.x

Wayman, J. L., & Alyea, L. (2000). *Picking the best biometric for your applications. National biometric test center collected works* (pp. 269–275). San Jose, CA: National Biometric Test Center.

Werts, C. E., & Linn, R. L. (1960). Path analysis: Psychological examples. *Psychological Bulletin*, *74*(2), 194–212.

Woodward, J. D. (2001). *Army biometric applications*. RAND Corporation.

Yogesh, M., & Malhotra, D. (1999). *Extending the Technology Acceptance Model to account for social influence*. Theoretical Bases and Empirical Validation.

Zanna, M. P., & Rempel, J. K. (1988). *Attitudes: A new look at an old concept*. New York: Cambridge University Press.

ENDNOTES

1. Euromonitor (2006) & CDSI Statistical Year book (2006)
2. UK Year book (2007) & SAMA statistics (2008)
3. The Saudi Communication and Information Technology Commission (2008)
4. www.privacyrights.org
5. http://www.computing.co.uk/computing/news/2221020/lords-say-crime-nust-reported
6. http://coeia.edu.sa/index.php/ar/
7. http://www.biometricgroup.com/
8. http://www.ubb.com/
9. Retrieved Feb 12, 2008, from http://www.surveysystem.com/ssformu.htm.
10. http://www.unicef.org/infobycountry/saudiarabia_statistics.html

This work was previously published in International Journal of E-Business Research Volume 6, Issue 3, edited by In Lee, pp. 34-55, copyright 2010 by IGI Publishing (an imprint of IGI Global).

Chapter 8
An Empirical Study of Predicting Hong Kong Consumers' Online Shopping Intentions:
Personal Hygiene Products

T. C. E. Cheng
The Hong Kong Polytechnic University, China

M. W. Chung
The Hong Kong Polytechnic University, China

ABSTRACT

The digital technologies that have made electronic commerce (EC) a reality have changed the landscape of operations management (OM). Past OM studies have focused on the relationship between information systems (IS) and OM strategy, however, there is a lack of investigation into the environmental factors and OM strategy in the EC setting. Moreover, little empirical research has been performed outside Europe and the United States. Thus, the authors' research fills this gap with a view of understanding the potential factors influencing Hong Kong consumers' online shopping intentions. The authors collected data from Facebook users via a Web-based survey and their research results support the previous literature and behavioral models in that perceived usefulness and perceived convenience and inconvenience are significantly related to consumers' online shopping intentions. Further, gender difference plays a role in predicting consumers' attitudes toward the positive features of online shopping, as well as predicting personal innovativeness toward information technology. In the collectivist culture of Hong Kong, the findings show that men are positively associated with the subjective norm. Unlike gender, education can only explain consumers' perceived usefulness.

DOI: 10.4018/978-1-61350-462-8.ch008

INTRODUCTION

The emergence of electronic commerce (EC) has transformed business operations from the physical marketplace to the virtual "marketspace" and re-shaped consumers' shopping habits. EC is regarded as any business transaction carried out on the World Wide Web or the Internet (Baber, 2005; Schniederjans & Cao, 2002). Digital technologies have substantially changed the daily routines of operations management (OM) whereby the application of computer technology helps automate business transactions and workflows on the Internet (Yang & Lester, 2005).

In this study we examined the customer market component of the business environment in order to understand the key factors that determine the success of electronic business operations. We investigated the business-to-consumers (B2C) market in Hong Kong. Hong Kong is an ideal place in Asia to examine EC and consumers' intentions of Internet shopping since the city is the leading business hub that has one of the best business operating environments in Asia and has earned recognition for its world-class IT infrastructure (Business Monitor International, 2008). Despite the popularity of the Internet, EC is still in its infancy in Asia (Cheng et al., 2006; Sin & Tse, 2002; Telecom Research Asia-Pacific, 2004). The EC penetration in Hong Kong is low and has grown at a decelerating rate (Telecom Research Asia-Pacific, 2004). The under-developed EC businesses in Hong Kong may lead to exceptionally large unrealized profits and economic benefits. The importance of EC to Hong Kong's economy and retailing in particular, calls for examination of the consumer market in order to determine the adoption and growth of EC. In addition, there is little empirical research performed outside Europe and the United States and studies of Asian consumers' online shopping intentions are limited (Teo & Lim, 2000).

Past OM studies have focused on the relationship between information systems (IS) and OM strategy. There is a lack of investigation into the environmental factors and OM strategy in the EC setting (Schniederjans & Cao, 2002). With a view to understanding the key factors that determine the success of EC operations, we conducted this study to look into the business environment and study the customer market component. Our purposes were to: 1) gain some insights into consumers' online buying intentions in the Hong Kong market, 2) ascertain the impacts of demographics on the predictors of online buying intention, and 3) understand the role that consumers' characteristics play in EC. In sum, we sought to understand the predictors of online shopping intention, as well as the impacts of demographics in a collectivist culture. We also attempted to link our research findings to OM practices in the EC context.

LITERATURE REVIEW AND RESEARCH MODEL

In order to have a good grasp of the online shopping phenomenon, we conducted an extensive review of the relevant literature. Based on the review, we identified the key variables (constructs), developed a research model, and formulated several hypotheses for testing.

Human Behavioural Theories

Human adoption behaviour has long been a popular topic in behavioural science. The Theory of Reasoned Action (TRA) and the Technology Acceptance Model (TAM) provide a well developed theoretical framework for understanding human adoption behaviour. TRA argues that both behavioural attitude (AD) and subjective norm (SN) affect behavioural intention (INT), which in turn affects actual behaviour (Ajzen & Fishbein, 1980). According to TAM, perceived usefulness (PU) and perceived ease of use (PEOU) are the determinants of IT adoption and usage (Davis, 1989), and both have a direct effect on intention

(Lu et al., 2009). Since these theories have been extensively used to study usage intention, they provide a suitable framework for exploring the relationships between Hong Kong consumers' characteristics and the factors affecting their online shopping intentions.

Demographics and Socioeconomic Characteristics

Demographics have been reported as significant variables influencing Internet usage and purchases (Park & Jun, 2003). Prior research has reported that computer adoption and use are strongly associated with male consumers. On the other hand, female consumers are perceived as less technical and less spontaneous (Dholakia & Chiang, 2003). Although in most cultures women are perceived as the primary shopper for the household and shopping is categorized as "female tasks" (Dholakia, 1999; Garbarino & Strahilevitz, 2004; South & Spitze, 1994), they are less likely to make purchases online and spend less even if they do (Van Slyke et al., 2002). Girard et al. (2003) found that more males prefer to shop online for search products while more females prefer to shop online for experience products.

Besides gender difference, education is a commonly studied predictor of Internet usage and shopping. Lightner (2003) showed that there is a significant relationship between consumers who possess a high level of education and Internet usage. Moreover, Girard et al. (2003) found that education is positively related to preference of shopping on the Internet for search products. Lee et al. (2007) observed that education is a significant predictor of online shopping.

Features of EC - Perceived Convenience and Perceived Risk

Positive and negative features of EC play a role in influencing e-shoppers' buying intentions. Convenience has been the most frequently cited reason for using online shopping (Chiang & Dholakia, 2003). However, Hong Kong consumers may perceive the convenience associated with online shopping differently because of the great dispersion of stores, which are close to households, and the long opening hours of shopping malls/stores in the city. Further, security and privacy concerns are by-products of EC, which negatively affect purchases made online (Cai & Cude, 2008; Lian & Lin, 2008; Park & Jun, 2003). Complementary to trust, perceived risk has been extensively studied in the application of EC as it is empirically important for the analysis of online shopping behaviour (Kerem, 2005).

Research Model and Hypotheses

Taking previous research findings and behavioural theories into consideration, we developed a research model and formulated a set of hypothesis for our study. We selected personal hygiene products as the search goods that consumers frequently purchase.

We distinguished consumers' AD towards the positive, as well as the negative, features of online shopping. The positives represent the benefits while the negatives reflect the barriers and risks associated with shopping on the Internet. We hypothesized the following:

H1: Positive features of online shopping (POS_AD) are positively related to consumers' intentions to shop online (INT).

H2: Negative features of online shopping (NEG_AD) are negatively related to consumers' intentions to shop online (INT).

Another variable involved in the TRA model that explains human behavioural intention is SN. Previous literature has produced mixed results from studying the relationship between SN and INT. Yu and Wu (2007) suggested that behavioural intention is affected by AD and SN. Further, Baker et al. (2007) affirmed that SN is a significant and positive determinant of intention to use technology. However, Wang et al. (2007) commented that consumers' SN has no influence on their online shopping intentions. Nonetheless, it is believed that consumers' purchases are often affected by social value, where status is conferred by others (Andrews et al., 2005). Moreover, in a collectivist culture like Hong Kong, members of the in-group (e.g., immediate family members and peers) are perceived as more trustworthy than out-group members (Lim et al., 2004). Since SN is the perception of social influence to perform a given behaviour (Pavlou & Fygenson, 2006), it was measured by the perception of whether other people important to the shopper perceive they should use online shopping. Thus, we hypothesized that SN has a positive relationship with consumers' Internet buying intentions.

H3: SN is positively related to consumers' INT.

In additional to the variables involved in the behavioural models, we also studied personal innovativeness towards information technology (PIIT) because online shopping involves the use of technology, which needs to gain end users' acceptance in order for it to be adopted (Rogers, 1995). PIIT determines an individual's technology usage behaviour, which has been shown to be positively related to AD towards new technologies (O'Cass & Fenech, 2003; Rogers, 1995). The findings on PIIT affecting usage of online shopping vary. Some studies found a positive relationship between PIIT and INT (Donthu & Garcia, 1999; Limayem et al., 2000). On the other hand, other studies observed that the relationship between the two is insignificant (Citrin et al., 2000; Sin & Tse, 2002). In our study, we proposed that PIIT is positively related to consumers' online shopping intentions.

H4: PIIT is positively related to consumers' INT.

In sum, we hypothesized that POS_AD, NEG_AD, SN, and PIIT are explanatory variables of INT. We used INT as the dependent variable instead of actual behaviour because measuring behavioural intention is more appropriate given that we adopted a survey-based research design (Cheng et al., 2006). In addition, we investigated the role that demographics and socioeconomic characteristics play in our model. Unlike previous research, instead of investigating how gender (GEN) and education (EDU) are related to INT, we employed them to explain POS_AD, NEG_AD, SN, and PIIT.

According to a study carried out by Harris Interactive, male shoppers use the Internet for shopping the most while women search information online but make purchases offline (Business Wire, 2000). Besides, online shopping is computer-based that involves the use of computer technology and related knowledge and female shoppers view shopping offline as a major source of relaxation (Dholakia & Chiang, 2003). Considering these, we hypothesized the following:

H5a: Compared with women, men have a more positive attitude towards the positive features of online shopping.

H5b: Compared with women, men have a more positive attitude towards the negative features of online shopping.

Besides AD, gender difference plays a role in SN and PIIT. According to Grabarino and Strahilevitz (2004), family members'/friends' recommendations have no significant effect on men, whereas such recommendations have a strong effect on women's willingness to buy online. Likewise, Andrews et al. (2005) suggested that the social value of talking about Web purchasing activities has an insignificant effect on men but women are more likely to use the Web for purchasing when they receive recommendations from friends. Hence, we hypothesized that women are influenced by recommendations from family members/friends to shop online.

H5c: Compared with men, women are more likely to be associated with SN.

Further, Andrews (2005) found that males are the dominant group that makes purchases on the Internet. In general, men show more interest and confidence in using computers while women have less self-confidence in using the computer and have a more difficult time navigating arbitrary websites (Yang & Lester, 2005). Given these findings, we proposed that men are positively associated with PIIT.

H5d: Compared to women, men are positively associated with PIIT.

Another variable of interest in our research model is EDU. Online shopping requires an individual's digital technology skills from managing hardware and software to navigating skills, such as searching, selecting, and processing information on the Web in order to make purchases (Cai & Cude, 2008). Thus, we hypothesized the following:

H6a: Compared with the less educated individuals, the more educated individuals have a more positive attitude towards the positive features of online shopping.
H6b: Compared with the less educated individuals, the more educated individuals have a more positive attitude towards the negative features of online shopping.

Finally, in a collectivist culture like Hong Kong, people tend to be affiliating closely with others and maintaining social bonds. Since SN relates to a person's perception of social pressures (Yu & Wu, 2007), this variable is perceived as cultural specific. Besides, education has been found to be negatively related to computer anxiety (Igbaria & Parsuraman, 1989; So et al., 2005). It follows that more educated individuals would feel more comfortable in using computers and the Internet. Thus, we formulated the following hypotheses:

H6c: EDU is not associated with SN.
H6d: Compared with the less educated individuals, the more educated individuals are positively associated with PIIT.

RESEARCH METHODOLOGY

In view of the importance of the customer market to business performance, we focused on examining consumers' online shopping intentions in this research. We adopted the survey research methodology and used a Web-based questionnaire to collect data.

Population and Sample

The population of this study consisted of Internet users in Hong Kong because in order to shop online, consumers need to know how to use computers and how to surf on the Internet (Yang & Lester, 2005). They are the people who would purchase personal hygiene products for themselves or their family members. Our sampling frame consisted of people who have access to the Internet and have a Facebook account. We chose to distribute our questionnaire via the Facebook social network because there has been a noted decrease in interactions between users and email services (Beh, 2008). Facebook is reported to have captured the majority of the share of Internet users, as well as being the leading social network in Hong Kong (Lau, 2008). Some top consumer brands such as McDonald's, Nike, and Samsung have realized the potential of reaching their target audience via Facebook and shifted their advertising from traditional channels to Facebook (Hunsberger, 2008). Moreover, an Australian lawyer was permitted by the court to file legal documents against a couple via Facebook (BBC News, 2008). Given Facebook's popularity and recognition, we considered distributing the questionnaire via Facebook appropriate. We drew a systematic random sample from Hong Kong's Facebook network. To achieve a 95% confidence level with a ±5% confident interval, our sample size should be 384, assuming 50% of the population would buy personal hygiene products for themselves/family members.

Research and Questionnaire Design

We employed an electronic survey for data collection. The design of the survey questionnaire followed Lumsden (2006). Questions were adapted from related previous research studies. Since English is the default language for the Web and is the predominant language of business, which is widely spoken in professional circles in Hong Kong (Economist Intelligence Unit, 2007; Ernst & Young, 2007), the questionnaire was developed in English and was translated into traditional Chinese. To eliminate discrepancies, the Chinese version was back translated to English. In addition to cross-checking, the questionnaire was proofread by a language expert before its use in a pilot test. The pilot test was conducted to determine if the questionnaire contained ambiguous sentences/words. Following that, three respondents were interviewed to seek improvements to the questionnaire. Pre-testing of the Web survey was carried out before the official launch.

The questionnaire consists of only scale-rating and multiple-choice questions, without any open-ended questions and requiring no written answers. We used the five-point Likert-type scale (1 = strongly disagree, ..., 5 = strongly agree) and an eleven-point scale, ranging from "absolutely no" to "definitely yes", when we asked respondents to indicate their intentions of purchasing personal hygiene items, such as shower gel/bar soap, shampoo, toothbrush, facial cleanser etc. Besides, we pledged to make a small donation for every completed questionnaire to a local charity as an "incentive" to encourage more respondents to participate in the survey. We employed the mySurvey tool, provided by The Hong Kong Polytechnic University. Since the Web address of the survey contained the domain name "polyu.edu.hk", using mySurvey would increase respondents' confidence in completing the questionnaire. All of the responses were collected anonymously via mySurvey.

A total of 930 invitations were sent to the sample of Facebook users. The bilingual invitation included an embedded universal resource locator (URL) linking respondents directly to the survey site. Reminders were sent to the sampled subjects to encourage replies. Follow-up interviews were performed with some randomly selected respondents to find out more about their views on online shopping.

Data Analysis

The data collected were analyzed using SPSS. None of the responses collected in the pilot test was included in the sample. Some of the answers were coded for SPSS analysis. We computed some descriptive statistics to generate descriptive information about the data collected, check the presence of any "unexpected" value, and examine the pattern of responses in each variable. We also checked whether non-response bias existed by performing the two-sample t-test (Armstrong and Overton, 1977). We randomly selected some questionnaire items to test for non-response bias. Then we compared the means generated from respondents who replied before and after the first reminder. We also performed the z-test to assess population-sample disparities and used the goodness-of-fit test to check whether there were significant differences between the sample and the population. For data reduction, we applied exploratory factor analysis (EFA). We performed the principal component factor analysis on the variables to reduce them into a smaller set of manageable factors. To investigate the relationships between Hong Kong consumers' characteristics and their online shopping intentions, we applied multiple regression analyses to test the hypotheses.

Results and Findings

We received a total of 430 responses, among which 13 responses were discarded because the respondents indicated that they were not living in Hong Kong. Thus, the total number of respondents analyzed was 417, comfortably exceeding the required sample size of 384. The response rate of the questionnaire was 46.23%. Among the valid responses, there were 45.41% males and 54.59% females. A total of 16 respondents took part in the follow-up interviews, for which the ratio of males to females was 5:3. There were 9.61% of the respondents at Form 5 or below level of education, 24.88% at post-secondary level, and 65.52% at bachelor's degree or above level. The male respondents were generally higher educated than the female respondents, which may be attributed to the younger ages of the latter. Over half of the respondents were between 25 and below 45 years' old by the end of 2008, accounting for 78.35% of the total. As for personal monthly income, the majority of the respondents earned HK$30,000 (about US$3,800) or more per month, accounting for 36.44% of the total. Furthermore, the majority of the respondents had never married (62.88%) while respondents who were married or living with a partner made up the second largest group (35.10%).

Across the entire sample, the respondents strongly agreed that it is more difficult to return a defective product after purchasing it online (Mean_{AD13} = 4.14). On the other hand, many of the respondents disagreed that people who are important to them think that they should shop online (Mean_{SN2} = 2.38). We also noted that a relatively high dispersion for AD11r (standard deviation = 1.202) and a low dispersion for AD17 (standard deviation = 0.840). These indicate that respondents' opinions on "online shopping provides easy access to information" were much more alike than they were when they commented on the appropriateness of online shopping.

Among the valid responses, only 32.85% used the Internet to purchase goods. Besides, over 70% of the respondents indicated that they may not consider purchasing via the Internet in the next six months. The major factors influencing their intentions were the reputation of the vendor, level of convenience, variety of goods/services, and price while ease of effort was considered as a minor factor.

Non-Response Bias and Population-Sample Disparities Analyses

To ensure the data were free from bias, we performed a non-response bias check by comparing the data collected before and after the first reminder. The number of responses collected before the first reminder was 244. All of the *p*-values from the two-sample *t*-test were greater than 0.05, indicating there were no significant differences among the responses collected before and after the first reminder. Then we performed a *z*-test ($z_{obs} > -1.96, p > 0.05$) and found that gender percentage in the sample and population had no significant differences. We also carried out a goodness-of-fit test on the dispersion of EDU ($Z_{obs} < 5.99, p > 0.05$). The results suggest that our observed frequencies would be identical to the expected frequencies. Therefore, we proceeded with further analyses.

Exploratory Factor Analyses

In order to identify the underlying factors that explain the pattern of correlations within the set of AD, PIIT, and SN variables, we performed two EFAs. The goal of performing EFA was to reduce the set of variables into smaller sets of common factors. Before we employed the principal components analysis to compare and contrast the common factors, we dropped AD12 because the wording in this item could not reveal whether respondents perceive that the absence of sales assistants in the online shopping environment poses a problem.

We considered any variable that had a loading greater than 0.3 on more than one factor not pure and removed it from the factor analysis (Lewis-Beck, 1994). Through purification, six items were removed. We then grouped 12 variables into four factors of AD, namely Perceived Usefulness (PU), Perceived Convenience (CONV), Perceived Inconvenience (INCONV), and Time, while PIIT and SN remained as individual factors. In addition, in order to cross check the number of components to retain among the variables, we adopted the eigenvalue larger-than-one criterion. According to the Kaiser's rule, we retained only those components that had eigenvalues greater 1 (Stevens, 1996). Both of the extraction sums of squared loadings and scree test suggested there were four components of AD while two components for PIIT and SN. We then performed reliability analysis on each of the factors to ensure internal consistency.

Multiple Regression Analysis

The first regression model examined the overall AD variables towards both positive and negative features of online shopping, as well as PIIT and SN, while the second regression model expanded the AD variables into their constituent factors generated from the EFA.

Model One

In the first model, we considered POS_AD, NEG_AD, PIIT, and SN as independent variables. Having validated all of the assumptions of the regression model, we studied the model findings. We found that 13.8% of the variation in the dependent variable was explained by the variations in the independent variables included in the model. Further, the model had an F ratio of 15.626 and a p-value < 0.05, indicating that the relationship between the predictors and dependent variable was significant. Since the model was meaningful, we proceeded with interpreting the results.

The independent variables, POS_AD, PIIT, and SN, showed a positive relationship, while NEG_AD showed a negative relationship, with the dependent variable as expected. The regression equation was established as follows:

$$INT = -1.821 + 1.444*POS\_AD + (-.286*NEG\_AD) +.128*SN +.154*PIIT + \varepsilon.$$

To further investigate the impacts of the attitudes towards positive and negative features on intention, we performed another multiple regression analysis. Involving the second order construct, we considered PU, CONV, INCONV, and TIME as independent variables on top of PIIT and SN in Model Two.

Model Two

In this model, the adjusted R square was 0.137 while the F ratio was 10.423 with a p-value < 0.05, indicating that 13.7% of the variation in the dependent variable was explained by the variations in the independent variables. The revised linear equation was:

$$INT = -1.26 +.992*PU +.425*CONV -.302*INCONV -.021*TIME +.135*PIIT +.095*SN + \varepsilon.$$

According to the p-values, there was a strongly significant relationship between PU and INT (p-value = 0.000), while CONV and INCONV had a weaker but significant relationship with INT (0.05 < p-value < 0.10). The results indicate that the positive features of online shopping appear to be more important than the negative features in determining consumers' online shopping intentions. Hence, H1 was fully supported, while H2 was partially supported. Nevertheless, all of the regression coefficients had the theoretically expected signs. According to the findings, PU, CONV, and INCONV are the determining factors for online shopping intention. Based on the sizes of the standardized regression coefficients, PU had the greatest impact (0.273), while INCONV had the least impact (-0.097) on consumers' online shopping intentions among the three significant variables. Since PIIT and SN had no significant relationships with INT, H3 and H4 were not supported.

To further test the hypotheses, we performed other multiple regression analyses by having GEN and EDU as explanatory variables with PU, CONV, INCONV, TIME, PIIT, and SN as dependent variables. GEN had a significant relationship with PU, CONV, PIIT, and SN, indicating that H5a and H5d were supported. On the other hand, H5c was inversely supported, while H5b was not supported. As for EDU, it showed a significant relationship with only one dependant variable, PU. Hence, H6c was supported and H6a was partially supported; but both H6b and H6d were not supported. GEN and EDU explained only 3.6% of the variance in PU, 0.8% in CONV, -0.2% in INCONV, -0.3% in TIME, 7.9% in PIIT, and 1.6% in SN.

DISCUSSIONS

That our sample consisted of younger and higher educated individuals are in line with Internet users' characteristics in many countries (Lee et al., 2007; Sin & Tse, 2002). The empirical findings reveal that Hong Kong consumers' intentions to shop online for personal hygiene products, an example of low involvement-search goods, tend to be low. Our regression results suggest that online shopping intention is greatly influenced by PU, while CONV and INCONV in the online shopping environment have weaker influences.

Attitude

That H1 was supported is consistent with previous studies (Chiang & Dholakia, 2003). Similar to convenience, perceived inconvenience also has a significant impact on intention but in an inverse relationship (H2). The item "it is more difficult to return a defective product after I purchased online" received the highest mean score (4.14), indicating that majority of the respondents agreed to this statement. It is plausible that this feature of inconvenience reduces consumers' intentions to shop online because it affects consumers' confidence in making online purchases. Besides, waiting for delivery also weakens the intentions because consumers cannot enjoy immediate possession of the purchase items.

Personal Innovativeness towards Information Technology and Subjective Norms

Our finding on PIIT (H4) conflicts with Citrin et al. (2000), who affirmed that innovativeness has a direct influence on Internet shopping adoption and with Lee et al. (2007), who proposed that intention to make purchases online is positively related to level of innovativeness. This conflicting finding can be associated with product characteristics. Lian and Lin (2008) observed that PIIT positively affects consumers' attitude towards buying "high cost, infrequently purchased and intangible" goods/services online. Since personal hygiene products fall into the category of "low outlay, frequently purchased and tangible" goods, PIIT is not an influential factor when consumers consider making purchases on the Internet (Lian & Lin, 2008). Another reason for this contradicting finding is cultural difference. Park and Jun (2003) performed a study of Koreans, who share a similar culture to the Chinese, and found that innovativeness has no significant relationship with Internet shopping behaviour. Our finding is fully consistent with the previous studies performed in Hong Kong (Siu & Cheng, 2001; Sin & Tse, 2002). In the consumer markets of Hong Kong and Korea, consumers are overwhelmed with advanced technologies. The diffusion of high-tech products in these markets is faster than in Western countries. Thus, the time for measuring individuals' receptivity towards technology and innovation is so short that it is impossible to significantly tell individuals' differences in PIIT.

Another variable that we expected to have a cultural impact was SN. It is believed that in a collectivist culture like Hong Kong, people tend to be affiliating closely with others and maintaining connectedness, thus they are more likely to be influenced by family members/friends (Hofstede et al., 2002; Park & Jun, 2003). But our research results did not support the belief that SN influences INT (H3). Nonetheless, the results are consistent with many prior studies (Bhattacherjee & Premkumar, 2004; Hsu et al., 2006; Siu & Cheng, 2001). Bhattacherjee and Premkumar (2004) suggested that the effect of normative influences (SN) is rather mediated by individuals' voluntariness. Voluntariness is formed by a combination of factors, such as individuals' prior online shopping experience, perceptions about online shopping, and personal characteristics. Another reason for the insignificance of SN could be the low adoption rate of Internet shopping in Hong Kong.

Gender

The demographics and socioeconomic variables considered in our study were gender and education. Between the two variables, gender was found to have a significant relationship with PU (H5a), CONV (H5a), SN (H5c), and PIIT (H5d). Men and women no doubt behave differently. For example, they differ in information processing, reaction to stimuli, and IT usage (Garbarino & Strahilevitz, 2004).

Since electronic shopping is computer and technology based, which is associated with masculinity, it is not surprising to note that men are positively associated with PU, CONV, and PIIT in regard to Internet shopping intention. Besides, many scholars have suggested that women primarily use the Internet for social connections, such as chatting online, while men use the Internet for practicability (Rodgers & Harris, 2003; Wrench et al., 2008; Yang & Lester, 2005). Furthermore, women have more negative affective associations for computer technology, spend lesser time online (which is consistent with our sample characteristics), and are less likely to buy online than men (Doolin et al., 2005; Kwak et al., 2002; Sanchez-Franco, 2006; Yang & Lester, 2005). As for the attitude towards the negative features of online shopping, it is understandable that the relationship is insignificant (H5b) because the measuring items are related to some external factors, such as returning defective products and waiting for delivery. Thus, both men and women may have similar opinions. Furthermore, our finding on SN goes in the opposite direction of our hypothesis (H5c). Solomon (1999) suggested that individuals may acquire social status from others through achieving a certain placement in a social hierarchy, such as being an online purchaser. In other words, the finding that males are associated with normative influences can be explained by the fact that men seek status and self-image affirmation since the Internet is associated with masculinity.

Education

Previous literature has stated that the level of education is positively related to online buying (Sin & Tse, 2002; Lee et al., 2007). According to our study, education plays a significant role in PU (H6a). The commonly used language on the Internet is English and an individual with a low level of education may need to skip websites that are in English. As noted by Shih (2008), the Chinese Internet users tend to access Chinese websites rather than English websites. Therefore, the variety and completeness of information accessed by this group of consumers may not be as much as the more educated group, who would feel more comfortable with using English. Another reason could be the person's occupation. In general, more educated individuals are more likely to use computers at work. This group of consumers will have greater confidence in navigating websites. As a result, more educated consumers would perceive a higher level of usefulness in using the Internet to make purchases.

On the other hand, contrary to our hypotheses, education is not related to any of CONV, INCONV, TIME, and PIIT. These findings are consistent with some prior studies that suggest that education is not a good predictor of online buying (Doolin, 2005; Kwak et al., 2002; Donthu & Garcia, 1999). Examples of items that are grouped under CONV and INCONV are product variety, information quality, defective products return, and wait for delivery, which are external factors that are unrelated to one's education level. As for TIME, Web surfing, i.e., clicking on links and browsing through Web pages, is not rocket science that requires much knowledge. Besides, with more and more people surfing the Web, Kwak et al. (2002) suggested that the Internet may have shifted from the stage of early adoption to the stage of growth. Cheng et al. (2006) commented that online systems are becoming more user-friendly; thus, the difficulty in using the Web should not be a concern nowadays. Hence, differences in education level may not have much an impact on a person's intention to shop online.

Other Findings

Our study generated some interesting insights into the two factors that are commonly regarded as reasons that consumers do not shop online - security and privacy of sensitive information (i.e., perceived risk). Scale items measuring these two constructs were dropped from the EFA due to their mixed loadings. The results imply that perceived risk may not be a concern to the respondents when they shop online, which contrast with many of the previous literature (Chang et al., 2005; Garbarino & Strahilevitz, 2004) but are consistent with Shih's (2008) findings. This can be explained by product characteristics. For low outlay, frequently purchased, and tangible products, consumers are less concerned about the security issue than for more expensive items (Lian & Lin, 2008). Besides, the telecommunications infrastructure and legal systems in Hong Kong are well developed. Therefore, consumers in Hong Kong have more confidence in making purchases online.

Implications for Academics

This research addresses consumers' online shopping intentions in Hong Kong, producing results that add to the existing stock of knowledge in the Asian market. The results of this study are culture specific and reflect whether the findings in Western countries are applicable to Hong Kong's context. An interesting finding on SN, the variable we expected would be influenced by a collectivist culture, reveals that cultural difference may not apply to this antecedent of intention. Since many of the existing literature studied luxury items, such as accessories and air tickets, our study contributes some new findings concerning consumers' online shopping intentions with regard to necessity items, i.e., personal hygiene products.

Implications for Practitioners

Given our empirical findings that the positive features of online shopping appear to be more influential than the negative features in shaping consumers' online shopping intentions, we suggest that marketers focus their efforts on developing and expanding the positive features that highlight online shopping's benefits, thus overriding the weaknesses.

In our research, over 50% of the respondents indicated that convenience and product variety are the factors that they would consider for online shopping. Product variety has been shown to have a significant relationship with Internet buying (Sin & Tse, 2002; Yang & Lester, 2005). Uniqueness of items is another motivational factor for consumers to shop online. Consumers getting information on the Web influences their purchasing behaviours, thus marketers should be cautious about the amount of information made available on the Web. Price also plays a role in consumers' online shopping intentions (Chang et al., 2005). Many of the interviewees expressed that they often look for value-buy on the Internet. Another concern that consumers have in regard to making purchases on the Internet is it could be arduous to return an expired, damaged or defective products. Besides, some individuals perceive waiting for delivery a problem. Hence, marketers need to pay attention to quality and logistics management in order to provide faster services to consumers.

In our study vendor reputation, composed of comments from other buyers, popularity of the e-shop, and credibility in on-time delivery and transaction accuracy, was ranked as the top factor by consumers in considering making purchases online. The above suggestions can help improve firms' services quality, which has shown to have a significant impact on corporate image and customer satisfaction (Cheng et al., 2008).

CONCLUSION, LIMITATIONS AND FUTURE RESEARCH

Our research findings provide insights into customer objectives in the business environment. It is of prime importance to understand what benefits the shoppers seek in the online shopping environment. Then firms need to align their OM strategy with customers' expectations and demand of EC in order to obtain business success. A well-designed customer Internet interface does not depend on the use of high technology. It should support customers' cultures and preferences.

Conclusions

1. In the Hong Kong market, consumers' attitude towards online shopping has a significant effect on their Internet buying intentions. Particularly, perceived usefulness has the strongest influence.
2. Gender difference was found to have a significant relationship with consumers' attitude towards the positive features of online shopping, SN and PIIT. On the contrary, education can only explain individuals' PU.
3. Based on our research findings, e-shoppers attach a great importance to the availability of stocks, lead time between placing an order and receiving the items, total costs incurred in purchasing over the Internet, and whether defective products can be returned with ease. Therefore, OM practitioners in the EC setting need to devote efforts to enhance forecasting and scheduling management, inventory management, quality management, and logistics management.

Limitations and Suggestions for Future Studies

Although this study provides relevant and interesting insights into understanding Hong Kong consumers' online shopping intentions, it is important to recognize some limitations associated with the study.

While our study covered many explanatory variables, many topics of EC were left for further investigations, such as the number of hours spent on the Internet, type of personality, as well as consumers' perceptions about different Web designs. Another limitation of our study was the use of multiple regression analysis of the common factors generated from EFA. Unlike structure equation modeling (SEM), a sophisticated multi-variate statistical tool, EFA and multiple regressions use fewer indicators and drop many valuable data from the analysis.

Besides, the findings from this study may not be readily generalized to the overall population of Internet users because we could only capture the responses from the individuals who voluntarily participated in the survey. This may exclude some potential e-shoppers who did not have time or did not like to respond to our survey.

The findings may not be entirely applicable to other places as Hong Kong has a unique historical background, and a unique cultural and business environment. Besides, since value perceptions and technology change from time to time, the applicability of the findings can be short-lived. Moreover, findings obtained from Hong Kong may not be applicable to other places such as the Chinese mainland, where the telecommunications infrastructure and legal systems are different. Future research may survey respondents from different countries and perform cross-cultural comparisons between the individualist and the collectivist cultures.

Finally, it may also be interesting to look into the linkages between Generation X's and Generation Y's online shopping behaviours. Future research should also study factors affecting continued intention as the success of EC depends on e-shoppers' willingness to re-visit the website and make recurring purchases. Since mobile commerce is the second wave of EC (Lu et al., 2005), it is also worthwhile to examine mobile commerce from the OM's perspective.

REFERENCES

Ajzen, I., & Fishbein, M. (1980). *Understanding Attitudes and Predicting Social Behavior*. Englewood Cliffs, NJ: Prentice-Hall.

Andrews, L., Kiel, G., Drennan, J., Boyle, M. V., & Weerawardena, J. (2005). Gendered perceptions of experiential value in using web-based retail channels. *European Journal of Marketing*, *41*(5), 640–658. doi:10.1108/03090560710737660

Armstrong, J. S., & Overton, T. S. (1977). Estimating Nonresponse Bias in Mail Surveys. *JMR, Journal of Marketing Research*, *14*, 396–402. doi:10.2307/3150783

Baber, Z. (2005). *CyberAsia: The Internet and society in Asia*. Boston: Brill.

Baker, E. W., Al-Gahtani, S. S., & Hubona, G. S. (2007). The effects of gender and age on new technology implementation in a developing country - testing the theory of planned behavior (TPB). *Information Technology & People*, *20*(4), 352–375. doi:10.1108/09593840710839798

Beh, C. (2008). Friendster still tops but Facebook catching up. *Marketing-interactive.com*. Retrieved October 14, 2008, from http://www.marketing-interactive.com/news/5287

Bhattacherjee, A., & Premkumar, G. (2004). Understanding changes in belief and attitude toward information technology usage: a theoretical model and longitudinal test. *Management Information Systems Quarterly*, *28*(2), 229–254.

Business Monitor International. (2008). *Hong Kong Telecommunications Report*. London: Business Monitor International.

Business Wire. (2000). What kind of dot-shopper are you? ebates.com survey first to uncover 'dot-shopping types' defining online purchasers. *BNET*. Retrieved March 13, 2008, from http://findarticles.com/p/articles/mi_m0EIN/is_2000_May_12/ai_62048317/print?tag=artBody;col1

Cai, Y., & Cude, B. J. (2008). Online shopping. In Xiao, J. J. (Ed.), *Handbook of Consumer Finance Research*. New York: Springer Publishing. doi:10.1007/978-0-387-75734-6_9

Chang, M. K., Cheung, W., & Lai, V. S. (2005). Literature derived reference models for the adoption of online shopping. *Information & Management*, *42*, 543–559. doi:10.1016/S0378-7206(04)00051-5

Chase, R. B., Jacobs, F. R., & Aquilano, N. J. (2006). *Operations Management for Competitive Advantage* (11th ed.). New York: McGraw-Hill.

Cheng, T. C. E., Lai, L. C. F., & Yeung, A. C. L. (2008). The driving forces of customer loyalty: a study of Internet service providers in Hong Kong. *International Journal of E-Business Research*, *4*(4), 26–42.

Cheng, T. C. E., Lam, D. Y. C., & Yeung, A. C. L. (2006). Adoption of internet banking: an empirical study in Hong Kong. *Decision Support Systems*, *42*, 1558–1572. doi:10.1016/j.dss.2006.01.002

Chiang, K. P., & Dholakia, R. R. (2003). Factors driving consumer intention to shop online: an empirical investigation. *Journal of Consumer Psychology*, *13*(1), 177–183. doi:10.1207/153276603768344898

Citrin, A. V., Sprott, D. E., Silverman, S. N., & Stem, D. E. (2000). Adoption of internet shopping: the role of consumer innovativeness. *Industrial Management & Data Systems*, *100*(7), 294–300. doi:10.1108/02635570010304806

Davis, F. D. (1989). Perceived usefulness, perceived ease of use, and user acceptance of information technology. *Management Information Systems Quarterly*, *13*(3), 319–340. doi:10.2307/249008

Dholakia, R. R. (1999). Going shopping: key determinants of shopping behaviors and motivations. *International Journal of Retail and Distribution Management*, *27*(4), 154–165. doi:10.1108/09590559910268499

Dholakia, R. R., & Chiang, K. P. (2003). Shoppers in cyberspace: are they from Venus or Mars and does it matter? *Journal of Consumer Psychology*, *13*(1), 171–176. doi:10.1207/153276603768344889

Donthu, N., & Garcia, A. (1999). The internet shopper. *Journal of Advertising Research*, *39*(3), 52–58.

Doolin, B., Dillon, S., Thompson, F., & Corner, J. L. (2005). Perceived risk, the Internet shopping experience and online purchasing behavior: a New Zealand perspective. *Journal of Global Information Management*, *13*(2), 66–68.

Economist Intelligence Unit. (2007). *The 2007 e-readiness rankings - Raising the bar*. Retrieved October 29, 2008 from http://a330.g.akamai.net/7/330/25828/20070420195432/graphics.eiu.com/files/ad_pdfs/2007Ereadiness_Ranking_WP.pdf

Ernst & Young. (2007). *Doing Business in Hong Kong*. Hong Kong: Ernst & Young.

Garbarino, E., & Strahilevitz, M. (2004). Gender differences in the perceived risk of buying online and the effects of receiving a site recommendation. *Journal of Business Research*, *57*, 768–775. doi:10.1016/S0148-2963(02)00363-6

Girard, T., Korgaonkar, P., & Silverblatt, R. (2003). Relationship of type of product, shopping orientations, and demographics with preference for shopping on the Internet. *Journal of Business and Psychology*, *18*(1), 101–120. doi:10.1023/A:1025087021768

Hofstede, G. J., Pedersen, P., & Hofstede, G. H. (2002). *Exploring Culture: Exercises, Stories, and Synthetic Cultures*. Yarmouth, ME: Intercultural Press.

Hsu, M. H., Yen, C. H., Chiu, C. M., & Chang, C. M. (2006). A longitudinal investigation of continued online shopping behavior: an extension of the theory of planned behavior. *International Journal of Human-Computer Studies*, *64*, 889–904. doi:10.1016/j.ijhcs.2006.04.004

Hunsberger, B. (2008). Nike tests Facebook. *The Oregonian*. Retrieved September 17, 2008, from http://www.oregonlive.com/business/index.ssf/2008/06/nike_tests_facebook.html

Igbaria, M., & Parsuraman, S. (1989). A path analytic study of individual characteristics, computer anxiety and attitudes toward microcomputers. *Journal of Management*, *15*(3), 373–388. doi:10.1177/014920638901500302

Kerem, T. (2005). *Advances in the Economics of Information Systems*. Hershey, PA: IGI Global.

Kwak, H., Fox, R. J., & Zinkhan, G. M. (2002). What products can be successfully promoted and sold via the Internet. *Journal of Advertising Research, 42*(1), 23–38.

Lau, A. (2008). Facebook tops social networks in Hong Kong. *Marketing-interactive.com*. Retrieved October 14, 2008, from http://www.marketing-interactive.com/news/5328

Lee, H. Y., Qu, H. L., & Kim, Y. S. (2007). A study of the impact of personal innovativeness on online travel shopping behavior – a case study of Korean travelers. *Tourism Management, 28*(3), 886–897. doi:10.1016/j.tourman.2006.04.013

Lewis-Beck, M. S. (1994). *Factor Analysis and Related Techniques*. London: Sage Publications.

Lian, J. W., & Lin, T. M. (2008). Effects of consumer characteristics on their acceptance of online shopping: comparison among different product types. *Computers in Human Behavior, 24*, 48–65. doi:10.1016/j.chb.2007.01.002

Lightner, N. J. (2003). What users want in e-commerce design: effects of age, education and income. *Ergonomics, 46*(1-3), 153–168. doi:10.1080/00140130303530

Lim, K. H., Leung, K., Sia, C. L., & Lee, M. K. (2004). Is e-commerce boundary-less? Effects of individualism-collectivism and uncertainty avoidance on Internet shopping. *Journal of International Business Studies, 35*, 545–559. doi:10.1057/palgrave.jibs.8400104

Limayem, M., Khalifa, M., & Frini, A. (2000). What makes consumers buy from internet? A longitudinal study of online shopping. *IEEE Transactions on Systems, Man, and Cybernetics, 30*(4), 421–432. doi:10.1109/3468.852436

Lu, Y., Zhou, T., & Wang, B. (2009). Exploring Chinese users' acceptance of instant messaging using the theory of planned behavior, the technology acceptance model, and the flow theory. *Computers in Human Behavior, 25*(1), 1–252. doi:10.1016/j.chb.2008.06.002

Lumsden, J. (2006). Online-questionnaire design guidelines. In Reynolds, R. A., Woods, R., & Baker, J. D. (Eds.), *Handbook of Research on Electronic Surveys and Measurements*. Hershey, PA: IGI Global.

News, B. B. C. (2008, December 16). Legal papers served via Facebook. *BBC* (Online). Retrieved December 17, 2008, from http://news.bbc.co.uk/2/hi/asia-pacific/7785004.stm

O'Cass, A., & Fenech, T. (2003). Web retailing adoption: exploring the nature of Internet users web retailing behaviour. *Journal of Retailing and Consumer Services, 10*(2), 81–94. doi:10.1016/S0969-6989(02)00004-8

Park, C., & Jun, J. K. (2003). A cross-cultural comparison of Internet buying behavior – effects of Internet usage, perceived risks, and innovativeness. *International Marketing Review, 20*(5), 534–553. doi:10.1108/02651330310498771

Pavlou, P. A., & Fygenson, M. (2006). Understandings and predicting electronic commerce adoption: An extension of the theory of planned behavior. *Management Information Systems Quarterly, 30*(1), 115–143.

Rodgers, S., & Harris, M. A. (2003). Gender and e-commerce: an exploratory study. *Journal of Advertising Research, 43*(3), 322–329.

Rogers, E. (1995). *Diffusion of Innovation* (4th ed.). New York: Free Press.

Rohm, A., & Swaminathan, V. (2004). A typology of online shoppers based on shopping motivations. *Journal of Business Research, 57*(7), 748–757. doi:10.1016/S0148-2963(02)00351-X

Sanchez-Franco, M. J. (2006). Exploring the influence of gender on the web usage via partial least squares. *Behaviour & Information Technology, 25*(1), 19–36. doi:10.1080/01449290500124536

Schniederjans, M. J., & Cao, Q. (2002). *e-Commerce Operations Management*. River Edge, NJ: World Scientific.

Shih, H. P. (2008). Continued use of a Chinese online portal: An empirical study. *Behaviour & Information Technology*, *27*(3), 201–209. doi:10.1080/01449290600802403

Sin, L., & Tse, A. (2002). Profiling internet shoppers in Hong Kong: demographic, psychographic, attitudinal and experiential factors. *Journal of International Consumer Marketing*, *15*(1), 7–28. doi:10.1300/J046v15n01_02

Siu, N. Y., & Cheng, M. M. (2001). A study of the expected adoption of online shopping - the case of Hong Kong. *Journal of International Consumer Marketing*, *13*(3), 87–106. doi:10.1300/J046v13n03_06

So, W. C., Wong, T. N., & Sculli, D. (2005). Factors affecting intentions to purchase via the Internet. *Industrial Management & Data Systems*, *105*(9), 1225–1244. doi:10.1108/02635570510633275

Solomon, M. (1999). The value of status and the status of value. In Holbrook, M. B. (Ed.), *Consumer Value: A Framework for Analysis and Research*. New York: Routledge. doi:10.4324/9780203010679.ch3

South, S. J., & Spitze, G. (1994). Housework in martial and non-marital households. *American Sociological Review*, *59*, 327–347. doi:10.2307/2095937

Stevens, J. (1996). *Applied Multivariate Statistics for the Social Sciences* (3rd ed.). Mahwah, NJ: Lawrence Elbaum Associates.

Telecom Research Asia-Pacific. (2004). *Telecommunications in Hong Kong*. Hong Kong: Telecom Research Asia-Pacific Limited.

Teo, T. S. H., & Lim, V. K. G. (2000). Gender differences in internet usage and task preferences. *Behaviour & Information Technology*, *19*(4), 283–295. doi:10.1080/01449290050086390

Van Slyke, C., Comunale, C. L., & Belanger, F. (2002). Gender differences in perceptions of web-based shopping. *Communications of the ACM*, *45*(8), 82–86. doi:10.1145/545151.545155

Vijayasarathy, L. R. (2004). Predicting consumer intentions to use on-line shopping: the case for an augmented technology acceptance model. *Information & Management*, *41*, 747–762. doi:10.1016/j.im.2003.08.011

Wang, M. S., Chen, C. C., Chang, S. C., & Yang, Y. H. (2007). Effects of online shopping attitudes, subjective norms and control beliefs on online shopping intentions: a test of the Theory of Planned Behaviour. *International Journal of Management*, *24*(2), 296–303.

Wrench, J. S., McCroskey, J. C., & Richmond, V. P. (2008). *Human Communication in Everyday Life*. Boston: Pearson Education.

Yang, B., & Lester, B. (2005). Gender differences in e-commerce. *Applied Economics*, *37*, 2077–2089. doi:10.1080/00036840500293292

Yu, T. K., & Wu, G. S. (2007). Determinants of Internet shopping behavior: an application of Reasoned Behaviour Theory. *International Journal of Management*, *24*(4), 744–764.

This work was previously published in International Journal of E-Business Research Volume 6, Issue 3, edited by In Lee, pp. 56-70, copyright 2010 by IGI Publishing (an imprint of IGI Global).

Chapter 9
Effects of Website Interactivity on E-Loyalty: A Social Exchange Perspective

Wen-Jang Jih
Middle Tennessee State University, USA

Su-Fang Lee
Overseas Chinese University, Taiwan

Yuan-Cheng Tsai
Overseas Chinese University, Taiwan

ABSTRACT

This study examines the impact of major considerations of commercial website design and management, website interactivity, the important performance factor, and customer loyalty. Meaningful interactions with existing and potential customers allow e-commerce firms to enhance customer loyalty by providing satisfactory customer experience. Customer loyalty is a composite construct, yet consists of multiple dimensions that do not necessarily contribute to repeat purchase in the same manner. Guided by the concept of social exchange in social psychology, this study investigates the causal effect of website interactivity on two conceptually distinctive types of website loyalty—transactional loyalty and relational loyalty. This paper finds positive effects of website loyalty on both types of loyalty, while revealing positive effects of transactional loyalty on relational loyalty and overall loyalty. Furthermore, relational loyalty is found to contribute more to overall loyalty than transactional loyalty. E-commerce firms are advised to offer attractive incentives in a systematical manner to gradually convert more temporary transactional loyalty to more long-term relational loyalty. Finally, firms must devise a portfolio of different marketing strategies to cater to the customers located on different points on transactional loyalty.

DOI: 10.4018/978-1-61350-462-8.ch009

BACKGROUND

As Internet technologies rapidly advance to become an essential element of innovative business models, website features have received much attention from both business academician and practitioners. Driven by various supply and demand factors, the adoption of Web-enabled tools or services has been tightly tied with and even become a core work practice in carrying out daily business operations and supporting managerial decision making in today's information and knowledge-intensive organizations. Increasingly, customers in all sectors are experiencing the widespread impact of the Internet technology in most major aspects of their life, including shopping and communications. To a certain degree, this impact is reflected in the quarterly survey conducted by the U. S. Census Bureau of the Department of Commerce to track retail e-commerce sales in the U. S. market, Internet retail sales have steadily grown from $25.8 billion in the year of 2000 to the estimated $147.6 billion in 2008. This robust growth trend reflects many strategic advantages that Internet retailing offers over the pure brick-and-mortar business model. From the business strategy point of view, these advantages include global market outreach, broader product lines, near real-time communications with customers, and customization of products or services to satisfy customer needs. Internet retailing has appeared to be embraced by an increasing number of consumers for its convenience and flexibility in serving the life style in the knowledge economy. For individual companies that sell primarily via their websites, however, there are also significant operational challenges that must be properly addressed, for example, the lack of customer trust and loyalty.

High customer loyalty is important for a seller's profitability. Acquiring a new customer generally costs many times that of selling to an existing customer. The cost of serving a customer throughout the customer service life cycle is higher during the early transactions than when the loyal customer returns for repeat purchase (Reichheld & Sasser, 1990). Customer loyalty to a business that sells online, or e-loyalty, is especially hard to come by since the Internet is a nearly perfect market due to convenient access to product and service information. The dramatic reduction of information asymmetry can result in fierce price competition and low brand loyalty. Enhancing customer loyalty through a superb website is, therefore, an important strategic goal of e-commerce vendors (Kumar & Shah, 2004).

A crucial design element for effective online marketing is the level of website interactivity. Website interactivity, among other design factors, contributes to the performance of commercial websites (Auger, 2005). In an economy that is characterized by collaboration both within and between business partners, website interactivity also provides the essential technological foundation for collaborative commerce (Hartono & Holsapple, 2004). For website users, a highly interactive website encourages in-depth communications with the company and other customers. Communicating directly with the company facilitates responsive customer service, and may even enable customization of the product or service they receive from the vendor. Perhaps more importantly is the conversation supported by the interactivity features of the website enable among the customers. Significant value can be created by involving customers in the product life cycle.

Given the critical role played by interactivity in commercial website design and the technologies available for online marketers (e.g., Web 2.0 technologies) to incorporate proper interactivity features in their websites, it is important for marketers to have an in-depth knowledge regarding the impact of interactivity on customers' loyalty to the website. The purpose of this empirical research is to address this issue.

In the remainder of the paper, we first review the literature in customer-facing system interactivity and customer loyalty. This is followed by a discussion of the issues related to research method. We then present the results of data analysis. In the last section, a summary of the research is provided along with discussion of the implications and suggestions for future research.

LITERATURE REVIEW

Interactivity

The issue of interactivity has been discussed in a variety of fields, including information systems, education, advertising, marketing, communication, and strategy. Most existing research on interactivity emphasize the two-way nature of the information flow between communicating parties (e.g., Auger, 2005; Burgoon et al., 2000; Ghose & Dou, 1998; Johnson et al., 2006). A high level of website interactivity generally is not just a welcomed usability feature that pleases website visitors, it is even considered a critical factor for commercial website to attract and retain online customers. Recent developments and applications of Web 2.0 technologies have added more significance to the notion of website interactivity. At the fundamental level, the appeal of Web 2.0 primarily lies in its capability to support collaboration of all scales. The interactive features of Web 2.0 technologies allow companies, large and small, to involve the knowledge workers both within and outside the company in the knowledge creation and application processes. The result, as well as the process itself, of mass collaboration can be beneficial for all participants: companies reap the fruit of innovation, and contributors of innovative ideas often receive monetary or non-monetary awards. Furthermore, the positive feeling of customers as participants of community-building activities may boost the level of brand loyalty of customers as they are personally involved in certain stages of the product life cycle (Ghose & Dou, 1998; Jarvenpaa & Todd, 1997). The significant impact of website interactivity on various aspects of website performance has also been investigated in other studies (Srinivasan et al., 2002; Deignton, 1996; Johnson et al., 2006). Our research seeks to validate this causal effect by empirically testing the following hypothesis (formulated in the alternative form):

H1: Website interactivity positively impacts customer loyalty.

Behavioral Loyalty and Attitudinal Loyalty

Customer loyalty is a compound concept. Early views of brand loyalty almost exclusively emphasized the customer behavior of repeat purchase over a period of time (Cunningham, 1956; Day, 1969; Newman & Werbel, 1973; Reynolds & Darden, 1974; Monroe & Guiltinan, 1975). More recent research, however, contended that customer loyalty and repeat purchase do not equal (Dick & Basu, 1994; Dekimpe et al., 1997). Jacoby and Kyner (1973), for example, have indicated loyal customers tend to purchase repetitively but those who purchase repetitively do not necessarily do so out of loyalty, such as the scenario of lacking an alternative offering. Blattberg and Neslin (1990) also pointed out in their theory of repeat purchase that customers may purchase repetitively out of loyalty or as a result of promotion campaign. If motivated primarily by promotional offering, they would stop purchasing once the promotion ends. Strictly speaking, there is no customer loyalty involved in this scenario. True customer loyalty is driven by internal factors. A meaningful measurement of brand loyalty is considered to encompass both behavioral and attitudinal aspects of the concept (Chaudhuri & Holbrook, 2001; Jacoby & Kyner, 1973; Lim & Razzaque, 1997; Dick & Basu, 1994; Gommans et al., 2001). Although behavioral loyalty and attitudinal loyalty are two conceptually distinctive concepts, they often mutually reinforce each other in strengthening overall customer loyalty (Rowley, 2000).

In light of the distinction of these two types of loyalty (behavioral and attitudinal) and supported by the theory of social exchange (Blau, 1964), we conducted this research to examine the causal effect of an important website feature, website interactivity, on the type of loyalty that is associated with economic exchange (transactional loyalty) and the other type of loyalty that is connected with social exchange (relational loyalty). We also investigated how website interactivity impacts the overall loyalty of online customers.

Transactional Loyalty and Relational Loyalty

Repeat purchasing purely motivated by external factors does not contribute to profitability, since the economic benefit may be offset by the expense of the motivating factors. Lacking the availability of such motivators, customers immediately turn to other vendors. Thus the benefit of behavioral loyalty is essentially short-term in nature and is not a vital source of competitive advantage (Reinarts & Kumar, 2002). Companies must identify effective ways to strengthen the motivating factors that build customer loyalty from within and seek to develop and nurture favorable attitude of their customers. Understanding the nature of customer attitude is therefore an important issue. For this purpose we turned to consumer behavior and social psychology for conceptual guidance.

In the Tricomponent Attitude Model proposed by Schiffman and Kanuk (1997), consumer attitude consists of three aspects: cognition, affect, and conation. The cognitive component represents a consumer's knowledge and perceptions (i.e., beliefs) about products and services. The affective component is determined by the consumer's emotions or feelings with respect to the product or service. This is an important aspect of the attitude formation process, since it often determines an individual's overall favorableness toward the product or service. Finally, conation expresses the likelihood that the consumer will act in a specific fashion with respect to the attitude object, i.e., the intention to purchase the product or service. The three attitude components can be used to explain customers' purchasing decision making. The decision making process begins with gathering information about the product or service under consideration cognition). The evaluation of the perception results in the emotional feeling for the product or service. A strong favorable emotion toward the product increases the likelihood of purchasing. The type of customer loyalty that is driven only by the cognitive component is as strong as that driven by the affect and conation aspects of the attitude. The former is called transactional loyalty, and the latter is relational loyalty. From the social psychological point of view, transactional loyalty is motivated by economic exchange, and relational loyalty is driven by social exchange (Blau, 1964). Economic exchange requires that all the two-way transfers of benefits between the transaction participants that is to be made now or in the future must be agreed upon at the time of sale. In contrast, social exchange "tends to engender feelings of personal obligation, gratitude, and trust (Blau, 1964, p. 94)." In other words, relational loyalty supported by social exchange tends to be stronger than transactional loyalty that is based on economic exchange.

Long-term customer loyalty results from a long-term experience of satisfaction with the product or service. Temporary and purely economically-driven behavioral loyalty, therefore, may develop into long-term, relational loyalty (Solomon, 1999). The conversion from transactional loyalty to relational loyalty occurs gradually. At any point in time, a customer is located somewhere on the continuum from extremely transactional to extremely relational. Businesses must analyze their customers regarding their positions on this continuum and develop marketing strategies to accommodate their unique conditions (Anderson & Narus, 1990). A "one-size-fits-all" marketing strategy does not produce maximal benefit of customer loyalty.

Two research hypotheses were formulated for the testing of the impact of website interactivity on transactional loyalty and relational loyalty. Two hypotheses were set up to allow for the testing of the effect of transactional loyalty on relational loyalty. One hypothesis tests the impact of relational loyalty on overall loyalty. All the hypotheses are listed in the alternative form.

H2: Website interactivity positively impact transactional loyalty.
H3: Website interactivity positively impacts relational loyalty.
H4: Transactional loyalty positively impacts relational loyalty.
H5: Transactional loyalty positively impacts overall customer loyalty.
H6: Relational loyalty positively impacts overall customer loyalty.

Data Gathering and Data Analysis

Our primary research data were collected in Taiwan in spring 2006. Ranked 3rd in Brown University's Global e-Government rankings and 17th in the Economist Intelligence Unit's e-readiness rankings as of the 3rd quarter of 2007, Taiwan has aggressively invested in all major aspects of e-commerce, including infrastructures construction, merchant and customer education, and applications development (www.find.org.tw). E-commerce is now a major thrust in Taiwan's economic development, especially in the sector of small businesses (Jih, 2004). Supported by the 3rd-generation wireless networks and personal handyphone systems (PHS), Taiwanese shoppers are well equipped and conveniently linked with the online merchandisers worldwide. Research findings obtained from Taiwanese data provide valuable clues for understanding online shoppers' behavior elsewhere.

The survey questionnaire questions for website interactivity are adapted from the "contact interactivity" measure in Srinivasan et al. (2002). The question items on loyalty are based on Chaudhuri and Holbrook (2001), Jacoby and Kyner (1973), and Dick and Basu (1994) and cover behavioral dimensions of recommendation, repeat visit, and word-spreading. The constructs are phrased to relate to the concepts of economic exchange and social exchange, respectively. Respondents were not asked to indicate the specific website they had in mind when filling out the questionnaire. The revised questionnaire after pretesting was e-mailed to a convenient sample of online shoppers. An amount of $10 NT dollars was donated to a charity agency for each completed and returned questionnaire. Among the 564 returned questionnaires, 163 were removed because they either have never shopped online or have shopped only once. Additional eleven responses were removed due to significant incompleteness of content leaving a total of 390 effective responses were used in the data analysis. Of these, 36% were males and 64% were females. Other demographical data of the subjects are presented in Table 1.

The data analysis procedure involves two steps to address the issues of reliability and validity of the measurement instrument (Anderson & Gerbing, 1988). The first step evaluates the measurement model, and the second step assesses the structural relationships among latent constructs. Specifically, the analysis process consists of the calculation of Cronbach's α values for the sets of questions designed to measure research constructs (reliability analysis), validity analysis, exploratory factor analysis, correlation analysis, first-order confirmatory factor analysis, and structural equation modeling.

Table 2 displays the covariance matrix of the measurement items. The results of first-order confirmatory factor analysis are summarized in Table 3. All Squared Multiple Correlation (SMC) coefficients are greater than 0.5. All composite reliabilities are greater than 0.7. And all average variances extracted are greater than 0.5. These results indicate acceptable reliabilities and convergent validities of the measurement instrument.

Table 1. Demographics (Number of subjects = 390)

Measure	Frequency	Percent
Gender:	139	35.6
Male	251	64.4
Female		
Age:	10	2.6
<= 19	235	60.3
20-29	96	24.6
30-39	42	10.8
40-49	6	1.5
50-59	1	.3
>= 60		
Education Level:	7	4.4
High School	83	21.3
Junior College	236	60.5
4-year College	54	13.8
Graduate School		
Marriage Status:	87	22.3
Married	303	77.7
Single		
Occupation:	126	32.3
Student	14	3.6
Information Industry	36	9.2
Retailing	8	2.1
Military and Police	3	.8
Publishing	7	1.8
Government	26	6.7
Education	54	13.8
Manufacturing	6	1.5
Factory	1	.3
Fishery, Forestry & Holticulture	77	19.7
Other Service Industries	3	.8
Homecaring	29	7.4
Retiree		
Total	390	100.0

The Cronbach α value of each of the construct measurements is greater than 0.7 and indicates acceptable reliabilities. In assessing the discriminant validity of a measurement instrument, Fornell and Larcker (1981) suggested comparing the average extracted variance with the correlation coefficient of the two constructs under consideration. Acceptable discriminant validity is witnessed when the average extracted variance of the two theoretical constructs is greater than their correlation coefficient. The values on the diagonal in Table 4 represent the average extracted variances of the four research constructs (0.563, 0.791, 0.693, and 0.784). These values are all greater than the rest of the values in the table representing the correlation coefficients of all pairs of the constructs investigated in this study, an indication of reasonable discriminant validity.

Findings

The causal effects of the research constructs are evaluated using structural equation modeling (SEM) analysis. SEM is employed as this study's purpose is to confirm the theoretical model, comprising four theoretical constructs (website interactivity, transactional loyalty, relational loyalty, and overall loyalty) (Figure 1, Table 5), that is formulated on basis of literature. The six causal effects evaluated, the standardized structural coefficients and the corresponding t-values are:

Website Interactivity → Overall Loyalty (0.402, 5.911)
Website Interactivity → Transactional Loyalty (0.379, 5.661)
Website Interactivity → Relational Loyalty (0.385, 5.631)
Transactional Loyalty → Overall Loyalty (0.134, 2.747)
Transactional Loyalty → Relational Loyalty (0.346, 6.202)
Relational Loyalty → Overall Loyalty (0.379, 6.337)

These results, along with the model fitness values (GFI = 0.958, AGFI = 0.932, NFI = 0.965, CFI = 0.981, and RMSEA = 0.056), reveal positive causal effects of website interactivity on transactional loyalty, relational loyalty, and overall loyalty (Bentler & Hu, 1995). Three other causal effects are also validated: Transactional loyalty on relational loyalty and overall loyalty, and relational loyalty on overall loyalty. It is worth noting the fact the structural coefficient (0.379) of the impact of relational loyalty on overall loyalty is greater than that of the impact of transactional loyalty on overall loyalty (0.134) indicates that relational

Table 2. Covariances matrix of the measurement items

	INT1	INT2	INT3	OL1	OL2	OL3	RL1	RL2	RL3	TL1	TL2	TL3
INT1	.746											
INT2	.314	.677										
INT3	.365	.302	.855									
OL1	.304	.251	.291	.695								
OL2	.310	.256	.297	.588	.691							
OL3	.259	.214	.248	.492	.501	.638						
RL1	.248	.205	.238	.386	.394	.329	.928					
RL2	.256	.211	.245	.398	.406	.339	.626	.975				
RL3	.269	.222	.258	.418	.426	.357	.657	.678	.934			
TL1	.212	.175	.204	.326	.333	.278	.348	.359	.377	1.033		
TL2	.222	.184	.213	.341	.348	.291	.365	.376	.395	.864	.971	
TL3	.180	.149	.173	.277	.283	.236	.296	.305	.321	.701	.734	.917

INT: Interactivity; OL: Overall Loyalty; RL: Relational Loyalty; TL: Transactional Loyalty

Figure 1. The validated theoretical model

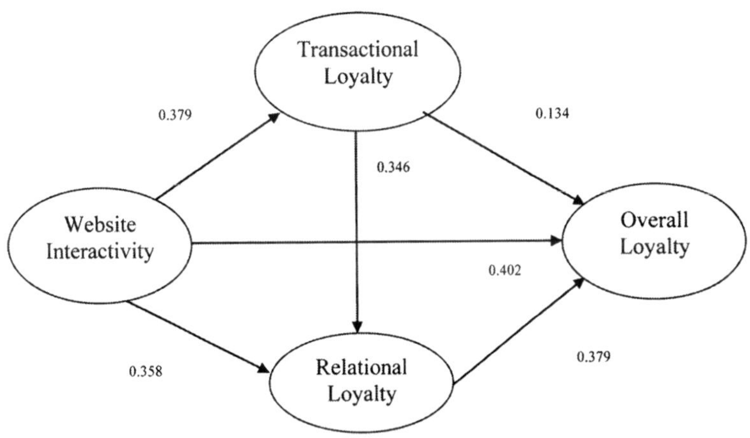

loyalty contributes more to the overall loyalty than transactional loyalty does. The t-values represent the significance level of these relationships ($p < 0.01$). The results of hypothesis testing are summarized as follows.

H1: Website interactivity positively impacts customer loyalty. (Accepted)
H2: Website interactivity positively impact transactional loyalty. (Accepted)
H3: Website interactivity positively impacts relational loyalty. (Accepted)
H4: Transactional loyalty positively impacts relational loyalty. (Accepted)
H5: Transactional loyalty positively impacts overall customer loyalty. (Accepted)
H6: Relational loyalty positively impacts overall customer loyalty. (Accepted)

Table 3. Reliabilities and convergent validities of model dimensions

Constructs	Measurement Items	Standardized Regression Weights	Cronbach's α	Squared Multiple Correlations	Average Variance Extracted	Composite Reliabilities
Website Interactivity	The website allows me to view products from different points of view	0.788	0.702	0.621	0.563	0.794
	The search capability of this website makes it easy to find the products	0.735		0.5421		
	The website is responsive to online customer inquiry	0.723		0.526		
Transactional Loyalty	I use this website because of the discount offers	0.798	0.915	0.637	0.791	0.919
	I use this website because it offers gift awards	0.974		0.948		
	I'll continue to use this website to accumulate bonus points.	0.888		0.789		
Relational Loyalty	I'll stick with this website even if there's something else better than this.	0.820	0.877	0.672	0.693	0.851
	When out of stock, I am willing to wait on back-order.	0.813		0.660		
	I like this website and will be switch to other websites easily.	0.865		0.748		
Overall Loyalty	Overall, I will continue to use this website	0.804	0.917	0.647	0.784	0.916
	Overall, I will recommend this website to relatives and friends	0.947		0.896		
	Overall, I will spread good words about this website	0.900		0.809		

Table 4. Discriminant validity assessment of research constructs

Constructs	Interactivity	Transactional Loyalty	Relational Loyalty	Overall Behavioral Loyalty
Website Interactivity	0.563			
Transactional Loyalty	0.102	0.791		
Relational Loyalty	0.158	0.206	0.693	
Overall Behavioral Loyalty	0.277	0.222	0.346	0.784

Table 5. Fitness of theoretical model

Causal Path	Expected Sign	Theoretical Model of Website Loyalty	
		Standardized Coefficient	t-value
Website Interactivity -> Overall Loyalty	+	0.402	5.911*
Website Interactivity -> Transactional Loyalty	+	0.379	5.661*
Website Interactivity -> Relational Loyalty	+	0.385	5.631*
Transactional Loyalty -> Overall Loyalty	+	0.134	2.747*
Transactional Loyalty -> Relational Loyalty	+	0.346	6.202*
Relational Loyalty -> Overall Loyalty	+	0.379	6.337*
Fitness Indices	Criteria		Fitness
GFI	> 0.9	0.958	Good
AGFI	> 0.9	0.932	Good
NFI	> 0.9	0.965	Good
CFI	> 0.95	0.981	Good
RMSEA	< 0.08	0.056	Good

* $p < 0.01$

CONCLUSION

Creative business practices often are enabled or facilitated by innovative technologies. In a paper presented at the MIT's 2007 Buck Weaver Award reception, Wind (2008) contended that the discipline of marketing must fundamentally change in order to keep up with the rapid change of technologies. One of the specific strategies he proposed is to replace the traditional notion of customer relationship management with the more realistic strategy of customer-managed relationship. In light of the rapid proliferation of Web 2.0 applications, the business thinking that positions customer influence at the center of decision-making functions would not sound like anything revolutionary. Website interactivity, among many other aspects of website design and management, represents a critical element of the enabling information infrastructure. Customer loyalty is an important performance indicator of marketing success. This study investigates the impact of website interactivity on customer loyalty. Two types of customer loyalty are examined as distinctive constructs to allow for better understanding of this composite variable.

In addition to confirming the findings of previous studies concerning the crucial role of interactivity in e-commerce such Auger (2005) and Ghose and Dou (1998), this research has verified the validity of characterizing customer in terms of their attitudinal components and distinguishing between the two types of customer loyalty. Since transactional loyalty is primarily economic in nature and does not necessarily result in repeat purchase in the long term and relational loyalty involves emotional identification, e-commerce companies must design their customer relationship management in light of these findings to strengthen their customer bases. Two practical implications of the research results are presented as follows.

The notions of transactional loyalty and relational loyalty can serve as a meaningful basis for customer segmentation. Given the importance of website interactivity in enhancing customer loyalty, e-commerce firms should interact with their online customers according to their different loyalty types. On the one hand, customers in the category of transactional loyalty, for example, tend to make purchasing decision in more rational way and expect economic benefits in return for their purchasing. Lacking a strong emotional bond with the vendor, these customers can only be acquired and kept by offering material benefits in the form of, for example, price discount, gift coupon, bonus point accumulation in return for free product or service, etc. On the other hand, relationally loyal customers have developed an emotional bondage with the vendor. They may rationally or emotionally act as loyal supporters or even strong defenders for the vendor. These customers are the most valuable assets and worth special attention to nourishing the relationship. Through mechanisms such as online community and customized or even personalized communications, these customers can effectively assist in new product or process design and testing. The first-hand market intelligence directly acquired from truly loyal customers provides valuable input for major strategy formulation.

It must be noted that customers are located on different points on the Transactional Loyalty – Relational Loyalty continuum. A customer may begin the journey completely driven by purely economic benefit. Over time and through consistently satisfactory experiences with the product or service, the customer gradually develops more trust for the company and becomes more relationally loyal. This trust-building is a dynamic process. Only through open and in-depth communications can the loyalty level of a particular customer be correctly identified. Furthermore, a key to strong customer relationship is free flow of information in both directions. Companies must make themselves as transparent as they can to make strongly loyal customer feel they truly belong to the community.

For e-commerce research, two research directions are suggested. Multidisciplinary in nature, e-commerce research prospers on innovative integration of frameworks originating in such fields as business strategy, marketing, consumer behavior, information technology, economics, and psychology. The combinations of factors relating people, technology, and business process represent a fertile field for systematic, theory-building inquiry. Collaborative design of new product or service participated by loyal customers, for example, has not received enough attention from academic researchers. The role of Web 2.0 services such as blogs in building fruitful online community also requires systematic investigation. Another direction for further research is to duplicate our study in different cultural settings. Findings from research conducted in multiple cultural settings contribute to better external validity of the resultant theory.

REFERENCES

Anderson, J. C., & Gerbing, D. W. (1988). Structural equation modeling in practice: a review and recommended two-step approach. *Psychological Bulletin*, *103*(3), 411–423. doi:10.1037/0033-2909.103.3.411

Anderson, J. C., & Narus, J. A. (1990). A Model of Distributor Firm and Manufacturer Firm Working Partnerships. *Journal of Marketing*, *54*, 42–58. doi:10.2307/1252172

Auger, P. (2005). The impact of interactivity and design sophistication on the performance of commercial websites for small businesses. *Journal of Small Business Management*, *43*(2), 119–137.

Bentler, P. M., & Hu, L. T. (1995). Evaluating model fit. In Hoyle, R. H. (Ed.), *Structural Eguation Modeling* (pp. 76–99). Thousand Oaks, CA: Sage Publications.

Blackman, J., & Crompton, L. (1991). The usefulness of selected variables for predicting activity loyalty. *Leisure Sciences*, *13*(3), 205–220. doi:10.1080/01490409109513138

Blattberg, R. C., & Neslin, S. A. (1990). *Sales Promotion-Concepts, Methods, and Strategies*. Englewood Cliffs, NJ: Prentice-Hall.

Blau, P. M. (1964). *Exchange and Power in Social Life*. New York: John Wiley & Sons, Inc.

Burgoon, J. K., Bonito, J. A., Bengtsson, B. Jr, Ramirez, A., Dunbar, N. E., & Miczo, N. (2000). Testing the Interactivity Model: Communication Processes, partner Assessments, and the Quality of Collaborative Work. *Journal of Management Information Systems*, *16*(3), 33–56.

Chaudhuri, A., & Holbrook, M. B. (2001). The Chain of Effects from Brand Trust and Brand Affect to Brand Performance: The Role of Brand Loyalty. *Journal of Marketing*, *65*(2), 81–93. doi:10.1509/jmkg.65.2.81.18255

Cunningham, R. M. (1956). Brand loyalty-what, where, how much? *Harvard Business Review*, *34*(1), 116–128.

Day, G. S. (1969). A two-dimensional concept of brand loyalty. *JMR, Journal of Marketing Research*, *9*, 29–36.

Deignton, J. (1996). The future of interactive marketing. *Harvard Business Review*, *74*, 151–160.

Dekimpe, M. G., Pierre, F., Srinath, G., Gary, L. L., & Christophe, V. B. (1997). Generalizing about trade show effectiveness: A cross-national comparison. *Journal of Marketing*, *61*(4), 55–64. doi:10.2307/1252086

Dick, A. S., & Basu, K. (1994). Customer Loyalty: Toward an Integrated Conceptual Framework. *Journal of the Academy of Marketing Science*, *22*(2), 99–113. doi:10.1177/0092070394222001

Fornell, C., & Larcker, D. F. (1981). Evaluating Structural Equations Models with Unobservable Variables and Measurement Error. *JMR, Journal of Marketing Research*, *18*, 39–50. doi:10.2307/3151312

Ghose, A., & Dou, W. (1998). Interactive Functions and Their Impacts on the Appeal of Internet Presence Sites. *Journal of Advertising Research*, *8*, 29–43.

Gommans, M., Krishan, K. S., & Scheddold, K. B. (2001). From Brand loyalty to e-loyalty: A conceptual framework. *Journal of Economic and Social Research*, *3*(1), 43–58.

Jacoby, J., & Kyner, D. B. (1973). Brand Loyalty vs. Repeat Purchasing Behavior. *JMR, Journal of Marketing Research*, *10*, 1–9. doi:10.2307/3149402

Jarvenpaa, S. L., & Todd, P. A. (1997). Is There a Future for Retailing on the Internet? In Peterson, R. A. (Ed.), *Electronic Marketing and the Consumer* (pp. 139–154). Thousand Oaks, CA: Sage.

Jih, W. J., & Lee, S. F. (2004). Relationship between Online Shoppers' Motivation and Life Style Indicators. *Journal of Computer Information Systems*, *44*(2), 65–73.

Johnson, G., Bruner, G. C. II, & Kumar, A. (2006). Interactivity and its facts revisited. *Journal of Advertising*, *35*(4), 35–52. doi:10.2753/JOA0091-3367350403

Kumar, V., & Shah, D. (2004). Building and Sustaining Profitable Customer Loyalty for the 21st Century. *Journal of Retailing*, *1*(80), 317–330. doi:10.1016/j.jretai.2004.10.007

Lim, K. S., & Razzaque, M. A. (1997). Brand Loyalty and Situational Effects: An Interactionist Perspective. *Journal of International Consumer Marketing*, *9*(4), 95–115. doi:10.1300/J046v09n04_06

Monroe, K. B., & Guiltinan, J. P. (1975). A Path-Analysis Exploration of Retail Patronage Influences. *Journal of Customer Research*, *2*, 19–28.

Newman, J., & Werbel, R. A. (1973). Multivariate analysis of brand loyalty for major household appliances. *Journal of Retailing*, *56*(1), 77–93.

Reichheld, A., & Sasser, W. E. Jr. (1990). Zero defection: quality comes to services. *Harvard Business Review*, *68*, 105–111.

Reinartz, W. J., & Kumar, V. (2002). The mismanagement of customer loyalty. *Harvard Business Review, 80*(7), 86.

Reynolds, F. D., & Darden, W. R. (1974). *Constructing Life Style and Psychograghics*. Chicago: AMA.

Rowley, J. E. (2000). Understanding loyalty and loyal: The customer perspective. In *Proceedings of the Academy of Marketing.*

Schiffman, L. G., & Kanuk, L. L. (1997). *Consumer Behavior* (6th ed., pp. 238–240). Upper Saddle River, NJ: Prentice-Hall.

Schultz, D. (2000). Customer/Brand Loyalty in an Interactive Marketplace. *Journal of Advertising Research, 40*(3), 41–53.

Solomon, M. R. (1999). *Consumer Behavior* (4th ed., pp. 208–209). Upper Saddle River, NJ: Prentice-Hall.

Srinivasan, S. S., Anderson, R., & Ponnavolu, K. (2002). Customer loyalty in E-commerce: An exploration of its antecedents and consequences. *Journal of Retailing, 78*(1), 41–50. doi:10.1016/S0022-4359(01)00065-3

Wind, Y. (2008). A Plan to Invent the Marketing We Need Today. *MIT Sloan Management Review, 49*(4), 21–28.

This work was previously published in International Journal of E-Business Research Volume 6, Issue 4, edited by In Lee, pp. 1-12, copyright 2010 by IGI Publishing (an imprint of IGI Global).

Chapter 10
From Virtual Organization to E-Business:
Transformational Structuration

James J. Lee
Seattle University, USA

Bandula Jayatilaka
State University of New York at Binghamton, USA

Ben B. Kim
Seattle University, USA

Ted E. Lee
University of Memphis, USA

Pairin Katerattanakul
Western Michigan University, USA

Soongoo Hong
Dong-A University, USA

ABSTRACT

This article shows how the technical hype of 1990s has been transformed into the e-business organizations at the beginning of the 21st century. The authors took an interpretive stance in this study, grounded theory, and investigated the ontology of virtual organization by the metaphorical analysis. The metaphorical analysis adopted in this study provides the analytical power to conceptualize the social structure of virtual organization in the context of structuration theory with the process of grounded theory. As the e-business structuration indicates in this study, virtualization, the metaphor of virtual organization is interpreted as flexible domination, where the rational relationships in power distribution with other e-business characterized temporal (virtual) bond with low switching costs.

DOI: 10.4018/978-1-61350-462-8.ch010

INTRODUCTION

The Internet revolution enabled multi-dimensional connectivity between organizations that helped in emergence of novel concepts in organizing and social relations. We are living in the Internet revolution that changes the fundamentals of communications and this revolutionary phenomenon parallels the industrial revolution in the 18th century, which changed the principles of manipulating our surroundings. The emergent ways of communications have tremendous impacts on society as well as on the research disciplines. In the early days of the Internet revolution, researchers heavily focused on the concept of virtual organization giving various definitions based on the organizational settings arising from the Internet. For example, in traditional sense, we metaphorically represent organization of a big and solid company as a building on the land. But, now it has evolved to be a Web site representing the presence of an organization regardless of its business type, such as samsung.com as a manufacturer or amazon.com as an e-store. We attempt to find theoretical explanations of this evolutionary phenomenon.

However, to endorse or refute the claims in the business literature, it has been challenging for researchers to find proper theories to explain this rapid changing trend driven by highly advanced information technologies (IT). What we have observed have proved that the use of information and communication technology is rapidly transforming societies. With advanced communication protocols embedded in IT, we are now able to separate what we do and know from everyday practices to information systems (IS). Information and communications technology enables not only human-to-human communications via networks, but also computer-to-computer communications without any direct human interactions.

With this in mind, we designed our research taking an interpretive approach with metaphorical analysis. Metaphorical analysis takes accounts, such as live and dead metaphors of IS area, into the organizational structure. As the term 'organization' itself is a dead metaphor, this study shows how the already dead metaphor has shaped pure-play e-business organization in virtual settings. This research shows an intelligent analogy using the power of metaphors to realize and extend the world surrounding us.

This study intended to conceptualize the ontological understanding of virtual organization, which was promising back in 1990s. Though the term 'virtual organization' is still being used, companies have turned their key words into terms like e-business, cloud computing, SaaS (software as a service), etc. This study mainly revealed the origins of virtual organization in 1990s using metaphors. This will show how the technical hype of 1990s has been transformed into the e-business organizations in the beginning of the 21st century. This study theorized virtual organization by analyzing and interpreting the literature from 1994 to 2001. Therefore, the theory developed in this study is not just about a reflection of current trend of e-business, but it entails the Internet revolution era where World Wide Web technology emerged since 1992. Then, it showed how the concepts are produced, reproduced, and transformed along with the advancement of information and communications technologies. To achieve this objective, we investigated the articles in this period that especially discussed 'virtual organization'.

VIRTUAL ORGANIZATION STRUCTURES

After Lakoff and Johnson (1980) established the academic foundation of metaphors, scholars have used them in business literature. As discussed in his latest publication, Lackoff (2001) declared that even mathematics originated from metaphors of our consciousness. As we accepted theorems and axioms in geometry as a given without any reservations (Reichenbach, 1958), our languages in academic world fundamentally started from

Figure 1. Structuration of virtual organization metaphors

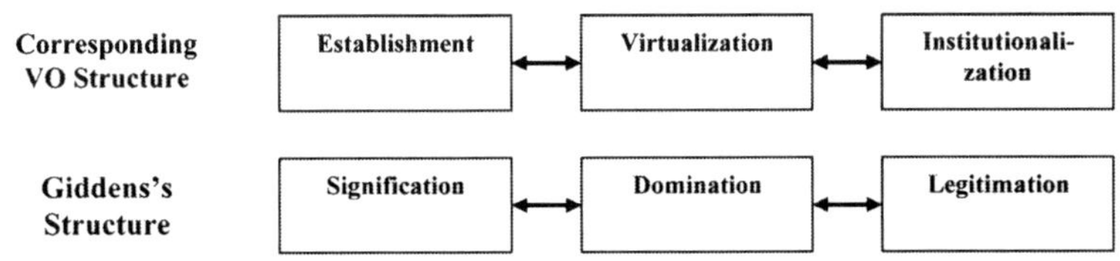

our concepts and these concepts are produced, reproduced, and transformed (Giddens, 1984) by our scholarly processes from insight, analogy, isomorphism, and finally to scientific model (Beer, 1984). The problem lies in how we use the metaphors to analyze and interpret the phenomenon a researcher wants to investigate. In organizational studies, metaphors have been used continuously (Palmer, 1996; Gibson et al., 2001; Pentland, 1995; Gioia et al., 1984; Garud et al., 1993; Lennie, 1999) after Morgan (2006) ignited the modern use of metaphors. There have been arguments on how to use metaphors in organization study and some even claimed the inappropriate use of metaphors in the field (Tinker, 1986; Bourgeois et al., 1983).

Having known the controversial aspect of metaphors, in this study we intended to focus on how to apply metaphorical analysis, not metaphor use appeared in the previous work (Schultze & Orlikowski, 2001). Therefore, this research uses induction (interpretive) rather than deduction (positivistic). Induction is appropriate because it reveals patterns from the accounts of rapid changes such as the industrial or the Internet revolutions. The Internet technology, including World Wide Web (WWW), has especially built the images of entities such as business organizations elated on the Web, which are the metaphors of organizations in the virtual spaces.

Since the early days of virtual organization, Davidow and Malone (1992), and Mowshowitz (1994) have made significant insights and contributions to the initial stage of virtual organization concept and they have discussed virtual organization as the future of organization. A good lens to conceptualize virtual organization is the use of social theory that can explain why social structures are evolving over time and show the implications of what have happened since the initial Web use in 1992. Structuration theory (Giddens, 1984) especially provides a way of thinking about social changes and will give an inductive explanation of e-business as a result. As shown in Figure 1 (Jayatilaka et al., 2007), we metaphorically represent virtual organization along with structuration based on Gidden's structuration theory.

Establishment is a structure in which a virtual organization is established to accomplish the shared goal of participating members. Organizations in virtual space become realizable or implementable because advanced information and communication technologies connect members. Virtual organization is often viewed as the extension of traditional IT in a way that supports and improves existing administration of members' activities and roles. Virtual organization's deployment of IT, therefore, focuses on the establishment of linkages and operations between business processes and members (Johnston & Carrico, 1988).

Virtualization is a structure in which the virtual organization grows into the new dimension. Virtual organization deploys IT in more proactive and integrated ways to manage conflict and encourage cooperation. Integrated IT can provide improved communication channels and linkages among members, and ultimately establish new relationships and SOPs (standard operation procedures) in its newly defined boundary. In this regard, virtual organization continues to ensure the strategic use of IT to achieve benefits in managing relationships among members (Johnston & Carrico, 1988).

Institutionalization is a structure in which social mechanisms develop. The virtual organization flourishes by establishing its own culture (e.g., shared belief and value system), norms, etc. Culture and norms are institutionalized within the virtual organization and enforced. In other words, members constitute social mechanisms and they are constituted by social mechanisms (Orlikowski et al., 2001). At this point, IT needs to evolve appropriately to facilitate, support, and capitalize on the strategies (Johnston & Carrico, 1988). Establishing guidelines that build relationships among members and produce a collaborative and cooperative working environment accomplishes this.

METAPHORICAL ANALYSIS

From a MIS research perspective, Schultze and Orlikowski (2001) proposed the metaphors of virtuality. Their study showed the insights of metaphors and identified how we treat the virtuality. However, it provided only static and a discontinuous framework of virtuality, which seemed difficult to adopt directly in an empirical study. Therefore, while adopting the approach of their work the authors of this study extended it with a broader view that would allow us to derive a generic set of metaphors that could explain the evolution of the virtual organization.

The procedures in our metaphorical analysis are an adoption of basic approaches of interpretive study. Schultze and Orlikowski (2001) used the six steps to analyze metaphors in their study based on Beer's method (Beer, 1984). Beer's accounts of scientific modeling (Beer, 1984; Tsoukas, 1991) explain the transformation of concepts when researchers conduct and build theories. However, the basic assumption of this model is our reductive thinking. Nonetheless, it discusses the role of metaphors thus it is important for authors to map Beer's method with the main strategy of this study, grounded theory (Glaser & Strauss, 1967). Grounded theory is basic theory building process thus it is based on researcher's inductive thinking (Charmaz, 2000). Since both methods start from the insights and end at scientific theory (grounded theory) and model (Beer's method), authors identify the gaps between the corresponding steps in each methodology due to different directions in its logics, induction vs. deduction. These gaps were further analyzed and used in metaphorical analysis as shown as 'Activities' in Table 1.

While main concepts and context of each step in metaphorical analysis are from Beer's method, how to analyze text data is from Grounded Theory (Strauss & Corbin, 1990). The basic approach of Grounded Theory is the *theoretical sampling* (Glaser, 1978; Glaser & Strauss, 1967). It is the main difference from the positivist's approach since it uses intentional data site selection. Authors decide which data will be analyzed and where they find interesting phenomenon. After initial data are collected then analyzing data immediately while additional data are being collecting. Thus, it uses parallel approach, i.e., conducting both data collection and analysis together. Main focus is to collect selected data which are important for conducting research. In this way, researchers can accumulate and build knowledge and theories. Data collection stage will be ended when researchers reach the point where no more new data is needed – theoretical saturation.

There are six steps in this study (Table 1). We applied these steps for both practitioners' and academic journals while prohibiting no interaction between activities in each set of data. Step 1 and 2 is mainly for data collection and the initial analyses of data (Open Coding: Strauss & Corbin, 1990). The example output of step 2 is shown in the Figure 2. The authors did not use any previous work or ideas of virtual organizations in this process. The concept of *epoche* in Phenomenology is applied to prevent any theoretical or conceptual interference from other assumptions (Sanders, 1982). Step 3 and 4 uses the notion of *comparative analysis* developed by Glaser and Strauss

Table 1. Procedures of metaphorical analysis

Step	Beer's Method	Activities	Grounded Theory
1	Insight	Find metaphoric contents in the articles. Try to find all of the metaphoric expressions contained in the articles, whether they appeared to be directly relevant to virtual organization or not.	Data Collection
2		Each instance of a metaphoric expression was captured in a list together with the page number on which the trope was located, terms of source domain used for metaphorical expressions, and a short comment explaining the meaning or significance that the researcher had attributed to it at the time of reading.	Open Coding
3	Perception	From multiple analyses, eliminate from initial lists those metaphoric expressions that are unrelated to virtual organizations and refine the terms of metaphorical expressions.	Open Coding
4	Analogy	Review the terms of metaphorical expressions and create categories to put similar attributes together. Focus attentions on the strategy in the texts – where authors contrast the "real" organization with the "virtual" one in order to construct a definition of virtual organizations.	Open Coding
5	Homomorphism	Analyze categories from the previous stage and discover the connections between each category using meaning of metaphors, entailment, affordance, and challenge.	Axial Coding
6	Isomorphism	Establish a set of metaphors defining, using the concepts of Opposition, Meaning of metaphors, Source, Entailment, Affordance, and Challenge.	Selective Coding

(1967). While examining the initial data, authors frequently exchange their understandings and ideas from initial efforts. Many times step 1 and 2 were revisited (comparative analysis in grounded theory) and refined. Step 5 involves deeper analysis of categories developed in the previous steps. First, relational diagram was developed with the considerations of each category's relationship and their characteristics where some categories (super-categories) incorporate other categories (sub-categories). This process gives the life to the separated and discontinued categories, making it one continuous process.

The emerging categories are further analyzed (step 6) with following four themes: meaning of metaphor, entailment, affordance, and challenge (Schultze & Orlikowski, 2001). *Meaning of metaphor* is the form of evolved core category among similar metaphors. It represents one of important aspects of virtual organization. *Entailment* is the transition and/or the connections between source and target domain. This dynamic dimension explains how conceptual analogy between target and source domain transcends to metaphorical coherence. Schultze and Orlikowski (2001) defined the notion of *affordance* as "the perceived and actual properties of things, primarily those fundamental properties that determine just how the thing could possibly be used." Then, they added the consequences caused from social, cultural, and historical construction of an object. Our interpretation of the notion of affordance is epistemology of virtual organization while entailment discourse the ontological transition of the perception of organization. However, there is no one metaphor that can explain virtual organization as a whole. *Challenges* should be articulated in the way how each metaphor has to be directed and related to other metaphors.

METAPHORS OF VIRTUAL ORGANIZATION

By following the steps 1 – 3 (cf., Table 1) we initially generated 555 terms from 59 articles in

Figure 2. Examples of metaphorical expressions, terms, and comments

Reference 116

Page 11, Paragraph 4, Sentence 1-2
"Many corporations are motivated to form alliances by marketing and manufacturing considerations. Some form joint ventures with foreign partners (or even competitors) simply to gain better coverage of international markets or to take advantage of reduced labor and delivery charges in other countries.

Term: alliances
Comments: Once again the term alliance is being used here to talk about business strategies and how to make businesses both domestically and nationally successful. "Alliances" are formed between different countries and also competitors so both can prosper. Forming these "alliances" can also help communication across countries/businesses because information will be more readily available.

Page 12, Paragraph 2, Sentence 1-3
"The new, ultra-mobile work forces, nicknamed "road-warriors" are message-intensive. They talk on the go and go where the action is. Road warriors need new tools as they go on their "infoquests" into the offices, factories, and homes of their clients."

Term: road-warriors
Comments: The employees that are always on the job, working for virtual organizations, are referred to here as "road-warriors". Like warriors themselves, they are successful with use of tools and they go to where the action is. They are constantly looking for and making new tools for the appropriate type of war and for them to be most undefeated. Workers for virtual organizations are constantly changing atmospheres, locations, and their type of tools used for the appropriate work.

IS journals, spanning from 1994 to 2001. After a reevaluation of all the terms, we eliminated irrelevant terms. Final categorization resulted in 16 metaphors. Based on the above 16 metaphors derived from the aforementioned literature, seven groups emerged: components, commitment, process, centrality, facility, culture, and norm. This section will discuss each metaphor in seven groups.

1. COMPONENTS

Structure

This metaphor describes the connections that exist among the various members of a virtual entity (i.e., an organization or a team). The source here is 'relationship' while the target is 'the interaction between companies and people in a virtual setting - structure.' The participants' level of connection depends on the degree of dependency and cooperation, whether temporary or permanent (Palmer & Speier, 1997).

Formation

This metaphor refers to the construction of an object as the result of a conscious act. The source here is 'architecture' while the target helps with the 'understanding of the shape and structure of the virtual organization – formation.' The main focus of this metaphor is placed upon the arrangements or spatial meaning of a building with how different components are placed with relative to the others providing the inhabitability.

2. COMMITMENT

Integration

Integration implies connectivity or continuity, and it enables associated objects (or objects tied to it), to move smoothly. The source domain is 'Wheel' and the target domain is the 'leader of the virtual organization that keeps the integrity – integration.' A wheel represents constant motion and nothing remains stationary. Things work in a cycle, with information moving from one person to another and to another and so on.

Coordination

This category contains a variety of words describing the growth of one thriving organism from various different non-living elements. The source here is Nature/Garden and the target is 'the overall ambience of the Virtual Organization – link.' Link is the key aspect of a virtual organization. Suppliers, distributors, and customers should coordinate. Once these components are coordinated and linked together, they thrive and produce at optimal level for virtual organization.

3. PROCESS

Functionality

This metaphor reinforces different parts that work collectively to stay alive. The source is the 'Human Body' and the target implies that 'the different parts of a virtual organization functioning separately yet in conjunction with each other – functionality.' Though we perceive each part of body independently, it affects the rest of the organization when one thing goes wrong. Likewise, the virtual organization will be successful when all areas are functioning and working towards one common goal.

Constitution

This metaphor represents a basic necessity for the survival of our bodies – an ingredient that gives us strength and nourishment. The source is 'Food' and the target comprises 'the things in a virtual organization that allow it to flourish or to put it simply, the things that nourish it – constitution.' With respect to virtual organizations, it describes the entities necessary at the foundation of the organization - the things that are needed to allow other things to work.

4. CENTRALITY

Fluidity

This metaphor comprises moving data, information, knowledge, and products and services from place to place. The source domain here is 'Transportation' and the target domain is 'how data, information, knowledge and products are transferred among members of a virtual organization or indicates lack of specificity with relative to space or location - fluidity'. This metaphor initiates the 'mobility' entrenched in a virtual organization, therefore the efficiency of organization is increased.

Flexibility

Water takes multiple forms and symbolizes the flexibility, which represents the various forms and structures, of a virtual organization. The source here is 'Water' and the target is 'the flexibility of the virtual organizational structure'. Because virtual organizations do not have to be confined to a particular form, it opens up a great opportunity to integrate with different companies as surroundings are changed.

5. FACILITY

Virtuality

This metaphor is staged performances that are targeted at capturing the attention of customers and perhaps portraying an image of an organization different from the one in existence. The source is 'Fashion/Theatre' and the target describes 'the model and structure of a virtual organization that attempts to create an impression - virtuality'. It describes how people must apprehend 'an appearance' of virtual organization that works together in an interwoven manner, enabling people from different locations and companies to work in coordination.

Power

This metaphor is the act or process of governing, especially the control and administration of public policy in a political unit. The source is 'Government' and the target describes 'how the virtual organization is run and who is involved in its running - power.' In a Virtual Organization, the manner in which the duties, responsibilities and jobs are distributed is critical. The metaphor helps in understanding how virtual power exercises in distributing duties and responsibilities in a particular fashion.

6. CULTURE

Practice

This metaphor shows a very high degree of cooperation among the participants in a party. Because there are rivals who compete against you, people in a same party work together to win over competitors, achieving a common purpose. This metaphor helps bring across the idea of cooperation, competition, and survival. The source here is 'War' and the target is 'any struggle needing cooperation that occurs when a group of people faces conflicts – practice'. This metaphor describes the way in which a virtual organization's employees deal with conflict. Their actions in the virtual organization directly affect the outcome of business strategy and business operations.

Transformation

This metaphor seeks to discover the meaning of all aspects and to understand the conditions for their existence. It shifts a paradigm of current structure constantly, being improved or developed to increase the efficiency of the world. The source here is 'Science/Nature/Technology' and the target explains 'the composition of virtual organizations and how the various aspects of the virtual organization are considered important in the business world – transformation.' The metaphor indicates how the virtual organization has "evolved" from more formal and traditional work settings.

Foundation

This metaphor is the totality of socially transmitted behavior patterns, arts, beliefs, institutions, and all other products of human work and thought. These patterns, traits, and products are considered to be the expression of a particular period, class, community, or society. The source is the 'Community and Culture' while the target is the 'ambience and social relations of the employees in virtual organization – foundation.' People work together with common interests that are within their organizational confines. This common interest allows them to form relationships, and form their own little community within the organization.

7. NORM

Institution

This metaphor is thought of as ways of communication aiming at developing one's comprehension and capacity. The source is Language/Literature and the target is 'the language of the virtual organization, the lingo used, and the way people learn how to function in the organization - institution.' It is important to have standard organizational procedures in a virtual organization. Through institutions, people in virtual organizations learn the ways to communicate in virtual settings.

Value

This metaphor relates to a belief system that a group or an individual holds about a certain deity.

Figure 3. Social framework of virtual organization

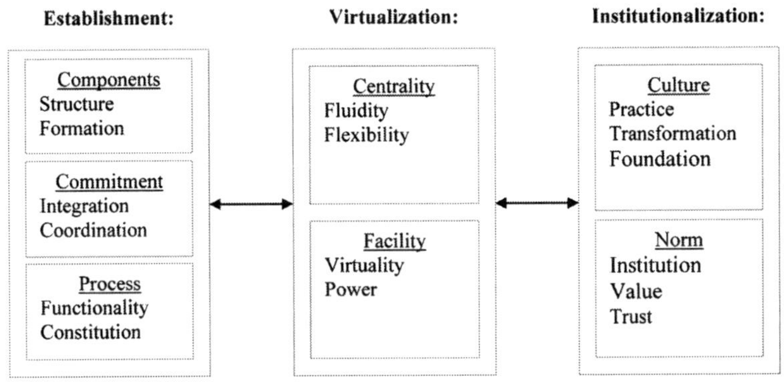

A belief system can be a powerful tool because it relates to impetus and motivation of members. The source is 'Religion' and the target is 'the belief system a virtual organization's employees live by - value.' This metaphor describes the way in which employees work in virtual organizations. It shapes what is required to be done, how it needs to be done, what is appropriate, and what is not.

Trust

This metaphor relates to an intangible confidence that a group or an individuals hold about a certain idea or faith. The source is the 'Faith or abstract entity' while the target is 'a trust system that virtual organization employees live by – trust.' Because virtual organizations do not have a particular shape or a physical structure, trust is a useful dimension or a descriptor. Organizations or members can rely on a trust system as a safeguard. This metaphor helps them build strategic alliances in value systems.

STRUCTURATION OF VIRTUAL ORGANIZATION

Though Figure 3 comes from metaphors of virtual organization, we organized groups of metaphors on the basis of the structuration theory (Giddens, 1984). However, each structure of signification, legitimation, and domination are not directly mapped into this framework. Rather, we conceptualized the relationships between each structure as shown in Figure 1. Each structure name from structuration theory is reproduced to reflect unique characteristics of the Internet environment.

The first structure is establishment. In this structure, the organization comes into existence as a formal organization. Members of virtual organization initiate its use, and identify with components, commitment, and processes. Components are basic properties of virtual organization while process explains how virtual organization works. Commitment gives roles to human users that open interfaces and interactions between human users and virtual organization. This structure includes the origin of virtual organization (Barner, 1996; Zajtchuk & Satava, 1997), the level of commitment and connection among members of a virtual entity, the degree of dependency (Mowshowitz, 1997) and temporary/permanent cooperation among members (Palmer & Speier, 1997), and the image of virtual organization due to a conscious acts of members (Barner, 1996). The other characteristic related to this structure is a usage of information and communication technology (ICT) to connect its business data and members to function in conjunction with each other (Bleecker, 1994), a usage of manpower and information in geographically distributed locations and the way to put these scattered resources together and create continuous functionality (Rayport & Sviokla, 1995), and the value-added activities of virtual organization (Gunn & Burroughs, 1996; Weiner, Nohria, Hickman, & Smith, 1997).

Virtual organization undergoes virtualization; becoming streamlined in terms of centrality and facility. Centrality determines the degree to which relations are guided by the formal hierarchy while facility measures the strength of the relation between individuals then facilitates the relationships (Tichy et al., 1979). This structure includes virtual organization's higher-level goals (Nosek & Shephard, 1995; Rayport & Sviokla, 1995) and members working together in distant locations (Davenport & Pearlson, 1998; Faucheux, 1997; Voss, 1996), members' loyalty due to its higher-level goal, a flexible structure that constantly changes to deal with changing offices, projects, fellow co-workers, time zones, etc. (Bleecker, 1994; Boudreau, Loch, Robey, & Straub, 1998; Weiner et al., 1997), and its flexible standards, schedules, and autonomy invested in members (Storey et al., 2000; Werther, 1999).

As virtual organization is stabilized, members go through a process of institutionalization developing new social mechanisms, building culture and norm. Culture and norm are entertainingly used to determine the limits of organizational boundaries or to be constituted by human intervention. This actually transcends the duality of structure by Giddens (1984). Culture, however, explains external rules, while norm express internal rules of virtual organization. This structure includes the concept of (virtual) team (Mowshowitz, 1997; O'Leary, Kuokka, & Plant, 1997; Storey, Straub, Stewart, & Welke, 2000) and formal and traditional work settings, its focus on the actions and relationships among members on a timely manner (Mowshowitz, 1997; Rayport & Sviokla, 1995; Storey et al., 2000), its capability to accommodate specific rules and business processes due to advanced IT (Chesbrough & Teece, 1995; Werther, 1999), its fluid transformation as business environment changes (Faucheux, 1997; Hardwick & Bolton, 1997), and its creation of particular culture based on the needs and evolving relationships among members (Barner, 1996; Davenport & Pearlson, 1998; Hardwick & Bolton, 1997; Mowshowitz, 1997).

E-BUSINESS: VIRTUAL ORGANIZATIONAL TRANSFORMATION

We took an interpretive stance in this study (use of grounded theory). Authors investigated the ontology of virtual organization through the metaphorical analysis. Theorized framework (Figure 3) in this study is produced by the Metaphors of virtual organization research published in the past, then reproduced by the Grounded Theory approach, then finally transformed to e-business implications using Structuration Theory.

The metaphorical analysis adopted in this study provides the analytical power to conceptualize the social structure of virtual organization in the context of structuration theory with the process of grounded theory. We identified the sets of systematic procedures of metaphorical analysis by comparing metaphors and grounded theory. Concepts from deductive thinking, which lead to a scientific model, and methodologies from inductive thinking, which lead to a scientific theory, were compared and gaps were found. Identified gaps are the basis of metaphorical analysis to conceptualize the phenomenon of e-business.

Numerous contributions can be identified, but among those, this study shows the new way of research approach with the triangulation of multiple methods. We believe that, similar to the structuration theory, our framework has the potential to bridge the different explanations based on different paradigms (Jayatilaka, Klein, & Lee, 2007). This type of research attitude is very necessary in rapidly changing environment. In addition to methodological contribution, the social framework of virtual organization of this study provides critical view of virtual organization which we continuously produce, reproduce, and transform in our consciousness.

The social framework discovered in this study showed the constitutions of making virtual organization in 1990s and it is still lived through the business presence on the Web. E-business, as a metaphor of physical organization, formalizes business processes in information systems through establishment, virtualization, and institutionalization (Figure 3). As the e-business structuration indicates in this study, virtualization, the heart of e-business, is the metaphor of centrality and facility. Virtualization can be interpreted as *flexible domination*, where the rational relationships in power distribution with other e-business characterized temporal (virtual) bond with low switching costs. Therefore, doing e-business faces different expectations from other business partners and customers, which can be explained in flexible domination.

REFERENCES

Barley, S. R., & Tolbert, P. S. (1997). Institutionalization and structuration: Studying the links between action and institution. *Organization Studies*, *18*(1), 93–117. doi:10.1177/017084069701800106

Barner, R. (1996, March/April). The new millennium workplace: Seven changes that challenge managers-workers. *The Futurist*, 14–18.

Beer, S. (1984). The viable system model: its provenance, development, methodology and pathology. *The Journal of the Operational Research Society*, *35*(1), 7–25.

Bleecker, S. E. (1994, March/April). The virtual organization. *The Futurist*, *28*, 9–14.

Boudreau, M., Loch, K., Robey, D., & Straub, D. (1998). Going global: Using information technology to advance the competitiveness of the virtual transitional organization. *The Academy of Management Executive*, *12*(4), 120–128.

Bourgeois, V. W., & Pinder, C. C. (1983). Contrasting philosophical perspectives in administrative science: A reply to Morgan. *Administrative Science Quarterly*, *28*, 608–613. doi:10.2307/2393012

Charmaz, K. (2000). Grounded Theory: Objectivist and Constructivist Methods. In Denzin, N. K., & Lincoln, Y. S. (Eds.), *Handbook of qualitative research*. Thousand Oaks, CA: Sage Publications.

Chesbrough, H., & Teece, D. (1995). When is virtual virtuous? Organizing for innovation. *Harvard Business Review*, 65–72.

Davenport, D., & Pearlson, K. (1998). Two cheers for the virtual office. *Sloan Management Review*, 51–68.

Davidow, W. H., & Malone, M. S. (1992). *The virtual corporation: structuring and revitalizing the corporation of the 21st Century*. New York: Harper Collins.

Faucheux, C. (1997). How virtual organizing is transforming management science. *Communications of the ACM*, *40*(9), 50–55. doi:10.1145/260750.260766

Garud, R., & Kotha, S. (1993). Using the brain as metaphor to model flexible production systems. *Academy of Management Review*, *19*(4), 671–698. doi:10.2307/258741

Gibson, C., & Zellmer-Bruhn, M. (2001). Metaphors and meaning: an intercultural analysis of the concept of teamwork. *Administrative Science Quarterly*, *46*(2), 274–303. doi:10.2307/2667088

Giddens, A. (1984). *The constitution of society*. Berkeley, CA: University of California Press.

Gioi, D., & Poole, P. (1984). Scripts in organizational behavior. *Academy of Management Review*, *9*(4), 449–459. doi:10.2307/258285

Glaser, B. G. (1978). *Theoretic Sensitivity*. Mill Valley, CA: The Sociology Press.

Glaser, B. G., & Strauss, A. L. (1967). *The discovery of grounded theory: strategies for qualitative research*. New York: Aldine Publishing Company.

Gunn, R., & Burroughs, M. (1996). Work spaces that work: Designing high-performance offices. *The Futurist*, 19–24.

Hardwick, M., & Bolton, R. (1997). The industrial virtual enterprise. *Communications of the ACM*, *40*(9), 59–60. doi:10.1145/260750.260770

Jayatilaka, B., Klein, H. K., & Lee, J. (2007). On Categorizing the IS Research literature - User Oriented Perspective. *Informing Science Journal*, *10*, 237–271.

Johnston, H. R., & Carrico, S. R. (1988). Developing capabilities to use information strategically. *Management Information Systems Quarterly*, *12*(1), 37–47. doi:10.2307/248801

Lakoff, G., & Johnson, M. (1980). *Metaphors we live by*. Chicago: University of Chicago Press.

Lakoff, G., & Nunez, R. E. (2001). *Where Mathematics Comes From: How the Embodied Mind Brings Mathematics into Being*. New York: Basic Books.

Lee, J., Jayatilaka, B., & Kwok, R. *(2002). Virtual Organizations: Duality of Human Identities in Consciousness and Entity. In* Proceedings of the Information Resources Management Association International Conference *(IRMA).*

Lennie, I. (1999). Managing metaphorically. *Studies in Cultures, Organizations and Societies*, *5*, 46–59. doi:10.1080/10245289908523520

Morgan, G. (2006). *Images of organization*. Thousand Oaks, CA: Sage.

Mowshowitz, A. (1994). Virtual organization: A vision of management in the information age. *The Information Society*, *10*(4), 267–288. doi:10.1080/01972243.1994.9960172

Mowshowitz, A. (1997). Virtual organization. *Communications of the ACM*, *40*(9), 30–37. doi:10.1145/260750.260759

Nosek, J., & Shephard, G. (1995). View point: Making groupware payoff-British model. *Communications of the ACM*, *38*(6), 11–13. doi:10.1145/213859.214784

O'Leary, D., Kuokka, D., & Plant, R. (1997). Artificial intelligence and virtual organizations. *Communications of the ACM*, *40*(1), 52–69. doi:10.1145/242857.242871

Orlikowski, W. J., & Robey, D. (1991). Information technology and the structuring of organizations. *Information Systems Research*, *2*(2), 143–169. doi:10.1287/isre.2.2.143

Palmer, I., & Dunford, R. (1996). Conflicting uses of methapors: reconceptualizing their use in the field of organizational change. *Academy of Management Review*, *21*(3), 691–717. doi:10.2307/258999

Palmer, J. W., & Speier, C. (1997). *A Typology of Virtual Organizations: An Empirical Study*. Retrieved from http://hsb.baylor.edu/ramsower/ais.ac.97/papers/palm_spe.htm

Pentland, B. (1995). Grammatical models of organizational processes. *Organization Science*, *6*(5), 541–556. doi:10.1287/orsc.6.5.541

Rayport, J., & Sviokla, J. (1995). Exploiting virtual value chain. *Harvard Business Review*, 75–85.

Reichenbach, H. (1958). *The philosophy of space & time*. New York: Dover Publications, Inc.

Sanders, P. (1982). Phenomenology: A new way of viewing organizational research. *Academy of Management Review*, *7*(3), 353–360. doi:10.2307/257327

Schultze, U., & Orlikowski, W. J. (2001). Metaphors of virtuality: shaping an emergent reality. *Information and Organization*, *11*(1), 45–77. doi:10.1016/S1471-7727(00)00003-8

Storey, V., Straub, D., Stewart, K., & Welke, R. (2000). A conceptual investigation of the E-commerce industry. *Communications of the ACM*, *43*(7), 117–124. doi:10.1145/341852.341871

Strauss, A., & Corbin, J. (1990). *Basics of Qualitative Research: Grounded Theory Procedures and Techniques*. Newbury Park, CA: Sage.

Tinker, T. (1986). Metaphor or reification: are radical humanists really libertarian anarchists? *Journal of Management Studies*, *23*(4), 363–384. doi:10.1111/j.1467-6486.1986.tb00423.x

Tsoukas, H. (1991). The missing link: a transformational view of metaphors in organizational science. *Academy of Management Review*, *16*(3), 566–585. doi:10.2307/258918

Voss, H. (1996). Virtual organizations: The future is now. *Strategy and Leadership*, *24*(4), 12–16. doi:10.1108/eb054559

Weiner, M., Nohria, N. E. R. G., Hickman, A., & Smith, H. (1997). Value networks: The future of the U.S. electric utility industry. *Sloan Management Review*, 21–34.

Werther, W. (1999, March/April). Structure-driven strategy and virtual organization design. *Business Horizon,* 13-18.

Zajtchuk, R., & Satava, R. (1997). Medical application of virtual reality. *Communications of the ACM*, *40*(9), 63–64. doi:10.1145/260750.260768

This work was previously published in International Journal of E-Business Research Volume 6, Issue 4, edited by In Lee, pp. 13-25, copyright 2010 by IGI Publishing (an imprint of IGI Global).

Chapter 11
Taking a Strategic Alliance Approach to Enhance M-Commerce Development:
Challenges and Strategies

Fang Zhao
American University of Sharjah, UAE

ABSTRACT

Current m-commerce business models show that m-commerce depends on complex networks of business relationships, which often comprise telecommunications service providers, mobile device makers, banking industry, Internet search engine providers, and various third-party value-adding companies. Due to the nature of m-commerce, the key to success in m-commerce lies predominantly in managing a network of alliances. This paper answers research questions, such as why do companies team up for m-commerce? What are the key challenges facing the alliances? How can companies address the challenges? What does the future hold for the study of strategic alliances including m-commerce alliances? This paper extends strategic alliance theories to the study of m-commerce alliances that are formulated in various cultural and national backgrounds. The authors examine both strategic and operational strategies for m-commerce alliances and discuss a wide range of issues in the formulation and implementation of m-commerce alliance strategy.

INTRODUCTION

Mobile commerce (m-commerce) has been increasingly gaining momentum in mobile payments, mobile ticketing, mobile coupons and mobile banking in the m-commerce marketplace. It is estimated that revenue from mobile content and services (excluding SMS) accounts for 7-10 percent of total mobile revenues worldwide while SMS accounts for a further 10 percent of total mobile data revenues ("Research and Markets", 2008). M-commerce and wireless Internet using wireless technologies such as wireless

DOI: 10.4018/978-1-61350-462-8.ch011

application protocol (WAP), Bluetooth, WML, and XML etc, represents the future of e-commerce and addresses the growing demand for mobility. Countries in the Asia Pacific regions are taking the lead in using mobile phones for m-payments which accounts for about 85 percent of consumers worldwide ("Research and Markets", 2008). In Asia, m-commerce has developed rapidly and become an efficient and established consumer buying channels. In 2007, the trade volume through m-commerce exceeded $6 billion in Japan. In the USA, there are more than 250 million wireless consumers with an m-commerce retail market of approximately over $500 million in 2008, according to Murphy (2008).

Today's digital world offers a borderless and virtual village where individuals, companies are interconnected and interdependent through a complex local and global network of relationships (transactions, partnerships, alliances, and even vertical integration). According to business network theories (Håkansson & Snehota, 1992; Batonda & Perry, 2003), it is this connectedness and interdependence through information technology including wireless technologies that drive the business world today. A glimpse of current m-commerce business models shows that m-commerce depends on complex networks of business relationships which often comprise telecommunications service providers, mobile device makers, banking industry, Internet search engine providers, and various third-party value-adding companies (Varshney, 2008). Due to the nature of m-commerce the key to success in m-commerce lies predominantly in managing the network of alliances.

A review of the current m-commerce literature finds that the majority of English language publications have focused primarily upon designing applications and technology support, location management and mobile transactions largely from the technology perspective (Varshney, 2008). For example, Read (2008) identifies mobile web issues and provides key advice to enable successful mobile web and data. Very few publications study the human side of issues such as the challenges and strategies in managing partnerships and alliances between companies in developing m-commerce. This paper addresses this gap based upon strategic alliance theories and explores the management issues in regard to a strategic alliance approach to the development of m-commerce taken by many companies to boost their m-commerce capacity.

This paper aims to answer the following research questions:

- Why do companies team up for m-commerce?
- What are the key challenges facing the alliances?
- How can companies address the challenges?
- Where does the future hold for the study of strategic alliances including m-commerce alliances?

As m-commerce alliance is a largely under-researched area in the literature, this paper delivers new insights and practical recommendations for corporate strategy and action. The paper starts with a review of a few influential cases of alliances formed internationally to promote m-commerce. After that, the paper discusses the key challenges and broad strategies in the formulation and implementation of strategic alliances for m-commerce before concluding it with directions for future research.

METHODOLOGY

This study draws upon case study design, taking strategic alliances as unit of analysis. Although it does not prelude a capacity for generation, case study can provide detailed illustration of issues and answer the questions of "how" and "why" (Coffey & Atkinson, 1996). The essence of a case study lies in its ability to illuminate actions and/or decisions, such as, why they were taken, how they were implemented, and with what result (Schramm, 1971; Yin, 2003). According to Yin (2003), studying multiple cases makes it possible to build a logical chain of evidence through cross-case analysis. As a result, multi-case studies and cross-case analysis help improve the validity of the study. This is why this study chose the case study methodology to answer the research questions: why do companies team up for m-commerce? What are the key challenges facing the alliances?

Sampling Procedure and Data Collection

Purposeful sampling was used to select the information-rich cases that are likely to manifest the phenomenon studied intensely (Patton, 2002). The cases of m-commerce strategic alliance selected for the present study should meet one or more criteria as follows:

- The case companies must be influential;
- The alliances take different forms;
- The alliances take place in various regions and continents in the world.

By meeting these criteria, the multi-case study is likely to generate strong evidence that captures the key drivers behind m-commerce strategic alliances across industries, countries and continents and the various challenges facing them.

Data Analysis

To maintain originality and precision, content analysis was employed in this study to process the original texts of documents and news release about m-commerce alliances published in the public domain. As a research technique, content analysis provides new insights, ensures a better understanding of particular phenomena, and informs practical actions (Krippendorff, 2004). To guard against superficial face validity of content analysis, this researcher gave greater weight to the analysis of latent content or meaning conveyed rather than merely an interpretation of the surface or literal contents (Berg, 1995).

MULTI-CASE STUDIES OF INTER-FIRM ALLIANCES IN M-COMMERCE

There has been a strong need of forming alliances for developing m-commerce. In early days, inter-firm alliances in m-commerce occurred most often between banking and insurance institutions and mobile operators with a focus on mobile banking and other financial applications which provided customers with mobile payment solutions. With rapid development of m-commerce, inter-firm alliances have gone beyond mobile operators and financial institutions. They extend to more operations areas to meet the diversity of emerging m-commerce services. The cost of building m-commerce infrastructure also requires collaboration among multiple players including wireless service providers, content providers and developers, access providers, aggregators, and etc (Varshney, 2009). The following cases give a glimpse of why and how companies teamed up for m-commerce.

Motorola and MasterCard

MasterCard International and Motorola teamed up for the purpose of improving the ability to manage purchases and financial transactions on-the-go using wireless Internet devices worldwide. Motorola and MasterCard extended their collaboration to research and development projects to assure interoperability between MasterCard's electronic developed next-generation m-commerce technologies, under the initiatives of Global Mobile Commerce Interoperability Group (GMCIG), of which both companies were key members. The alliance also included cooperation on a number of other key initiatives to promote the benefits of m-commerce to consumers through joint marketing activities. The collaboration was a critical step in the delivery of e-commerce capabilities to the mobile consumer, thereby enabling m-commerce. (Motorola, MasterCard Promote Mobile E-Commerce, 2000). Motorola's alliance with MasterCard has enabled Motorola to work closely with MasterCard's 22,000 member financial institutions. As a result, the alliance has helped Motorola realize continued improvements in the capabilities and functionality of its existing wireless Internet devices and platforms.

Motorola and Union Mobile Pay

Most recently Motorola has announced an alliance with Union Mobile Pay (UMPay), a leading mobile payment company established by China Mobile Communications Corporation (CMCC). In collaboration with the UMPay, Motorola is able to provide Chinese consumers with a comprehensive m-commerce solution for financial and retail merchandising needs. The solution consisting of an M-Wallet and back-end management platform is Motorola's first m-commerce platform launched in China. In 2007 UMPay's mobile payment platform reached 50 million registered users. The management of UMPay sees the alliance as an opportunity to strengthen its business and growth. When interviewed, Zhang Bin, the Managing Director of UMPay made the following comments:

We see tremendous value in our relationship with Motorola to further strengthen our position in the market.... With Motorola's expertise and experience in M-Commerce, we can create new solutions and services that address the needs of customers, bringing easy to use, secure and reliable services to them. (Motorola Enhances Mobile Commerce Solutions, 2009)

Motorola and CIECC

Motorola has entered another alliance with China International Electronic Commerce Center (CIECC), an agency operating under China's Ministry of Commerce responsible for the development of electronic information projects. This new alliance aims to instigate a m-commerce laboratory in Beijing. The laboratory will allow Motorola and its partners to develop an ecosystem, and pre-integrate and validate the service prior to commercial deployments in China. It has been reported that the m-commerce laboratory is anticipated to interconnect industry, government and academia to facilitate certain aspects in the m-commerce environment such as research and commercialization of mobile applications, mobile payment and security, and enterprise mobile commerce application (Motorola & CIECC in an agreement to launch m-commerce laboratory, 2008).

Nokia and Google

Like its competitor Motorola, Nokia formed an alliance with Google (Nokia, 2008). The alliance integrates Google's popular search engine with the Nokia Search application which will eventually make Google search available to Nokia's customers in over 100 countries around the world, serving mobile owners who speak more than 40 languages. Nokia Search, now with Google, offers fast and easy access to web information from the handset. In addition, Nokia Search makes it possible for users to search content on their device and use local search engines for a complete search experience.

Kasikornbank and Venyon

Kasikornbank, the fourth largest commercial bank in Thailand has entered into a partnership with Venyon, a Finland-based Near Field Communication (NFC) mobile payment service provider. The partnership will provide customers with the facility of making contactless payments when purchasing goods by waving their mobile handset over a contactless reader at the point of sale (Venyon, Thailand bank, 2009).

T-Mobile and Yahoo!

T-Mobile USA, Inc. and Yahoo! Inc. have teamed up too. According to an agreement, Yahoo! oneSearch™ powers mobile Internet searches through T-Mobile's new web2goSM service. Web2go provides a better Web browsing experience on mobile phone by making it easier to view and navigate the Web through a customizable home page, a simplified shopping and download experience, and Yahoo!'s mobile search experience. Yahoo! oneSearch provides customers who search through T-Mobile's web2go with access to what they are looking for - from news, financial information, weather conditions, Flickr™ photos and Web images, to Web and mobile web sites. Yahoo! oneSearch will also display relevant ringtones, wallpaper, games and other content available in the T-Mobile Store® within search results (Yahoo, 2008).

Zain and Citigroup and Standard Chartered

On the African continent side, Zain, a leading emerging market player in the field of telecommunications, is partnering with leading international and regional banks including Citigroup and Standard Chartered and launches a m-commerce service called Zap. Zap is considered the most comprehensive service enabling customers to access their money faster and more conveniently. The service will allow Zain customers to use their mobile phone to:

- Pay for goods/services/bills;
- Send/Receive money to friends, family, and bank accounts;
- Manage their bank accounts;
- Withdraw cash;
- Top up their own airtime account or top up someone else's; and
- Send airtime to Zain customers in East Africa.

The service will be available to over 100 million people in Kenya, Tanzania and Uganda (Oberthur Technologies, 2009).

Bell Mobility, Rogers Wireless and TELUS Mobility

In Canada, the three biggest wireless market players – Bell Mobility, Rogers Wireless and TELUS Mobility joined together to work on a solution to achieve inter-carrier interoperability. In the current m-commerce markets, closed solution offerings are still prevailing. Users of m-commerce have to be either a customer of a particular operator or a given financial institution. This alliance heralds a significant progress in overcome the barrier to m-commerce development (Recently the Three Biggest Operators, 2007).

As seen in the above cases, partnerships and alliances such as those between financial institutions and telecommunication industries are common and indispensable in running m-commerce business. The above cases show that companies resorted to alliance strategy to leverage the much needed resources from alliances and to develop and sustain their business. It is reported that about 20-50 per cent of corporate value has been generated from corporate partnerships (Bamford et al., 2003). Study show that cross-firm and cross-industry alliances have been increasing steadily and dramatically, and that competition tends to be between alliances rather than between individual firms (Zhao, 2006). In m-commerce, there has been fierce competition between bank-telecommunication alliances. For example, Bank of New Zealand and Vodafone versus ASB Bank (New Zealand) and Telecom (Banks and Telcos Vie for Position, 2004).

In summary, as shown in the above cases, the key drivers and expected benefits of inter-firm alliances in m-commerce include achieving competitive advantage, attaining much needed resources, sharing risks, and sharing knowledge and expertise.

CHALLENGES AND STRATEGIES – CROSS-CASE ANALYSIS

The various alliances reviewed in this paper indicate some patterns of m-commerce alliances which take advantage of core competency of each partner to generate a win-win outcome for the alliances. However, the viability and sustainability of the alliances remains questionable due to several inherited challenges as follows. This section discusses the challenges and suggests strategies to deal with the challenges.

A Strategic Alliance Approach

M-commerce requires "strong partner network" (Recently the Three Biggest Operators, 2007). This is a key challenge to successful inter-firm alliances. This researcher posits that it is important for firms to take a strategic alliance approach when seeking collaboration with other firms. Different from unplanned, casual and short-term oriented partnership, strategic alliance is generally defined as a long-term cooperative arrangement at the strategic level between firms to improve their competitive position and performance by sharing resources and risks. Lu and Choy (2004, p. 288) maintain that "fundamentally, strategic alliances are an externally oriented business approach emphasizing on improving business performance through the linkages in a supply network".

Strategic alliances can be limited to one or two functional areas or in a much broader scale. An alliance strategy that focuses on only one element of the businesses such as research and development (R & D), sales or distribution is termed a functional alliance strategy. The alliance between Yahoo! and Nokia is an example of strategic alliances with a clearly defined scope of cooperation, in which each company has agreed to a "hands-off" approach to the other's core competencies. Motorola's strategic alliance with China International Electronic Commerce Centre that focuses on R & D collaboration is another example of functional alliance strategy. There can be a broader scope of cooperation between participating firms in an alliance who agree to perform together at multiple stages of the process by which goods or services are brought to the market: product development, marketing, sales, distribution, etc. This is referred to as a comprehensive alliance strategy. The alliance between Motorola and MasterCard described above falls into this category. By taking a comprehensive alliance strategy, participating firms can combine and consolidate their strengths and resources to get a competitive advantage. However, integrating the different operating procedures of several participating firms over a broad range of functional areas can be a great challenge to strategic alliances. This kind of comprehensive alliance is still a rare breed in the current m-commerce partnerships.

Strategy Planning and Implementation for M-Commerce Alliances

It is found that most of the m-commerce alliances studied in this paper resulted from a well-designed strategy planning process. For example, Motorola has been systematically pursuing its global alliance strategy with multiple key players in m-commerce for years (Motorola, 2009), which contributes to its success and steady growth in global alliance and business.

However, strategic planning and implementation is a very challenging and formidable task given the multiple partners and industries involved in m-commerce. When formulating an alliance strategy, some key questions should be answered as follows:

- With whom should we partner or network?
- What are the key criteria to use to select our strategic alliances?
- What is the scope of cooperation through the alliances?
- What is our operating model for the alliances including revenue sharing model?
- What form or structure will our alliance take?
- What are the potential pitfalls of alliances and how can we address them?

To answer these questions, most strategists agree that it is important to undertake the external environment analysis before the internal organizational analysis in order to reduce possible biases (Mintzberg et al., 1995). The competitive analysis helps organizations to compare their strengths and weaknesses with those of their competitors in order to devise an effective strategy for gaining a competitive edge. To survive and sustain in today's volatile marketplace, organizations must be able to seize and quickly exploit opportunities for m-commerce. They must have a well-defined mission, a clear vision and an unambiguous understanding of how they intend to achieve their organizational goals. If a strategic alliance is a necessary means to help achieve the goals, the formulation of an effective strategy becomes paramount.

Alliance strategy has little value in practice if it stands alone. The synergy of alliance strategy with other strategies of an organization should also be a consideration in the strategy formulation process. First, the alliance strategy should be an integral part of m-commerce strategy if an organization decides to jointly take up m-commerce business. Second, the alliance strategy should have an interface with other organizational strategies, including corporate strategy, business strategy and functional strategy including marketing strategy, information systems strategy, operations strategy and R & D strategy. Thirdly, while a m-commerce alliance strategy is embedded in organizational strategies, it should be clearly defined and articulated.

Too often, companies enter into an alliance without giving considerable thought to strategic purpose and the alliance implementation strategy. Implementation of alliance intent is often far more complicated and difficult than striking a partnership deal. Even the best alliance strategy can fail if management does not implement it and evaluate the results of implementation properly (Zhao, 2006). "Corporate strategy is an organization process, in many ways inseparable from the structure, behaviour, and culture of the company in which it takes place" (Mintzberg et al., 1995, p. 67).

Selection of Alliance Partners

As shown in the case of Motorola's multiple alliances with other companies, m-commerce often involves multiple partners, such as, telecommunications (i.e., wireless) service providers, mobile device makers, banking industry, Internet search engine providers, and various third-party value-adding companies and requires effective collaboration among them. Selection of partners is another key challenge to successful alliances. The m-commerce alliances studied in this paper have one thing in common, that is, core competency. For example, in the cases of T-Mobile and Yahoo!, and Nokia and Google, each partner has its unique strength and leadership in its field which is likely to sustain the alliance by adding complimentary services and technology to the alliance.

Given the nature of strategic alliances, selection of partnering firms in an alliance should not only take into account the need for a long-term cooperative relationship at the strategic level, but the need to have strategic flexibility in order to modify the alliance when the environment changes, or to exit the alliance when the environment no longer requires one. Understanding each organization's dynamics is crucial to the selection as it will help you utilize it to your own advantage.

The crucial criteria for selecting the right alliances/partners are (i) complementarity and compatibility, (ii) core competency, (iii) electronic and technological capability, (iv) financial commitment, and (v) strong leadership and commitment. In the context of m-commerce, the key questions to be answered when selecting your potential alliances are:

1. Are the skills, capabilities and resources of the prospective partners complementary to those of our organization for a m-commerce alliance?
2. Are the prospective partners compatible to our organization in terms of management style and organizational culture? Can we get along with them?
3. How well do the prospective partners manage other partnerships (track record of partnerships)?
4. Do the prospective partners have a distinctive strength (core competency) that can be leveraged to gain a competitive advantage in the very competitive and dynamic m-commerce market?
5. Are the prospective partners ready to link up their systems with ours and other partners' in the alliances? For example, in the case of the alliance between the three biggest wireless market players in Canada. Their alliance is based upon a common goal, namely, achieving inter-carrier interoperability.
6. Are the prospective partners prepared to invest in the alliance? Do they have a fiscal future?
7. Do the prospective partners have a strong leadership team or leader?
8. Most importantly, will the prospective partners be committed to the alliance?

If the answers to the first three questions are no, it does not necessarily mean that you should refrain from entering into an alliance. The issue is how to approach these differences and deal with the challenges jointly from the outset to increase the chances of success in the strategic alliance.

A Balanced Portfolio and Network Management

As shown in the case of Motorola's alliances with MasterCard, UMPay, and CIECC in this paper, developing a balanced portfolio approach to alliances is a popular strategy. A balanced alliance portfolio has several advantages. It can provide a variety of complementary partnerships to strengthen a company's network position through increasing and expanding its variety of products and services. It serves as a buffer to mitigate the damage caused by the loss of one alliance if the partnership goes sour, and it can also consolidate the independent status of your company by not over-relying on one alliance. The portfolio approach may increase the bargaining power of a company in negotiating prices or stabilizing the prices and better services. Other advantages of the approach may include more market opportunities, expanded customer base brought in by partners, reduction of competition, and gaining access to more social capital. Social capital is often viewed as the sum of the actual and potential resources derived from a network of alliances possessed both by individuals and organizations. However, there is a need to standardize the processes for selection of portfolio partners to ensure transparency and consistency.

The success of a strategic alliance depends not only on how well a firm cooperates with its direct partners but also on how well its partners cooperate with its partners' partners. From a network perspective, there are three key components in network management: activities, actors, and resources. Activities refer to the commercial, technical and administrative functions of individual firms. In this respect one of the key tasks of network management is to create links between firms to facilitate these functions. Actors refer to both institutions (e.g., suppliers, distributors, customers, transport and logistics providers and other intermediaries) and individuals (e.g., managers of the institutions). Network management bonds the actors together and aligns them in the web of organizational relationships. Resources often include tangible manpower, equipment and financial and production capacity, in addition to intangible resources of knowledge, organizational learning, market image, innovative capabilities and patent rights. Network management concerns the control of resources which, in turn, defines the actors and their activities (Halldorsson et al., 2003). This network perspective helps better demonstrate the nature of network management and the key contingencies of network coordination.

In terms of the governance structure of network management, there are generally two popular forms of inter-firm governance in the literature that can be adopted for m-commerce strategic alliances: unilateral governance or hierarchy, and cooperative or bilateral governance. Unilateral governance or hierarchy is an authority structure in which one partner exercises a decisive power by imposing decisions on participating firms, and develops rules and instructions for them to follow. The dominating partner is responsible for the setup of the structure and the corresponding costs. However, these costs will only be borne if the partnership is intended not to be short-term oriented. In contrast to the one-party dominating structure, cooperative or bilateral governance represents a flexible cooperative mechanism which is based upon bargaining and negotiation between partners. The structure often involves the reach of an agreement or contract which must fulfill the interests of the cooperative parties and ensure that partners maintain decisive influence over the process and outcomes of partnership (Albers et al., 2003). The two governance structures and modes are distinctive but in reality, they are not mutually exclusive. In order to motivate and eventually enhance the performance of partners, many partnerships take on a combined governance mode containing both hierarchy and cooperative structures. Given the flexibility and dynamic nature of m-commerce, managers should give priority to the cooperative or bilateral governance mode because it allows for a multitude of alternative arrangements and horizontal communication and interfaces.

Cross-Cultural Management

The m-commerce alliances discussed in this paper are cross-industry and mostly cross national borders. How to manage the differences in organizational culture and national culture remains a challenging question to every alliance partner. Culture can be defined as the shared assumptions, beliefs, values and norms that characterize an organization. Each organization has its own culture developed from its own particular experiences, its own role and the way its owners or managers get things done (Hellard, 1995). A company's decision-making style and processes often affect the way inter-firm alliances interact and the degree to which interactive exchanges are made. Adaptive and open organizations are more conducive to cultural change. Davis and Spekman (2004, p. 202) suggest that a win-win orientation be promoted amongst partners in a strategic alliance because this "can be reflected in the psychological contract that shapes expectations of how partners will interact". The win-win orientation can help to foster a supportive organizational culture, encourage partner-like behavior and maximize the mutual benefits that accrue from strategic alliances. One of the key factors in any failed new business operation or partnership is associated with organizational culture. Differences in culture and resistance to cultural changes affect partners' ability to achieve alliance advantage. This applies to m-commerce alliances too.

In addition to cultural differences at organizational level, most of m-commerce alliances in this paper are international, such as the alliance between Kasikornbank and Venyon, and Zain's alliance with Citigroup and Standard Chartered. The alliances inevitably encounter the culture barriers caused by various cultural differences between nations as acknowledged by Hofstede's study (2001). Miroshinik (2002) pointed out that the first major contributor to problems and failures of international business is cultural differences. Similarly, Dickson, Den Hartog, and Mitchelson (2003) explained that different cultural environments require different managerial behaviors, strategies and structures. In a cross-cultural context, alliance partners are increasingly confronted with the need to understand how people from different cultures view them and interpret their actions (Yukl, 2002). According to a new perspective in cross-culture study, culture intelligence is an effective way to manage the difference. The key components of culture intelligence include (1) culture awareness; (2) motivation of culturally adapting one's behavioural skills or actions; and (3) aptitude to determine where new behaviours are needed and how to execute them effectively. Furthermore, effective cross-culture communication helps improve cross-culture intelligence.

In addition to culture differences, there are also unavoidable legal, political and economic differences in cross-border m-commerce alliances. Managing cultural and system change necessary to resolve differences across organizations and across nations is therefore a critical issue that managers of international alliance must negotiate. The legal, political and economic issues and challenges are beyond the scope of this paper.

CONCLUSION AND FUTURE RESEARCH

In the current global financial crisis, building m-commerce capacity is seen as another critical means for businesses to survive through cutting costs and boosting online operations (Wolf, 2009). With the popularity of smart phones such as iPhones, m-commerce is becoming one of the most important businesses in 2009. It is estimated that the global market for m-commerce should reach US$88 billion by 2009 (M-commerce, 2009). M-commerce can hardly work on its own without a network of alliances involving multiple players as shown in the cases discussed in this paper.

Inter-firm m-commerce alliances are complex undertakings as they involve not only inter-firm management but also cross-industry management as well as international management in many cases as shown in this paper (Dodourova, 2009). The complexity requires effective strategies to deal with the challenges in forming and managing m-commerce alliances.

This paper reviews and explores alliances in the context of m-commerce that are currently formulated in various cultural and national backgrounds. The paper deals with both strategic and operational strategies for m-commerce alliances and discusses critically a wide range of issues in the formulation and implementation of m-commerce alliance strategy.

There is a general belief in the current literature that e-business including m-commerce allows a lesser degree of strategic planning because of the uncertainties and lack of knowledge associated with the dynamics of e-businesses and m-commerce. However, the quick demise of a myriad of Internet firms challenges a rush-in mentality and indicates the importance of strategic planning and management. Strategic approach is an approach to dealing with change and turbulent environments particularly in the globalization of markets. It encourages and does not adverse change.

Based upon this study, the areas for future research in strategic alliances including m-commerce alliances could include:

- The complex, dynamic (opportunistic and instable) and sophisticated structures (often with confusing nomenclature) under which strategic alliances in m-commerce operate;

- The specific nature and characteristics of the operations of each typology of alliances in different industry sectors and in different country settings;
- The impact of specific m-commerce technological innovations on specific functional areas of strategic alliance interfaces;
- Identification of m-commerce networks and relationships that are most likely to change, and how they change, and;
- Investigation of the projected differences in developing m-commerce alliance strategies from an international perspective, in terms of business nature, size, typology, dimension, governance structure and management, international regulations.

It is this researcher's view based upon the past and present experiences that the business environment in the next decade will intensively include the management of inter-firm strategic alliances including m-commerce strategic alliances.

REFERENCES

Albers, S., Gehring, M., & Heuermann, C. (2003). A configurational approach to supply chain governance. In Seuring, S., Muller, M., Goldbach, M., & Schneidewind, U. (Eds.), *Strategy and organization in supply chains* (pp. 100–113). New York: Physica-Verlag.

Bamford, J. D., Gomes-Casseres, B., & Robinson, M. S. (2003). *Mastering alliance strategy: a comprehensive guide to design, management, and organization*. San Francisco, CA: John Wiley & Sons.

Banks and Telcos Vie for Position (p. 9). (2004, December 6). Wellington, New Zealand: The Dominion Post.

Batonda, G., & Perry, C. (2003). Approaches to relationship development processes in inter-firm networks. *European Journal of Marketing, 37*(10), 1457–1484. doi:10.1108/03090560310487194

Berg, B. L. (1995). *Qualitative Research Methods for the Social Sciences*. Boston: Allyn and Bacon.

Coffey, A., & Atkinson, P. (1996). *Making sense of qualitative data: complementary research strategies*. Thousand Oaks, CA: Sage Publications.

Davis, E. W., & Spekman, R. E. (2004). *The Extended enterprise: gaining competitive advantage through collaborative supply chains*. Upper Saddle River, NJ: Prentice Hall.

Dickson, M. W., Den Hartog, D. N., & Mitchelson, J. K. (2003). Research on leadership in a cross-cultural context: Making progress, and raising new questions. *The Leadership Quarterly, 14*(6), 729–768. doi:10.1016/j.leaqua.2003.09.002

Dodourova, M. (2009). Alliances as strategic tools: a cross-industry study of partnership planning, formation and success. *Management Decision, 47*(5). doi:10.1108/00251740910960150

Håkansson, H., & Snehota, I. (1992). Analyzing business relationships. In *Developing Relationships in Business Networks*. London: Routledge.

Halldorsson, A., Skjott-Larsen, T., & Kotzab, H. (2003). Inter-organizational theories behind supply chain management – discussion and applications. In Seuring, S., Muller, M., Goldbach, M., & Schneidewind, U. (Eds.), *Strategy and organization in supply chains* (pp. 32–46). New York: Physica-Verlag.

Hellard, R. B. (1995). *Project partnering: principle and practice*. London: Thomas Telford Publications. doi:10.1680/pppap.20436

Hofstede, G. (2001). *Culture's consequences: Comparing values, behaviors, institutions, and organizations across nations* (2nd ed.). Thousand Oaks, CA: Sage.

Krippendorff, K. (2004). *Content Analysis: An Introduction to Its Methodology*. Thousand Oaks, CA: Sage.

Lu, D., & Choy, T. K. L. (2004). Competitive supply chain development through strategic alliance and dynamic relationship. In *Proceedings of Supply Chain Management and Information System 2004,* Hong Kong (pp. 287-296).

M-commerce. (2009). Retrieved March 10, 2009, from http://gsmworld.com/technology/services/m-commerce.htm#nav-6

Mintzberg, H., Quinn, J. B., & Voyer, J. (1995). *The strategy process*. Englewood Cliffs, NJ: Prentice-Hall.

Miroshnik, V. (2002). Culture and international management: a review. *Journal of Management Development*, *21*(7), 521–544. doi:10.1108/02621710210434647

Motorola & CIECC in an agreement to launch m-commerce laboratory. (2008, December 9). *Wireless Federation*. Retrieved December 27, 2008, from http://wirelessfederation.com/news/motorola-ciecc-in-an-agreement-to-launch-m-commerce-laboratory/

Motorola. (2009). *Alliances*. Retrieved March 17, 2009, from http://www.symbol.com/alliances

Motorola enhances mobile commerce solutions for consumers. (2008, December 29). *Wireless News*. Retrieved February 2, 2009, from http://proquest.umi.com.ezproxy.aus.edu/pqdlink?Ver=1&Exp=02-18- 2014&FMT=7&DID=1615270131&RQT=309&cfc=1

Motorola, MasterCard Promote Mobile E-Commerce. (2000, July 18). *Internet Retailer*. Retrieved April 16, 2008, from http://www.internetretailer.com/internet/marketing- conference/64034-motorola-mastercard-promote-mobile-e-commerce.html

Murphy, S. (2008). The view from the top. *Chain Store Age*, *84*(9), 58.

Nokia. (2008). *Nokia and Google to Offer Google Search to Nokia customers worldwide*. Retrieved April 14, 2008, from http://www.nokia.com/A4136001?newsid=1190585

Oberthur Technologies; Zain Group Selects Oberthur Technologies' End-to-End Security Platform for Africa's Biggest Mobile Commerce Service. (2009, March 3). *Science Letter*, 3882.

Patton, M. Q. (2002). *Qualitative research and evaluaion methods*. Thousand Oaks, CA: Sage Publication, Inc.

Read, B. (2008). Mobile CRM: "Are we there yet? *Customer Interaction Solutions*, *27*(5), 5.

Recently the Three Biggest Operators in the Canadian Wireless Market, Bell Mobility, Rogers Wireless and TELUS Mobility. (2007, June 11). *Business Wire*. Retrieved March 12, 2009, from http://www.researchandmarkets.com/reports/c59365

Research and Markets Adds Report: 'Mobile Commerce: Prospects for Payments, Ticketing, Coupons and Banking 2008-2013'. (2008, November 22). *Wireless News*. Retrieved March 23, 2009, from http://proquest.umi.com.ezproxy.aus.edu/pqdweb?index=4&did=1650795911&SrchMode=2&sid=1&Fmt=3&VInst=PROD&VType=PQD&RQT=309&VName=PQD&TS=1235586003&clientId=19323

Schramm, W. (1971). *Notes on case studies of instructional media projects*. Washington, DC: Academy for Educational Development.

Varshney, U. (2008, November/December). Business models for mobile commerce service: requirements, design, and the future. *IT Pro*, 48-55.

Venyon, Thailand bank cooperate on contactless mobile payments. (2009, February 10). *Contactless News*. Retrieved March 10, 2009, from http://www.contactlessnews.com/2009/02/10/venyon-thailand-bank-cooperate-on- contactless-mobile-payments

Wolf, A. (2009). Surviving 2009: cost –cutting, e-commerce considered critical. *Twice*, *24*(1), 74.

Yahoo. (2008). *Yahoo! oneSearch™ Launches as the Mobile Search Service on T-Mobile USA's New Web2goSM Service*. Retrieved January 11, 2009, from http://yhoo.client.shareholder.com/press/releasedetail-mobile.cfm?ReleaseID=350074

Yin, R. K. (2003). *Case Study Research: design and methods*. Thousand Oaks, CA: Sage Publications.

Zhao, F. (2006). *Maximize Business Profits through E-partnerships*. Hershey, PA: IGI Global.

This work was previously published in International Journal of E-Business Research Volume 6, Issue 4, edited by In Lee, pp. 26-37, copyright 2010 by IGI Publishing (an imprint of IGI Global).

Chapter 12
The Impact of the Internet on Marketing Strategy:
Revisiting Early Predictions

Kaan Varnali
Istanbul Bilgi University, Turkey

ABSTRACT

This conceptual paper revisits and reviews early milestone literature of influential authors, who discuss the impact of the Internet on major forces that shape a company's marketing strategy to assess whether their futurist predictions and conceptual claims turned out to be true or were later falsified by empirical research or market experience. In particular, the impact of the Internet on customers, design of products/services, pricing, marketing communications, intermediaries, publics, and social/cultural environment is explored. Throughout the paper, related under-researched and unsettled areas are discussed and future research avenues are suggested.

INTRODUCTION

Marketing practitioners and academics have devoted considerable effort to identify the effect of the Internet on marketing. Early pioneering scholarly works envisioned the Internet as a driver so significant that it would transform even the most basic premises of the business environment. For Hoffman and Novak (1996) "The Web combines elements from a variety of traditional media, yet is more than the sum of the parts" (p. 55). Peterson, Balasubramanian, and Bronnenberg (1997) stated that "The Internet can be considered a market discontinuity because it represents a shift in any of the market forces or their interrelationships that can not be predicted by a continuation of historical trends and that, if it occurs, can dramatically affect the performance of a firm or an industry"

DOI: 10.4018/978-1-61350-462-8.ch012

(p. 330). Achrol and Kotler (1999) described the growth of e-commerce as the single biggest opportunity and threat facing almost every industry. These pioneering articles regarded the Internet as a major driver of change for marketing, in other words as a new technology introducing novel dimensions that offer a new basis of competition or way of doing business. Throughout the years, academic knowledge about the impact of the Internet and other emerging technologies on marketing has accumulated and due to the unique characteristics of the virtual world and its huge business potential a distinct research stream has emerged drawing academic interest from both IS and marketing disciplines.

This paper revisits and reviews early articles of influential authors about the impact of the Internet on the major forces that shape a company's marketing strategy. The foremost purpose of this study is to assess whether the futurist predictions included in these milestone articles did actually turned out to be true or were later falsified by empirical research or market experiences. It is important to note that the purpose of this article is not to judge the value of these visionary works of art. These articles dutifully served their purpose as courageous attempts to shed light on an undiscovered intellectual territory on which academic understanding was lagging behind the developments occurring within the business environment. The birth and the explosion of the Internet was so rapid that only after the burst of the dot.com bubble the academia could catch up to provide theoretical explanations and normative prescriptions. Therefore, acknowledging their fundamental academic value, this article aims to present a systematic analysis of the early predictions on how the Internet will change business environment and its subsequent impact on marketing strategy. In order to establish a coherent flow of arguments the framework shown in Figure 1 is used categorize the topics to be investigated. The figure presents a grand summary of the marketing process and the forces shaping a company's marketing strategy. Throughout the paper the academic knowledge on the impact of the Internet on each of those forces is explored, and future research avenues are suggested.

CONCEPTUAL BACKGROUND AND DISCUSSION

Sheth and Sisodia (1999) have classified marketing's lawlike generalizations into four key categories as location-centric, time-centric, market-centric and competition-centric. Location-centric ones focus on the physical separation between buyers and sellers. Time-centric ones focus on modeling same or similar patterns of product and market evolution over time. Market-centric ones focus on dividing the consumer market into well-defined and relatively homogeneous markets. And finally, competition-centric ones focus on competitors and intensive competition. The authors argued that the Internet will most strongly influence location-centric concepts, which were to be fundamentally altered by the evolution of the Internet. Much has changed since 1999. The penetration rate of the Internet significantly increased all over the world, connection speed and quality have improved, interactivity and visual aesthetics of user interfaces have been signifi-

Figure 1. Components of marketing strategy

cantly enhanced and the industry has accumulated tremendous experience and know-how regarding online consumer behavior. These developments lead to radical changes not only on location-centric concepts but on the entire context of marketing. Furthermore, the proliferation of mobile technology started to create unprecedented opportunities for marketers to reach existing and potential customers anytime anywhere with a great precision. The impact of the Internet on the traditional time cycles of marketing, segmentation and targeting, and the creation of totally personalized marketing mixes create an entirely new environment for competition.

The Impact of the Internet on Customers

The investigation of the impact of the Internet on customers has started following the commercialization of the Internet in early 90s. Initial conceptual works focused on shifts in communication patterns and consumer interaction through the online medium and with the digital content (Hoffman & Novak, 1996; Trevino & Webster, 1992). Later, together with the rapidly increasing penetration rate of the Internet, the business potential of the Internet exploded. Consequently, the significance of understanding online consumer behavior became paramount. Theories of Reasoned Action, Planned Behavior, Diffusion of Innovations, Expectation-Confirmation, and Technology Acceptance Model have been repeatedly used in the study of online consumer behavior (Jarvenpaa, Tractinsky, & Vitale, 2000; Koufaris, 2002; Lin & Lu, 2000). These theories were typically used to explain website usage patterns, online purchase intentions, online purchase and loyalty behaviors. In their thorough literature review of online consumer behavior, Cheung, Chan, and Limayem (2005) arrived at the conclusion that classical theories such as TRA and related theories constitute only a starting point in understanding online consumer behavior; therefore, future research should base their research more on irrational behavior models.

The first and the most important impact of the Internet on customers that consequently affect the world of marketing is the significant improvement in consumer power. Importance of consumer power is introduced to marketing by Day and Aaker (1970). They proposed that consumer rights such as the right to choose, the right to safety, the right to be informed and the right to be heard affect how consumers view themselves in the marketplace and how companies relate to them. The thesis that the Internet will cause a decisive increase in consumer power has been supported by many authors (e.g., Bakos, 1997; Hoffman & Novak, 1997; Murphy, 2000; Pitt, Berthon, Watson, & Zinkhan, 2002; Reisch, 2001; Rha & Widdows, 2002; Van Raaij, 1998). Market experience showed that the Internet, allowing consumers to easily access accurate and up-to-date information about products, companies and legal regulations from a vast variety of sources, and interact with others with almost no cost and physical effort, significantly enhanced consumers' ability to reduce his or her lack of information through information gathering. Search engines allow consumers to search with no cost and compare competing products easily through many perspectives. Also online communities allow customers to spread their compliments and complaints more effectively and efficiently than it were ever possible before. These communities allow consumers to freely share information about alternative value propositions with other consumers, forming their own expectations about quality and reducing dependence on suppliers' communications. In some cases, these communities facilitate formation of collective buying groups which give consumers the ability to negotiate better terms with suppliers (Pires, Stanton, & Rita, 2006). The speed and ease of information sharing is even more improved with the convenience and ubiquity provided by the mobile medium. The Internet have transformed information scarcity into "information democracy" (Sawhney &

Kotler, 2001) or "transparency" (Deshpande, 2002). These functions of the Internet gave the control of market knowledge, once controlled by suppliers, to the consumers which ultimately altered power dynamics in favor of consumers (Urban, 2004).

Kucuk and Krishnamurthy (2007) investigated this power shift towards customers and concluded that it happens along four dimensions: technologic, economic, social and legal. Technologic sources of power stem from superior mobility, ability to participate, and track and control transactions. Economic sources of power stem from the access to better value and markets and ability to construct economic value from C2C markets. Social sources of power stem from increased connectivity to communities and experts. And legal sources of power stem from access to governmental legal sources, consumerist support and activists. Internet, being a platform in which all nodes are equal in status, seems to perform as a mechanism that equalizes the power between companies, consumers and all other agents in the marketplace. As a result, the Internet facilitates choice and risk reduction which in turn enables consumers to exercise increasing control over their consumption activities (Harrison, Waite, & Hunter, 2006).

The Impact of the Internet on Product/Service Design

Web-based interaction with customers broadened differentiation opportunities for companies and enabled ultimate level of customization of services which is popularized by the term customerization. The concept of customerization is quite different than mass customization. In early 90s, increasing operational flexibility and divergent customer needs resulted in extreme differentiation of products, called mass customization (Pine, 1993). Mass customization was about producing huge variety of product versions at the price of standardized mass produced alternatives (Hart, 1996). However, mass customization approach has a serious fall-back. It often resulted in mass confusion due to the fact that consumers do not often have enough expertise and capability to define their preferences effectively when faced with increased variety (Huffman & Kahn, 1998). To overcome this problem, companies started to adopt one-to-one basis customer relationship management and micro level personalization of services (Peppers, Rogers, & Dorf, 1999) and the Internet provided the necessary medium for such interaction. More recent developments in the Internet technology, especially the introduction of the mobile Internet, have enabled firms to further enhance their one-to-one based relationship with customers and this ultimate form of differentiation of total consumption experience is called as customerization (Wind & Rangaswamy, 2001). The most recent form of customerization involves co-creation of value with customers. Today, interactive online systems allow customers to actively participate in the value creation process by choosing attributes, components, prices and delivery options. Firms no longer create value for the customer; rather, they create value with the customer and incorporate the customer into their value-creation chains (Prahalad & Ramaswamy, 2004). Within co-creation of value practices the company provides complementary knowledge, resources and tools to the customers and lets them to design/configure their own products and services. This new logic presupposes a longer term relationship between customers and suppliers, which creates opportunities for acquiring more knowledge, thus making the service provider better able to adapt to the customer and to provide higher quality (Wikstrom, 1996). Consumers now act as the designer, as well as public relations executive regarding problems caused by the services provided by the firm, and the main critic source for everything that the firm introduces to the market (Krishnamurthy, 2001). This situation demonstrates the birth of the customer age, and the end of product age (Kucuk & Krishnamurthy, 2007).

While a stream of research concentrated on consumers' new role as product designers, critique sources and public relations executives, another concentrated on the suitability of different product categories for online selling. Peterson, Balasubramanian, and Bronnenberg (1997) stated that the suitability of the Internet for marketing depends on the characteristics of the products and services being marketed. Alba et al. (1997) suggested that product characteristics together with situation and consumer characteristics will influence the evaluation and selection of the proper retailing channel. Phau and Poon (2000) pointed out that some limitations of the Internet as a retail channel restrict the products that are sellable online. This perspective suggests that some attributes of products are perceived to be better delivered online or offline. More recently, Levin, Levin, and Weller (2005) arrived at the same conclusion and suggested that predominance of specific product attributes influence online/offline shopping preference.

A literature review of "e-potentials of products research stream" reveals seventeen product characteristics that may have an influence on a product's suitability for online retailing. These product characteristics are: tangibility, tactility, information intensity, efficiency-of-delivery, importance of instant gratification, differentiability, level of retailer added-value (level of standardization), importance of owning, difficulty of finding, having a trusted brand name, price sensitivity, price, margin, involvement level, pleasantness of shopping experience in offline retailer, gift orientation, frequency of purchase, perishability, and personal nature of the product (Burke, 2002; Feather, 2002; Levin, Levin, & Weller, 2005; Liang & Huang, 1998; Phau & Poon, 2000; Toker & Varnali, 2009; Vijayasarathy, 2002).

Prediction of how the portfolio of products or services exchanged over the Internet will change remains to be an important research agenda. Although studies that examine the suitability of different product categories for online shopping are abundant, the rapid change in consumer perceptions about the online medium and improving technologies continuously change the range of products and services that are suitable for online sales. Additionally, as the mobile handsets proliferate in terms of usability and graphical interface, the mobile medium is on the verge of transforming into the most convenient platform available for electronic commerce. The study of design and optimization of products and services for mobile commerce is a fruitful avenue for research. Consequently, up-to-date understanding of the portfolio of products or services suitable for online shopping requires continual research.

The Impact of the Internet on Price

Early studies predicted that the great degree of price transparency caused by the Internet would eventually increase standardization of prices across borders, or at least cause narrower price differentials (Hamill, 1997; Samiee, 1998) and several studies provided empirical support for this thesis (e.g., Brown & Goolsbee, 2002;Scott Morton, Zettelmayer, & Silva-Risso, 2001; Strader & Shaw, 1999). The logic was straightforward: the freedom of information inherent in the Internet would result in numerous price reductions, which would be more apparent where products and services are hard to differentiate. However, contrary to expectations, the price dispersion on the posted prices remained surprisingly high (Clay, Krishnan, Wolff, & Fernandes, 2002; Lee & Gosain, 2002; Pan, Ratchford, & Shankar, 2003), and online retailers with the lower prices did not make the most sales. Lehmann (2003) investigated online prices for holiday packages and found that price dispersion is higher online for lower quality locations, while lower for higher quality locations. Brynjolfsson and Smith (2000) attributed this dispersion in prices to heterogeneity in consumer awareness and heterogeneity in retailer trust. Research on the impact of the Internet on price levels still remains inconclusive and hence more causal research is required on this subject.

The early theories regarding the impact of the Internet on price rested on the assumption that the Internet enables consumers to make more informed decisions more conveniently and with more efficient price-clearing mechanisms than was previously possible (Hamill, 1997; Peterson et al., 1997; Samiee, 1998). However, there are studies showing that a great majority online shoppers do not act solely based on price comparisons and still make ill-informed decisions (Johnson, Moe, Fader, Bellmann, & Lohse, 2004; Lal & Sarvary, 1999). It has been shown that consumers tend to buy at the firstly visited online shop (Rezabakhsh et al., 2006), and the search intensity of many consumers decreases as their experience increases (Suri, Long, & Monroe, 2003). More recently, Spann and Tellis (2006) studied the extent to which decisions that are made in an online name-your-own-price auction are rational. Their results showed that a large number of consumers do not exhibit rational decision making. This finding indicates that the Internet does not eliminate or lower consumers' irrational decisions, as predicted by the existing literature, and consumers still apply heuristics and decision shortcuts. Therefore, researchers and practitioners should be careful when relying solely on models of behavior that assume strict rationality, because such models may not reflect reality and may recommend suboptimal pricing strategies (Spann & Tellis, 2006). Future research should continue investigating the extent to which the Internet facilitates informed and rational decisions.

Another impact of the Internet on price is the transformation of the consumer from a passive price taker into a price maker (Kauffmann & Wang, 2001; Pitt et al., 2001). The interactivity and immediacy provided by the Internet gave birth to unprecedented opportunities for dynamic pricing strategies such as real-time auctions, name-your-own-price formats and group-buying (Wind & Mahajan, 2002). These pricing formats enable consumers to exert direct influence on posted prices both at the individual and collective levels, and consequently boost consumer power.

The use of the mobile medium in dissemination of marketing-related information has an inverse effect on pricing when compared with the use of conventional PC-based Internet. As opposed to the PC-based Internet which increases price transparency, the mobile medium provides exceptional price differentiation opportunities for marketers. Since mobile devices are typically used by a sole individual and are very personal in nature, they provide an ultimate tool of one-to-one communication with target consumers. The mobile Internet allows showing different prices and information to different people at different locations through delivery of highly personalized information at the most effective time and place with great precision. The impact of such mobile differential pricing strategies on perceptions, feelings and attitudes of customers has yet to be researched and provides fruitful research opportunities.

The Impact of the Internet on Customer Communication

The power of the Internet as a communication channel for all kinds of information is well understood by the marketers today. The information-based, technological and continuous communication opportunities hosted by the Internet make it a fruitful environment for advertising and marketing communications. Global advertising costs, as a barrier to entry are significantly reduced as the Internet makes it possible to reach a global audience more cheaply. At its birth, the Internet represented a major long-run threat to the established patterns of inter-media rivalry. The Internet emerged as an adaptive, hybrid medium with respect to audience addressability, audience control and contractual flexibility. Now, it became a potential substitute or complement for all major categories of existing media. The exact nature and time of its influence on existing media in different parts of the world depends on the penetration rate of the Internet, consumer demand for information, development of alternative advertising formats, and its appeal to advertisers in various product/service categories (Silk, Klein, & Bernt, 2001).

Considerable academic effort is spent on exploring the effectiveness of websites as advertising platforms (Chatterjee, Hoffman, & Novak, 2003; Pieters, Rosbergen, &Wedel, 1999; Tan, 1999). It is found that, the push method of promotion, bombarding the target market with advertising messages, is not very effective on the Internet. It may even mean brand suicide in mobile advertising. The literature unanimously agrees on the importance of the prior explicit permission of the consumer for the acceptance and success of a marketing message (Barnes & Scornavacca, 2004; Carroll et al., 2007; Godin, 1999). Prior permission is even more critical in the mobile context because in all other marketing channels, consumers may simply choose to ignore or get away from marketing effort if they are not interested in it or do not like it, which is not possible when the message is delivered to the personal mobile handset of an individual. Permission-based mobile marketing requires that consumers have to 'opt in' before they receive mobile advertising messages of any kind, have the ability to control timing and frequency of message delivery and the content of the message, and have the option to 'opt out' at any stage.

One strategic implication of the Internet on promotion strategy is through electronic word of mouth (eWOM). Internet provides the opportunity for consumers to offer their own consumption-related advice by engaging in eWOM. Although not explicitly named as eWOM, the concept first appeared in the pioneering work of Hoffman and Novak (1996). They defined two types of behavior in which consumers engage during time spent in a computer mediated environment. "Goal-directed" behavior is differentiated from "Experiential" behavior in terms of extrinsic versus intrinsic motivation, instrumental versus ritualized orientation, situational versus enduring involvement, directed versus non-directed search, and goal directed versus navigational search, respectively. During experiential behavior opinion leaders may be exposed to new product knowledge and disseminate this knowledge by word-of-mouth. Therefore, marketers are advised that they should not only focus on goal directed behavior but also on experiential behavior as well, in order to control the affects of eWOM.

Given that online communication is directed to multiple individuals, available to other consumers for an indefinite period of time, and anonymous, eWOM is more enduring, easier and faster to spread. The importance of eWOM becomes paramount in managing and strengthening brands in the online environment (McWilliam, 2000). Customer participation and customer to customer interaction can be so essential that the favorableness and message number on a discussion forum may have a significant impact on brand evaluations and attitudes towards the online platform (Chiou & Cheng, 2003). The major difference of online environment with respect to offline environment is that in the offline environment there are many cues a consumer can use to evaluate a brand, however in the online environment, these cues are not present. If the customer is not viewing the corporate website, he/she has to rely exclusively on eWOM. Concern for other customers, positive self enhancement, social benefits, economic incentives and advice seeking are motivators for visiting eWOM platforms (Hennig-Thurau et al., 2004). The actions of online marketers suggest that they have understood the importance of virtual communities and viral effects, as they make investments in virtual community infrastructure designed to motivate people to spend time at their websites. On the other hand eWOM and virtual communities provide a fruitful avenue for research for marketing academicians. Impact of virtual communities on life-styles of younger generations who grew-up with the Internet, the impact on the importance of offline communication channels for those who extensively engage in virtual communities, convergence of virtual identity with real-life identity and its impact on consumer perceptions are among many possible research issues waiting for further investigation.

One of the hot topics in the Internet research, that has been around since the commercialization the Internet in early 90s, is the tactics to convert website visitors into shoppers. The article of Schlosser, White and Lloyd (2006) adequately summarizes this research stream and extends its theoretical knowledge base. They found out that instead of serving a purely aesthetic function, investments in Web site design signal that a firm's ability can be trusted. In turn, ability beliefs are found to be the most significant driver of online purchase intentions for visitors engaging in prepurchase search. They suggest that even when privacy/security statements improve beliefs about a firm's trustworthiness, they are still relatively less effective in converting visitors into buyers. Thus, although investing in security and providing privacy/security statements are necessary, they are not sufficient in increasing searchers' willingness to buy online and less effective than website design in converting visitors into buyers. Golden rules of success for online selling and Internet marketing will always be welcomed by marketing practitioners.

The Impact of the Internet on Intermediaries

The literature on impact of the Internet on marketing intermediaries is initiated by Sheth and Sisodia (1999). They state that the major impact of Internet is felt on location-centric concepts of marketing. The borderless nature of the Internet creates the opportunity to sell over the Web in a standardized way and subsequently eliminates many natural barriers to entry. E-Commerce brought the shopping environment into customer's home. The major impact of the Internet on traditional marketing intermediaries is caused by the resulted transparency in ordering and inventory process. The Internet provides a direct connection between end-users and producers which reduces the importance of traditional intermediaries. As predicted by Eid and Trueman (2002), such intermediaries had to create added value to the system through collection, interpretation and dissemination of vast amount of information to survive.

Although the Internet is not yet considered as a significant threat to traditional retailers in general or seems to be replacing them in the near future, market experience reveals that it had proved to be tremendously powerful in selling certain product and service categories. The major problem of the Internet as a retail channel is the lack of pleasurable experiences, face-to-face social interaction, and personal consultation that define experiences in bricks-and-mortar shopping (Barlow et al., 2004; Burke, 2002; Feather, 2002; Levin, Levin, & Weller, 2005). Recent articles attempt to offer partial solutions to the impersonal nature of Internet. One such article suggests the use of avatars (Holzwarth, Janiszewski, & Neumann, 2006). Avatars are virtual characters that can be used as company representatives. Attractive avatars are persuasive because of their likeability, whereas expert avatars are persuasive because of their credibility. Future research should investigate the extent to which the developments in 3-D technologies and virtual reality can satisfy such interactive needs of customers during online shopping.

The Impact of the Internet on Publics

The anonymity, uncensored nature and power to reach multiple individuals, make the Internet a new avenue for political action. Having these attributes, the Internet presents outstanding opportunities for individuals to criticize, defame and tease organizations efficiently and effectively (Bailey, 1999). The Internet enables individuals to communicate their arguments better, organize more effectively, and recruit others to their cause more easily. Customer activist groups such as

Greenpeace and PETA effectively use such sites to rally campaigns against policies and actions of specific companies.

The Internet's power in influencing masses resulted in increased degree of corporate accountability, which has wide implications for the manner in which business is conducted today (Melewar & Smith, 2000). In order to avoid unanticipated blows to their reputation, companies should constantly monitor the Internet and virtual social platforms for aggressive content that could have negative implications for both their global reputation and credibility. Companies should be ready to provide satisfactory responses to all accusations being disseminated through online social content because unchallenged statements may be perceived as truths. As it is being used by the activists, the Internet is being used by the companies as a tool of image building either through the corporate website or by disseminating information using virtual communities and online social networking platforms (Harrison-Walker, 2001; Hennig-Thurau et al., 2004). Drivers of success for such viral marketing campaigns launched via virtual social communities offers vast research opportunities.

The Impact of the Internet on Social /Cultural Environment

As the penetration rate of the Internet increased and technology advanced, kinds of activities individuals engage in the Internet varied as well. Today, an average Internet user may interact and play games with his or her friends, purchase almost all kinds of products and services through e-stores, keep online diaries, engage in word of mouth, find mates and establish friendships, basically perform all social activities over the Internet except for the ones that require physical contact. When there are massive numbers of individuals get together and interact, they eventually develop norms of interaction. That is how the Internet has built itself a culture based on norms of collaboration, cooperation, and a willingness to share resources. Later, reciprocity has become the norm that defined interactions online (Mathwick, 2002). Norms of reciprocity imposes a duty to repay or at least be grateful for benefits received from others (Gouldner, 1960). As in traditional communities, this moral code introduced a sense of duty to the community as a whole and to the individual members of that community as well. The result is a social consciousness, in other words shared set of norms which define what is right and wrong, appropriate and inappropriate (Muniz & O'Guinn, 2001). These norms are shared among the Internet users from all over the world regardless of their fundamental culture. Further, through time, Internet users have developed a language that consists of jargon and symbols which are commonly understood by Internet users from all nationalities. With the help of its shared culture and language, the Internet became an effective tool for marketers in enhancing meaning transfer between national cultures. Although the Internet helps reducing cultural barriers, it does not destroy the presence of fundamental cultural and language differences (Melewar & Smith, 2003). Future research should investigate the extent to which Internet culture masks the fundamental culture of younger generations of developed countries. This investigation might allow further understanding of the magnitude of the impact of Internet on marketing context.

MANAGERIAL IMPLICATIONS

Much of the literature has predicted a technology-led revolution in which power will shift to customers, and industries will have to redesign themselves accordingly. Marketing, being the function or philosophy dealing with the customer constituency, will prominently increase its importance and influence within organizations together with this power shift towards the consumers. The Internet has changed the context of business so drastically that now organizations have to deal with every individual customer with constant care. The Internet's world wide penetration rate is rapidly increasing together with the population of technology-confident consumers. It has already become indispensable for many consumers. Today, there is almost no product or service that is not being marketed on the Internet. In 2001, Porter (2001) suggested that companies have no choice but to deploy Internet technology if they want to remain competitive. Nearly a decade has passed and market place developments showed that Porter was right. Today, the question is not whether to employ the Internet technology or not, but it is how to best utilize it and the discussions provided throughout this article should provide the necessary guidance in this respect.

In recent years, the explosion in the number and popularity of the social networking sites such as Facebook, Twitter, MySpace and FriendFeed, caused drastic shifts in traditional socialization practices. The success of social networking sites shows that people, especially the youth, loves sharing social media content. They share videos, pictures, stories, jokes, comments, critics, games, articles, profiles, preferences and even advertisements with their peers through these online platforms. People even enjoy broadcasting their real-time whereabouts and statuses. Furthermore, widgets downloaded to mobile handsets further facilitate this process. It is expected that virtually every online social network application is going to have a mobile component over the next few years. A typical social networking site allows a user to build and maintain a network of friends for social or professional interaction. With the current software that is available, interactions within social networks are not limited to exchanging simple messages on a 1-to-1 basis but are constantly evolving towards the sophisticated interactions of virtual communities. Proliferation and wide spread adoption of these social networking sites have two important implications for marketing. First, by allowing the tracking of WOM, along with new member sign-ups, these sites enable marketers to establish a concrete and measurable link between observed WOM activity and customer acquisition. Such data can be used to quantify WOM effects and to incorporate them into the planning of marketing communication strategies (Trusov, Bucklin, & Pauwels, 2009). Second, these sites drive the convergence of virtual identity and market space with real-life identity and market place. Although this new environment presents exceptional opportunities such as viral marketing, availability of rich customer data and delivery of highly personalized messages to target customers, it also presents new challenges for marketers. Now that the expertise levels are higher and people got sick of spamming, consumers are more concerned about their privacy and are becoming more and more selective regarding online marketing initiatives. The analysis presented in this article showed that not all the early theories about the impact of the Internet on marketing strategy are applicable today. As the technology evolves, consumers' perceptions, attitudes and expertise levels evolve too. Therefore, practitioners should continually review and revise their best-practices.

CONCLUSION

As a tribute to pioneering works focusing on the impact of the Internet on the context of marketing, this article has revisited and reviewed those visionary works of arts which planted the seeds of a distinct research stream. This article presented a systematic analysis of the early predictions on how the Internet will change the business environment and its subsequent impact on marketing strategy and compared early conceptual claims with more recent empiric findings and market experience. Throughout the paper the academic knowledge on the impact of the Internet on each of the forces shaping a company's marketing strategy is explored, and related future research avenues were suggested.

Unlike late 90s, when online presence itself was considered as an innovation for many companies, today the technological basis of the Internet became more robust, and the population of technology-confident, to some degree technology-dependent consumers consist much of the economically developed societies, we are coming towards the end of the first generation of online customers (Barlow et al., 2004). Latest advances in information communication technologies allow an online experience that is engaging, interactive and memorable. In developed economies and some developing countries, a considerable portion of younger consumers spend more time in social networking sites than they do in real-life social events. Future research should acknowledge the fact that human-computer interaction is perceived differently by the new generation who grew up with computers and the Internet. All existing online consumer behavior models should be constantly reviewed and revised if necessary. It is the academicians' duty to shed light onto the pathway of the organizations in today's fierce and global economy, and find ways for them to be gentle with the world and society while taming the beast called technology.

ACKNOWLEDGMENT

I would like to express my sincere appreciation to the anonymous reviewers who commented on a previous draft.

REFERENCES

Achrol, R. S., & Kotler, P. (1999). Marketing in the network economy. *Journal of Marketing*, *63*, 146–163. doi:10.2307/1252108

Alba, J., Lynch, J., Weitz, B., Janiszewski, C., Lutz, R., Sawyer, A., & Wood, S. (1997). Interactive Home Shopping: Consumer, Retailer, and Manufacturer Incentives to Participate in Electronic Marketplaces. *Journal of Marketing*, *61*(3), 38–53. doi:10.2307/1251788

Bailey, L. (1999, November). The triumph of the individual. *New Statesmen*, 1.

Bakos, J. Y. (1997). Reducing buyer search costs: Implications for electronic marketplaces. *Management Science*, *43*, 1676–1692. doi:10.1287/mnsc.43.12.1676

Barnes, S. J., & Scornavacca, E. (2004). Mobile marketing: the role of permission and acceptance. *International Journal of Mobile Communications*, *2*(2), 128–139. doi:10.1504/IJMC.2004.004663

Brown, J. R., & Goolsbee, A. (2002). Does the Internet make markets more competitive? Evidence from the life insurance industry. *The Journal of Political Economy*, *110*, 481–507. doi:10.1086/339714

Brynjolfsson, E., & Smith, M. D. (2000). Frictionless Commerce? A comparison of Internet and conventional retailers. *Management Science*, *46*(4), 563–585. doi:10.1287/mnsc.46.4.563.12061

Burke, R. R. (2002). Technology and the customer interface: What consumers want in the physical and virtual store. *Academy of Marketing Science Journal*, *30*(4), 411–432. doi:10.1177/009207002236914

Carroll, A., Barnes, S. J., Scornavacca, E., & Fletcher, K. (2007). Consumer perceptions and attitudes towards SMS advertising: recent evidence from New Zealand. *International Journal of Advertising*, *26*(1), 79–98.

Chatterjee, P., Hoffman, D. L., & Novak, T. P. (2003). Modeling the Clickstream: Implications for Web-based Advertising Efforts. *Marketing Science*, *22*(4), 520–541. doi:10.1287/mksc.22.4.520.24906

Cheung, C. M. K., Chan, G. W. W., & Limayem, M. (2005). A critical review of online consumer behavior: Empiric research. *Journal of Electronic Commerce in Organizations*, *3*(4), 1–19.

Chiou, J., & Cheng, C. (2003). Should a Company Have Message Boards on its Web Sites? *Journal of Interactive Marketing*, *17*(3), 50–61. doi:10.1002/dir.10059

Clay, K., Krishnan, R., Wolff, E., & Fernandes, D. (2002). Retail strategies on the Web: Price and non-price competition in the online book industry. *The Journal of Industrial Economics*, *50*, 351–367.

Day, G. S., & Aaker, D. A. (1970). A guide to consumerism. *Journal of Marketing*, *34*, 12–19. doi:10.2307/1249814

Deshpande, R. (2002). Performance companies. *International Journal of Medical Marketing*, *2*(3), 225–232. doi:10.1057/palgrave.jmm.5040080

Eid, R., & Trueman, M. (2002). The Internet: New international marketing issues. *Management Research News*, *25*(12), 54–67. doi:10.1108/01409170210783278

Feather, F. (2002). *Future consumer.com*. Toronto: Warwick Publishing.

Godin, S. (1999). *Permission Marketing: Turning Strangers into Friends, and Friends into Customers*. New York: Simon and Schuster.

Gouldner, A. W. (1960). The norm of reciprocity: A preliminary statement. *American Sociological Review*, *25*(2), 161–178. doi:10.2307/2092623

Hamill, J. (1997). The Internet and international marketing. *International Marketing Review*, *14*(5), 300–323. doi:10.1108/02651339710184280

Harrison, T., Waite, K., & Hunter, G. L. (2006). The internet, information and empowerment. *European Journal of Marketing*, *40*(9/10), 972–993. doi:10.1108/03090560610680961

Harrison-Walker, L. J. (2001). E-Complaining: A content analysis of an Internet complaint forum. *Journal of Services Marketing*, *15*(5), 397–412. doi:10.1108/EUM0000000005657

Hart, C. W. (1996). Made to Order. *Marketing Management*, *5*(2), 12–22.

Hennig-Thurau, T., Gwinner, K. P., Walsh, G., & Gremler, D. D. (2004). Electronic word-of-mouth platforms: What motivates consumers to articulate themselves on the Internet. *Journal of Interactive Marketing*, *18*(1), 38–52. doi:10.1002/dir.10073

Hoffman, D. L., & Novak, T. (1996). Marketing in Hypermedia Computer-Mediated Environments: Conceptual Foundation. *Journal of Marketing*, *60*, 50–68. doi:10.2307/1251841

Hoffman, D. L., & Novak, T. P. (1997). A new marketing paradigm for electronic commerce. *The Information Society*, *13*(1), 43–54. doi:10.1080/019722497129278

Holzwarth, M., Janiszewski, C., & Neumann, M. M. (2006). The Influence of Avatars on Online Consumer Shopping Behavior. *Journal of Marketing*, *70*(4), 19–36. doi:10.1509/jmkg.70.4.19

Huffman, C., & Kahn, B. E. (1998). Variety for Sale: Mass Customization or Mass Confusion? *Journal of Retailing*, *74*, 491–513. doi:10.1016/S0022-4359(99)80105-5

Jarvenpaa, S. L., Tractinsky, N., & Vitale, M. (2000). Consumer reactions to electronic shopping on the World Wide Web. *International Journal of Electronic Commerce*, *1*(2), 59–88.

Johnson, E. J., Moe, W., Fader, P., Bellmann, S., & Lohse, J. (2004). On the depth and dynamics of online search behavior. *Management Science*, *50*, 299–308. doi:10.1287/mnsc.1040.0194

Kauffmann, R. J., & Wang, B. (2001). New buyers' arrival under dynamic pricing market microstructure: The case of group-buying discounts on the Internet. *Journal of Management Information Systems*, *18*, 157–188.

Koufaris, M. (2002). Applying the Technology Acceptance Model and Flow Theory to online consumer behavior. *Information Systems Research, 13*(2), 205–223. doi:10.1287/isre.13.2.205.83

Krishnamurthy, S. (2001). A comprehensive analysis of permission marketing. *Journal of Computer-Mediated Communication, 6*(2).

Lal, R., & Sarvary, M. (1999). When and how is the Internet likely to decrease price competition? *Marketing Science, 18*, 485–503. doi:10.1287/mksc.18.4.485

Lee, Z., & Gosain, S. (2002). A longitudinal price comparison for music cds in electronics and brick-and-mortar markets: Pricing strategies in emergent electronic commerce. *The Journal of Business Strategy, 19*, 55–71.

Lehmann, E. E. (2003). Pricing behaviour on the Web: Evidence from online travel agencies. *Empirica, 30*, 379–396. doi:10.1023/B:EMPI.0000005240.01953.e0

Levin, A. M., Levin, I. P., & Weller, J. A. (2005). A multi-Attribute Analysis of Preferences for Online and Offline Shopping: Differences across Products, Consumers, and Shopping Stages. *Journal of Electronic Commerce Research, 6*(4), 281–290.

Liang, T. P., & Huang, J. S. (1998). An empiric study of consumer acceptance of products in electronic markets: A Transactional Cost Model. *Decision Support Systems, 24*(1), 29–43. doi:10.1016/S0167-9236(98)00061-X

Lin, J. C., & Lu, H. (2000). Towards an understanding of the behavioral intentions to use a website. *International Journal of Information Management, 20*(3), 197–208. doi:10.1016/S0268-4012(00)00005-0

Mathwick, C. (2002). Understanding the Online Consumer: A Typology of Online Relational Norms and Behavior. *Journal of Interactive Marketing, 16*(1), 40–55. doi:10.1002/dir.10003

McWilliam, G. (2000). Building Stronger Brands through Online Communities. *Sloan Management Review, 41*(3), 43–54.

Melewar, T. C., & Smith, N. (2003). The Internet revolution: some global marketing implications. *Marketing Intelligence & Planning, 21*(6), 363–369. doi:10.1108/02634500310499220

Morton, S. F., Zettelmeyer, F., & Silva-Risso, J. (2001). Internet car retailing. *The Journal of Industrial Economics, 49*, 501–519.

Muniz, A., & O'Guinn, T. C. (2001). Brand Community. *The Journal of Consumer Research, 58*, 20–38.

Murphy, T. (2000). *Web rules: How the Internet is changing the way consumers make choices.* Chicago: Dearborn.

Pan, X., Ratchford, B. T., & Shankar, V. (2003). The evolution of price dispersion in Internet retail markets. *Advances in Applied Microeconomics, 12*, 85–105. doi:10.1016/S0278-0984(03)12004-4

Peppers, D., Rogers, M., & Dorf, B. (1999). Is Your Company Ready for One-to-One Marketing? *Harvard Business Review, 77*, 151–160.

Peterson, R. A., Balasubramanian, S., & Bronnenberg, B. J. (1997). Exploring the Implications of the Internet for Consumer Marketing. *Journal of the Academy of Marketing Science, 25*(4), 329–346. doi:10.1177/0092070397254005

Phau, I., & Poon, S. M. (2000). Factors influencing types of products and services purchased over the Internet. *Journal of Internet Research, 10*(2), 102–113. doi:10.1108/10662240010322894

Pieters, R., Rosbergen, E., & Wedel, M. (1999). Visual attention to repeated print advertising: A test of scanpath theory. *JMR, Journal of Marketing Research, 36*, 424–438. doi:10.2307/3151998

Pine, B. J. (1993). *Mass Customization: The New Frontier in Business Competition*. Boston: Harvard Business School Press.

Pires, G. D., Stanton, J., & Rita, P. (2006). The internet, consumer empowerment and marketing strategies. *European Journal of Marketing, 40*(9/10), 936–949. doi:10.1108/03090560610680943

Pitt, L. F., Berthon, P., Watson, R. T., & Ewing, M. (2001). Pricing strategy and the Net. *Business Horizons, 44*(2), 45–54. doi:10.1016/S0007-6813(01)80022-7

Pitt, L. F., Berthon, P., Watson, R. T., & Zinkhan, G. M. (2002). The Internet and the birth of real consumer power. *Business Horizons, 45*(4), 7–14. doi:10.1016/S0007-6813(02)00220-3

Porter, M. E. (2001, March). Strategy and the Internet. *Harvard Business Review*, 63.

Prahalad, C. K., & Ramaswamy, V. (2004). Co-creation Experiences: The Next Practice in Value Creation. *Journal of Interactive Marketing, 18*(3), 5–14. doi:10.1002/dir.20015

Reisch, L. A. (2001). The Internet and sustainable consumption: Perspectives on a Janus face. *Journal of Consumer Policy, 24*, 251–286. doi:10.1023/A:1013977509623

Rezabakhsh, B., Bornemann, D., Hansen, U., & Schrader, U. (2006). Consumer Power: A Comparison of the Old Economy and the Internet Economy. *Journal of Consumer Policy, 29*, 3–36. doi:10.1007/s10603-005-3307-7

Rha, J.-Y., & Widdows, R. (2002). The Internet and the consumer: Countervailing power revisited. *Prometheus, 20*, 107–118. doi:10.1080/08109020210137493

Samiee, S. (1998). The Internet and international marketing: is there a fit? *Journal of Interactive Marketing, 12*(4), 5–21. doi:10.1002/(SICI)1520-6653(199823)12:4<5::AID-DIR2>3.0.CO;2-5

Sawhney, M., & Kotler, P. (2001). Marketing in the age of information democracy. In Iacobucci, D. (Ed.), *Kellogg on Marketing* (pp. 386–409). New York: Wiley.

Schlosser, A. E., White, T. B., & Lloyd, S. M. (2006). Converting Web Site Visitors into Buyers: How Web Site Investment Increases Consumer Trusting Beliefs and Online Purchase Intentions. *Journal of Marketing, 70*(2), 133–148. doi:10.1509/jmkg.70.2.133

Sheth, J. N., & Sisodia, R. S. (1999). Revisiting Marketing's Lawlike Generalizations. *Journal of the Academy of Marketing Science, 27*(1), 71–87. doi:10.1177/0092070399271006

Silk, A. J., Klein, L. R., & Bernt, E. R. (2001). The Emerging Position of the Internet as an Advertising Medium. *NETNOMICS: Economic Research and Electronic Networking, 3*(2), 129–148. doi:10.1023/A:1011455329830

Spann, M., & Tellis, G. J. (2006). Does the Internet Promote Better Consumer Decisions? The Case of Name-Your-Own-Price Auctions. *Journal of Marketing, 70*(1), 65–78. doi:10.1509/jmkg.2006.70.1.65

Strader, T. J., & Shaw, M. J. (1999). Consumer cost differences for traditional and Internet markets. *Internet Research: Electronic Networking Applications and Policy, 9*, 82–92. doi:10.1108/10662249910264819

Suri, R., Long, M., & Monroe, K. B. (2003). The impact of the Internet and consumer motivation on evaluation of prices. *Journal of Business Research, 56*, 379–390. doi:10.1016/S0148-2963(01)00228-4

Tan, S. J. (1999). Strategies for reducing consumers' risk aversion in Internet shopping. *Journal of Consumer Marketing, 16*(2), 163–180. doi:10.1108/07363769910260515

Toker, A., & Varnali, K. (2009). A Framework for E-Potential Assessment in Online Markets. *Journal of Internet Business*, *6*, 81–110.

Trevino, L. K., & Webster, J. (1992). Flow in Computer-Mediated Communication. *Communication Research*, *19*(5), 539–573. doi:10.1177/009365092019005001

Trusov, M., Bucklin, E. B., & Pauwels, K. (2009). Effects of word-of-mouth versus traditional marketing: findings from an Internet social networking site. *Journal of Marketing*, *73*, 90–102. doi:10.1509/jmkg.73.5.90

Urban, L. G. (2004). The emerging era of customer advocacy. *Sloan Management Review*, *45*(2), 77–82.

Van Raaij, W. F. (1998). Interactive communication: Consumer power and initiative. *Journal of Marketing Communications*, *4*, 1–8. doi:10.1080/135272698345843

Vijayasarathy, L. R. (2002). Product characteristics and Internet shopping intentions. *Journal of Internet Research*, *12*(5), 441–426.

Wikstrom, S. (1996). The customer as co-producer. *European Journal of Marketing*, *30*(4), 6. doi:10.1108/03090569610118803

Wind, Y., & Mahajan, V. (2002). Convergence marketing. *Journal of Interactive Marketing*, *16*(2), 64–79. doi:10.1002/dir.10009

Wind, Y., & Rangaswamy, A. (2001). Customerization: The next revolution in mass customization. *Journal of Interactive Marketing*, *15*(1), 13–32. doi:10.1002/1520-6653(200124)15:1<13::AID-DIR1001>3.0.CO;2-#

This work was previously published in International Journal of E-Business Research Volume 6, Issue 4, edited by In Lee, pp. 38-51, copyright 2010 by IGI Publishing (an imprint of IGI Global).

Chapter 13
Coping with Manufacturers' Dilemma in the E-Commerce Era:
A Relational Model and a Strategic Framework

Xiaolin Li
Towson University, USA

ABSTRACT

From a channel conflict perspective, this paper proposes a Relational Model identifying key determinants of manufacturers' intentions to establish an online direct sales channel (ODSC). The paper also develops a Strategic Framework as a guideline for manufacturers to formulate their Internet channel strategies. Unlike most other studies on channel conflicts that focus on channel management after the manufacturer has already established its ODSC, this study provides a comprehensive strategic analysis of channel conflicts based on a pre-ODSC context, which prevents unnecessary losses from channel conflicts before they occur.

INTRODUCTION

In recent year, most manufacturers have adopted the Internet for business purposes, but such adoption is still limited primarily to the gathering of business information, advertising (Fisher, Craig, & Bentley, 2007) and product search (Kula & Tatoglu, 2003). Few of them have taken advantage of the Internet's disimmediation function promised for manufacturers.

What is preventing many manufacturers from selling directly online and pursuing the evidently promising online business opportunities? The explanation may be found in a dilemma that

DOI: 10.4018/978-1-61350-462-8.ch013

many of those manufacturers share: to build an online direct sales channel (ODSC) or to rely on traditional resellers (a general term in this paper referring to wholesalers, distributors, and retailers) to distribute products. Choice of either side of the dilemma seems risky for the manufacturers. If they do not sell their products directly over the Internet, people will go to those who do, and if they do sell their products directly on the Internet, then their distributors may drop them and carry products from manufacturers who do not compete with them (Wilson, 1998).

On one side of the dilemma, manufacturers may benefit from multi-channel distribution strategies in many ways (Webb, 2002): First, multi-channel distribution strategies enable manufacturers to more effectively customize products to meet various customer needs and shopping preferences and thus reach more market segments. Second, a single channel is usually optimal for specific goods but only a multi-channel strategy can maximize the effectiveness of distribution of a manufacturer's whole production line. Third, manufacturers with excess capacity can take advantage of additional distribution outlets to sell their excess goods. The Internet has made multi-channel strategies even more desirable because of its potential to reduce costs, extend market reach, and customize products and services to meet customers' needs.

On the other side of the dilemma, the introduction of a new technology into the distribution channel potentially increases the likelihood of channel conflict (Rosenberg, 1974). The emergence of the Internet and its increasing popularity as a sales channel has made channel conflicts one of the most serious challenges that manufacturers face. When a manufacturer builds its own online direct-sales channel (ODSC) to sell products directly to end customers, it immediately turns itself into its resellers' competitor. The resellers may perceive the ODSC as a potential competitive threat to their profitability and may choose to retaliate by refusing to distribute or by reducing its distribution efforts for the manufacturer's products.

While Internet and e-commerce technologies have accelerated the use of multichannel model among manufacturers (Mukhopadhyay, Yao, & Yue, 2008), recent research has suggested that the existence of the online direct sales channel along with the traditional sales channel frequently causes channel conflicts (Eyuboglu & Kahadayi, 2005; Mukhopadhyay, Yao, & Yue, 2008; Rosenbloom, 2007; Yan & Pei, 2009; Zhou, Zhuang, & Yip, 2007). Although many manufacturers have adopted the relationship marketing philosophy in recent years in hope of creating long-term relationships between its traditional channel and the online direct sales channel (Zhou, Zhuang, & Yip, 2007), there is still a possibility of channel conflicts because the manufacturer and its traditional reseller instinctively seek for their own interests (Gudonavičienė, Alijošienė, & Aukščionis, 2008). As Rosenbloom (2007) suggests, while integrating online direct sales channels with conventional sales channels to create a "seamless" experience for customer is ideal in theory, it is difficult in practice.

Many of the key issues related to channel conflicts remain crucial today (Alberto, 2005; Zhou, Zhuang, & Yip, 2007). Appropriately resolving these issues will have substantial impacts on manufacturers' distribution performance, relationships with business partners, and the satisfaction of consumers' needs (Gudonavičienė, Alijošienė, & Aukščionis, 2008).

Being aware of the importance of these issues, many researchers have conducted research on channel conflicts. However, most of the extant research focus on channel management after the manufacturer has already established its ODSC. Very little has been studied on the factors that affect manufacturers' strategies prior to their establishment of an ODSC. Unfortunately, after a channel conflict occurs, damages to the manufacturer are already done and may never be repaired. Emotions may run high for a long period of time between the manufacturer and the reseller. A long-standing strategic relationship with a distribution partner, which often constitutes an important part of the manufacturer's competitive advantage, may already be damaged beyond repair or destroyed entirely. The findings from such studies, therefore, may be of limited usefulness.

To avoid unnecessary loss of business opportunities from an ODSC or unanticipated channel conflicts, manufacturers need to identify possible reactions of current resellers, analyze the potential sources of such reactions, recognize the possible nature of the conflict process, and then formulate a strategy based upon values gained and risks faced prior to establishing an ODSC. Building upon behavioral theories including the Theory of Reasoned Actions (TRA), Expected Utility Theory (EUT), and Prospect Theory (PT), this paper proposes a Relational Model, which maps the relationships between manufacturers' intentions to establish ODSC and critical factors underlying such intentions. Based on the Relational Model, the paper also develops a Strategic Framework, which can serve as a guideline for manufacturers to formulate their Internet channel strategies.

The Objectives of the study include:

- To establish a Relational Model mapping the relationships between key determinants of manufacturers' intentions and their distribution strategy prior to their initiation of an ODSC.
- To establish a Strategic Framework based upon the Relational Model. The Strategic Model serves as a guideline for manufacturers to formulate their channel strategies.
- To formulate a series of propositions that may be useful for future research.
- To present the academic and managerial implications of the study.

The rest of the paper is structured as following: Next section provides a literature review of channel conflicts and distribution strategies. Emphasis will be placed on literature about channel conflicts with the involvement of the ODSC. Then I develop the Relational Framework, which models a synergic TRA-EUT-PT view of determinants of manufacturers' intention to establish an ODSC. After that I present the Strategic Framework. The final section concludes the paper with implications of the study and direction for future research.

LITERATURE REVIEW

A distribution channel may be defined as producers and various middlemen joined to perform activities necessary to move goods from points of production to points of consumption (Rosenberg, 1974). It is one of the most critical factors for a manufacturer's success. Possessing strong distribution channels increases barriers for entrants and thus may constitute a significant competitive advantage. The channel conflict refers to the situation in which one channel member of a distribution network perceives another channel member's behavior as a barrier to achieving its business goals. This is perhaps one of the most serious concerns for manufacturers as they add e-commerce.

The channel conflict received enormous attention during the 1970s and 1980s (Frazier, 1999). In the 1990s, the channel conflict received relatively less attention because of the emphasis of the relationship-marketing paradigm, which is a strategy emphasizing the importance of relationship building between different members of the channel. Recently, however, with the emergence and growth of e-commerce, coupled with the trend toward multi-channel distribution, the channel conflict has regained attention and become a hot topic again (Eyuboglu & Kahadayi, 2005; Falk et al., 2007; Rosenbloom, 2007; Mukhopadhyay, Yao, & Yue, 2008; Yan & Pei, 2009; Zhou, Zhuang, & Yip, 2007).

Generally speaking, the channel conflict literature has focused primarily on three areas: (1) the effect, positive or negative, of channel conflicts on involved firms, (2) factors that contribute to channel conflicts, and (3) strategies to manage channel conflicts. The first area pertains to the positive (functional) or negative (dysfunctional) effects of channel conflicts. Many researchers (e.g., Stern, 1996) see channel conflicts as dysfunctional because it is opponent-centered and may cause reduction of channel efficiency, increase of channel cost, and emotional disruption of alliance between firms. However, others argue that channel conflicts also have positive or functional effects. For instance, channel conflicts can potentially propel changes and it can be used to evaluate management performance (Anderson & Narus, 1990; Rosenberg, 1974). Overall, previous studies view channel conflicts as damaging or even devastating to the firms involved.

Another frequently examined area of channel conflicts is the factors contributing to channel conflicts. Webb (2002) summarizes three primary factors that contribute to channel conflicts: goal incompatibility, domain dissensus, and differing perceptions of reality. Others identified more specific factors including online pricing, lack of trust (Steinfield et al., 2002), and channel power (Lee et al., 2003). While virtually every study has identified some unique factors contributing to channel conflicts, very few of them have integrated the factors into a conceptual model, which can be used as a guideline for manufacturers to manage channel conflicts.

In addition, other studies focus on manufacturers' strategies to ease or manage channel conflicts. Lee et al. (2003) prescribed market differentiation, product differentiation, intermediary support, conflict avoidance, channel absorption, information sharing, and profit sharing as possible strategies for coping with channel conflicts. Bucklin et al. (1997) cited Charles Schwab's establishment of an online channel to satisfy latent consumer demand for the "do-it-yourself" customer segment, and Kendall-Jackson's selling of brands of wine online different from those carried by its retail channels, as examples of differentiation strategy. Steinfield et al. (2002) presented click-and-mortar or synergy as an effective strategy to manage channel conflicts. Most of these strategies are based upon a post-ODSC context, where the manufacturers already built their ODSCs, often without strategic analysis prior to initiating such ODSCs. While those strategies may be of help to the manufacturers in controlling damages from channel conflicts, they are of little usefulness in preventing or avoiding unnecessary and unanticipated channel conflicts.

Unlike the case with most previous studies, the models proposed in this paper are based upon a pre-ODSC context. Building upon previous studies of factors contributing to channel conflicts and strategies to manage channel conflicts, we have developed a Relational Model that identifies key factors underlying a manufacturer's intention to pursue an ODSC initiative. Based on the Relational Model we then develop a Strategic Model, which can serve as a guideline for manufacturers to formulate their channel strategies.

THEORECTICAL DEVELOPMENT AND THE RELATIONAL MODEL

The Relational Model proposed in this paper draw support from various behavioral theories, particularly the Theory of Reasoned Actions (TRA), Expected Utility Theory (EUT), and the Prospect Theory (PT). Although these theories emphasize different aspects of organizational behaviors, they have several features in common: (1) they are all theories of decision-making under risky alternatives; (2) they all assume that decision makers (individuals and organizations) are self-interested; and finally (3) they all assume that decision makers are rational, that is, they plan for their actions through reasoning processes to maximize utilities and minimize risks. As noted earlier, whether to establish an ODSC is a manufacturer's dilemma of risky alternatives; moreover, it is reasonable to assume manufacturers are self-interested and rational. Hence, these theories are all appropriate to the ODSC issue.

Theory of Reasoned Actions

The Theory of Reasoned Action (TRA) and its later derivative, Theory of Planned Behavior (TPB), are concerned with the determinants of people's (or organization's) intentions to perform particular behaviors. Such determinants include motivational factors, which indicate how much effort the decision maker would like to make to perform the behavior, and non-motivational factors, such as the availability of opportunities and resources to perform the behavior.

Ajzen was among the earliest researchers to study TRA and TPB (Ajzen, 1991). He proposed three independent antecedents of intentions to perform a given behavior: (1) attitude toward the behavior, which refers to the decision maker's favorable or unfavorable evaluation or appraisal of a given behavior; (2) subject norms, which refers to the perceived social pressure to perform or not to perform the behavior; and (3) perceived behavior control, which refers to the level of efforts (time, money, efforts, etc.) available to perform the behavior. The more favorable the attitude and the subject's norms toward the behavior and the stronger the perceived behavioral control, the stronger would be the decision maker's intention to perform the behavior in question.

Based on such theory, the antecedents of a manufacturer's intention to establish an ODSC can be grouped in three factors (Figure 1): (1) the manufacturer's attitude toward ODSC on its potential to bring gains or losses; (2) social influences or pressures from other social parties such as the manufacturer's suppliers, clients and competitors; and (3) the availability of opportunities or resources to establish an ODSC, including the manufacturer's current IT infrastructure, IT smarts, and budgetary support for an ODSC. The more favorable an attitude it has toward ODSC, the more pressure the company faces from other parties, and the more resources it has available, the more likely the manufacturer will intend to establish an ODSC.

Figure 1. Determinants of a manufacturer's intention to establish ODSC: The TRA View

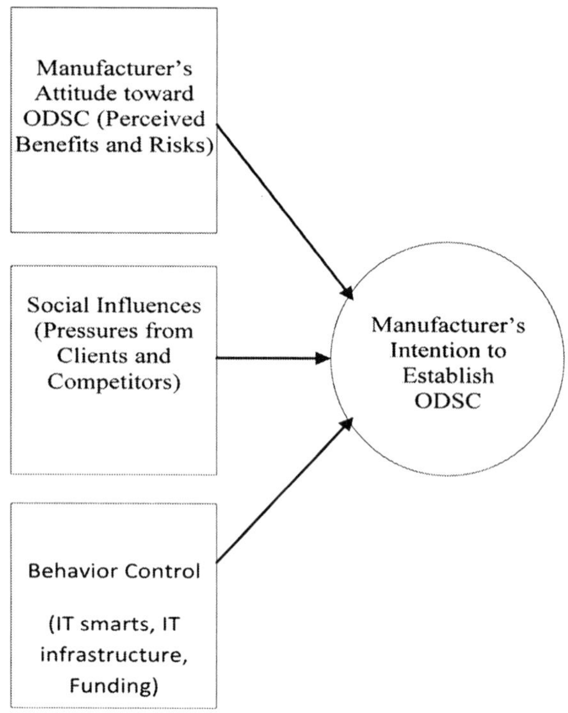

3.2 Expected Utility Theory

Expected Utility theory (EUT) can be traced back to the eighteenth century when Gabriel Cramer and Danniel Bernoulli worked separately to explain the Petersburg Paradox (Schoemaker, 1982). The EUT theory has been widely used for modeling decision-making under risk (Tamura, 2005; Schoemaker, 1982) and has played a central role in theories of measurable utility (Schoemaker, 1982). It has been frequently used prescriptively in management science (particularly decision analysis) and predictively in finance and economics (Schoemaker, 1982).

The core of the EUT concerns choices among risky alternatives. According to the EUT, if xi denotes the outcome vector and p_i denotes the probability associated with the outcome, then people will try to maximize:

$$\sum_{i=1}^{n} F(p_i)U(x_i)$$

Essentially, the EUT predicts or prescribes that people would combine outcomes and probabilities multiplicatively to make choices among risky alternatives. According to this theory, a manufacturer's intention to establish an ODSC is determined by its perceived utility gains from online sales or from maintaining conventional resellers and its perceived utility losses from missed online sale opportunities or from retaliation of resellers (Figure 2). The synthesis of the two factors, taking probabilities into account, determines the manufacturer's intention to establish an ODSC.

Prospect Theory

Prospect Theory (PT) was originally developed by Kahneman and Tversky (1979) and validated by many later studies (e.g., Chou, Chou, & Ko, 2009). The theory was developed as an extension to the EUT to solve some paradoxes unexplained by the EUT. PT has been found to be able to resolve some well-known issues associated with EUT, such as inconsistency of risky utility (Abdellaouia, Barrios, & Wakker, 2007) and descriptive deficiencies (Abellan-Perpiñan, Bleichrodt, & Pinto-Prades, 2009).

As in EUT, people make choices among risky alternatives based on gains and losses in PT. However, as opposed to EUT, in Kahneman and Tversky's Prospect Theory gains and losses are relative to a reference point (usually the status quo or one's current assets) rather than the final assets. In other words, value is treated as a function of two factors: (1) the asset position that serves as a reference point and (2) the magnitude of change (positive or negative) from the reference point (Kahneman & Tversky, 1979).

PT also replaces probabilities in EUT with a new variable named decision weights. Although Kahneman and Tversky (1979) argue that decision weights are different from probabilities, they are essentially similar, both referring to the decision maker's subjective probabilities inferred from preferences. The PT view of determinants of a manufacturer's intention to establish an ODSC (Figure 3) is similar to the EUT view with slight differences in names of terms. But with regard to the weighting function required to model precautionary decisions in prospect theory, risks tend to be exaggerated through the overweighting of

Figure 2. Determinants of a Manufacturer's Intention to Establish ODSC: The EUT View

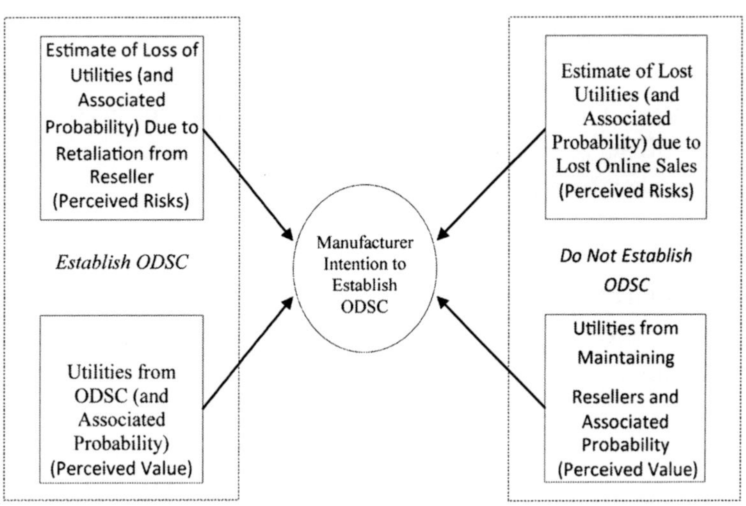

Figure 3. Determinants of a Manufacturer's Intention to Establish ODSC: The PT View

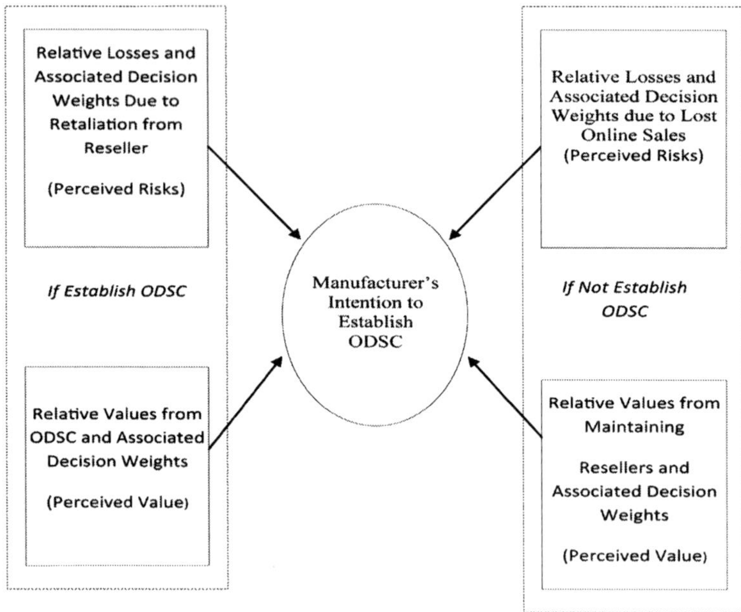

small, medium-sized, and moderately large probabilities (Kusev et al., 2009).

Although each of the three theories, TRA, EUT, and PT, has its own merits and limitations, they have something in common; that is, perceived benefits (utilities, values or gains) and perceived risks (losses) are consistently emphasized across all three theories. Future empirical studies should be cautious while selecting the theory to address particular issues related to decisions under risky alternatives.

Based on the above discussion, we propose,

- **Proposition 1:** The manufacturer's intention to establish an ODSC will be positively related to the foreseen benefits from an ODSC
- **Proposition 2:** The manufacturer's intention to establish an ODSC will be negatively related to the perceived risks from intermediary retaliation.

In addition to perceived benefits and risks, the availability of resources (Cragg & King, 1993; Lee, 2004) and social pressures such as the pressures from clients, suppliers, and competitors (Nielsen, Høst, & Mols, 2005) also contribute to the manufacturer's intention to initiate an ODSC. Therefore, we also propose:

- **Proposition 3:** The manufacturer's intention to establish an ODSC will be positively related to the availability of resources required.
- **Proposition 4:** The manufacturer's intention to establish an ODSC will be positively related to its perceived level of social influences.

As analyzed through the TRA, EUT, and PT views of the manufacturer's intention to establish an ODSC, the primary factors affecting a manufacturer's intention lie in the manufacturer's perceived benefits and risks of an ODSC, social influences, and availability of resources to establish an ODSC.

Further investigation into the determinants of these factors is useful in formulating a model that includes indirect factors useful for strategic channel analysis.

Among those factors, the availability of resources such as financial abilities and personnel competences essentially determines the manufacturer's ability to build an ODSC. Thus it directly affects the manufacturer's intention to build an ODSC. Social influences refer to the pressures a manufacturer faces from various parties to establish an ODSC. For instance, it may face pressure from a major competitor to sell directly online, or it may be compelled by some key clients to establish an ODSC for easy transactions. Sometimes, the manufacturer may intend to establish an ODSC simply because other manufacturers in the industry have done so. Social influences and the availability of resources directly affect the manufacturer's intention to initiate an ODSC; that is why they are included in the Relational Model. However, because the determinants of those two factors have little relevance to channel conflicts, analysis of such determinants is beyond the scope of this study.

Determinants of Perceived Risks

In extant research, perceived risks have frequently been found to be a critical antecedent of the behavioral intention to adopt an e-business technology. For example, Jih, Wong, and Chang (2005) found that perceived risks impacts the adoption of Internet banking services. However, little research attention has been focused on the determinants of perceived risks. The following section examines the factors that are likely to influence a manufacturer's perception of risks from a channel management perspective.

Threat to the Reseller's Profitability

As noted earlier, the manufacturer's major risk in the establishment of an ODSC is from possible retaliation of the reseller. The reseller will retaliate only if it views the manufacturer's move toward an ODSC as a serious threat to its profitability. The manufacturer should look for mechanisms to initiate a potentially profitable online direct sales channel without jeopardizing its current reselling channels. For this reason, it should analyze the relative importance of and its contribution to each existing reseller's revenues and profits. Through this analysis, the manufacturer would be better able to assess the likely reaction of its resellers and to develop appropriate channel strategies accordingly (Nair & Pleasance, 2005). Based on this discussion we propose:

- **Proposition 5:** The level of risk of an ODSC estimated by the manufacturer will be positively related to the perceived threat that the ODSC will pose to the reseller's profitability.

The perceived threat of the manufacturer's ODSC initiative to the reseller's profitability depends on two important variables: (1) the similarity of marketing mix between the planned ODSC and the reseller's sale channel and (2) the possibility of synergy between the ODSC and the reseller's sale channel. The marketing mix variables that strongly affect perceived threat to the reseller's profitability mainly include product (products sold on the ODSC), place (geographic location, or more generally, market segment served by the ODSC), and price (the pricing of products on the ODSC).

When the manufacturer's ODSC serves a different market segment or when it sells different products or sells similar products to different market segments, it will have limited impact on the reseller's profitability. Being aware of the sensitivity of products and market segment served, some manufacturers have tried to differentiate their ODSC from their resellers' sales channels in those aspects. For instance, through its ODSC Kendall-Jackson offers brands of wines that are rarely carried by its retail channels, which sell more popular brands such as Vintner's Reserve and Grand Reserve (Bucklin et al., 1997).

Price erosion on the Internet is among the biggest concerns of the reseller. Research has found that ODSC pricing is the single most critical issue over which most Internet channel conflict is generated (Webb, 2002). A general rule is that while ODSC is expected to distribute different products to different market segments from the reseller's sales channel, it is also always expected to price its products higher than the reseller's channel. Based on the analysis above, we propose:

- **Proposition 5a (1):** The perceived level of threat of the ODSC to the reseller's profitability will be positively related to the predicted similarity of products sold on the ODSC and on the reseller's sales channel.
- **Proposition 5a (2):** The perceived level of threat of the ODSC to the reseller's profitability will be positively related to the predicted similarity of market segments served by the ODSC and by the reseller's sales channel.
- **Proposition 5a (3):** The perceived level of threat of an ODSC to the reseller's profitability will be negatively related to the perceived level of the manufacturer's pricing of products on the ODSC as compared to on the reseller's sales channel.

One additional variable affecting the perceived threat to the reseller's profitability is the likelihood of synergy. The manufacturer and the reseller may pursue channel integration to improve profitability through synergy effects that may not be available to those that treat their channels independently (Anderson et al., 1997; Steinfield et al., 2002). For instance, they can divide channel roles so that different channel members focus on different functions in the value delivery channel (Bucklin et al., 1997). The benefits from synergies include lower costs, differentiation through value-added services, improved trust, geographic and market extension (Steinfield et al., 2001). The manufacturer's ODSC may also have spillover effects, generating increased purchases in the reseller's physical channel (Ward, 2001). The potential increase of overall profitability through increased quality of service and sales because of synergy effects may reduce or even eliminate the threat to the reseller's profitability. Thus, we propose,

- **Proposition 5b:** The perceived level of threat of a manufacturer's ODSC to its reseller's profitability will be negatively related to the level of potential for synergy between the ODSC and the reseller's sale channel.

The Manufacturer's Reliance on the Reseller

The manufacturer's reliance on the reseller is reflected in three aspects: (1) the percentage of volume of products distributed through the reseller, (2) the availability of alternative distribution channels; (3) the costs (time, financial and human resources) required to replace the original reseller with a new distribution channel (Lee et al., 2003).

Overall, traditional manufacturers usually rely heavily on the physical (as opposed to the Internet) distribution network to sell their products. Such a distribution network is also one of their most important competitive advantages. Giving up such a distribution channel constitutes a high risk of loss for them. However, a startup manufacturer who currently uses the physical distribution network but does not gain a strong competitive edge from such use may see it worthwhile seizing opportunities on the online sales channel. Since such companies have little to lose when the reseller retaliates, retaliation from the reseller may be of little use, therefore, such retaliation will be perceived as less likely to occur.

- **Proposition 6:** The manufacturer's perceived likelihood of reseller retaliation will be positively related to its perceived level of reliance on the reseller.

Relative Market Power of the Reseller

Channel power, which is a channel member's ability to control the decision variables of the marketing strategies of other channel members, is the most frequently cited cause of channel conflicts (Lee et al., 2003). As noted above, firms are rational and self-interested. When a reseller has superior channel power over the manufacturer it will be more likely to retaliate against a manufacturer's establishment of an ODSC in order to prevent others from following suit. As Rosenberg (1974) points out, compared with the manufacturer, the larger and the more powerful the reseller, the more likely a complaint can be magnified into a major conflict issue that affects the manufacturer's distribution system.

On the other hand, if the reseller has less market power than the manufacturer, it may be more tolerant of the manufacturer's ODSC initiative. For example, Kodak is a well-established manufacturer and it has superior channel power over its resellers. The consumer buys a Kodak camera from a physical store usually because of the brand of the product, not because of the name of the reseller. Essentially, Kodak rather than the reseller owns the customers. Therefore, retaliation from a reseller will pose very limited risk to Kodak. The ineffectiveness of such retaliation discourages the reseller from actually carrying out any retaliation. Therefore, we propose,

- **Proposition 7:** The manufacturer's perceived likelihood of reseller retaliation will be positively related to the relative channel power of the reseller.

Determinants of Perceived Benefits

Feasibility of Products to Sell on ODSC and Estimated Volumes

Although virtually every product can be sold online, not all products are equal on the Internet. The characteristics of the online sales channel (e.g., easy for information search and comparison but lack physical interactions) determine that some products with specific attributes are more suitable to be sold on the ODSC than others. The following list a few product classifications that may be linked to how effective a product can be sold on an ODSC.

- **Search Goods vs. Experience Goods:** Search goods are those marked by attributes that can be obtained prior to purchase while experience goods are those marked with attributes that cannot be known until the purchase and use of the product (Nelson, 1970). Experience goods, such as wine or automobiles, which need to be experienced (tasted or test-driven) to know whether they are good, may not be very suited to be sold on the ODSC unless a "virtual experience" can be created (Klein, 1998).
- **Durable vs. Nondurable Goods:** Whether a product is durable may also impact its online potential (Norton & Norton, 1988; Rhee, Riggins, & Kim, 2009). Nondurable goods such as grocery products that are purchased routinely tend to be more suited for a physical store while nondurable goods such as stereo systems that last for a long time and usually cost much may have a promising online market.
- **Digitized Goods vs. Nondigitized Goods:** the online channel is very convenient for transporting (downloading) digital products such as software, music, and video. These have become the most popular prod-

ucts online. According to the December 2002 monthly Online Retail Index (NRF/Forester, 2002), music, software, and videos are among the top online items while non-digitized products are relatively less popular online.

Other attributes of products that are likely to be linked to their online success (failure) include information intensity (Palmer & Griffith, 1998), the frequency and magnitude of purchase (Rhee, Riggins, & Kim, 2009), and asset specificity and complexity of the product descriptions (Malone, Yates, & Benjamin, 1987). The manufacturer must analyze the major attributes of its products and determine whether its products are promising to sell online, which in turn, should be taken into account when forecasting its sales volume on the ODSC and the potential values of initiating an ODSC. Based on the discussion above, we propose,

- **Proposition 8:** The foreseen benefits of a manufacturer's initiative in ODSC will be positively related to the perceived level of feasibility of its products to sell online.
- **Proposition 9:** The foreseen benefits of a manufacturer's initiative in ODSC will be positively related to its predicted sales volume through an ODS.

The Reseller's Effectiveness in Selling Products Online

One factor that determines the manufacturer's perceived value of an ODSC is whether its conventional resellers have effectively established a web presence for its products. If the answer is yes and the reseller is already working effectively in digging gold out of the web and/or through the physical distribution channel for the manufacturer, then the manufacturer will have no reason for selling products directly online itself. The perceived value of establishing an ODSC will also be relatively low. However, if the reseller is not effective in selling the manufacturer's products, then the manufacturer will be motivated to establish its own ODSC (Modahl, 2000). A manufacturer will take such action particularly when it is not satisfied with the reseller's efforts to sell its products but does not have sufficient negotiating power to force the reseller to build better web presence for its products.

When the reseller is not working as effectively as the manufacturer expect and the manufacturer does have a strong negotiating power, then the manufacturer may be motivated to adopt an online direct sale channel as an effective tool to motivate the retailer to improve its retail services (Yan & Pei, 2009). Based on the above analysis, we propose:

- **Proposition: 10:** The manufacturer's foreseen benefits of building its own ODSC will be negatively related to the reseller's ability to sell the manufacturer's products online.

The Relational Model

Synthesizing all the analyses above, I develop an integrative Relational Model that maps the relationships between the manufacturer's intention to establish an ODSC and key determinants underlying such intention (Figure 4).

The framework can serve as guideline for addressing some of the key issues related to channel conflicts (Bucklin, 1997): 1) are the online channel and traditional offline channel serving the same market (same products same consumers)? 2) Is the deteriorating profitability of one griping player (reseller, for instance) the result of another channel's encroachment? 3) Will the decline of one channel (either online channel or offline channel) necessarily harm the manufacturer's profits?

Figure 4. Relational Model of Manufacturer's Intentions to Establish ODSC

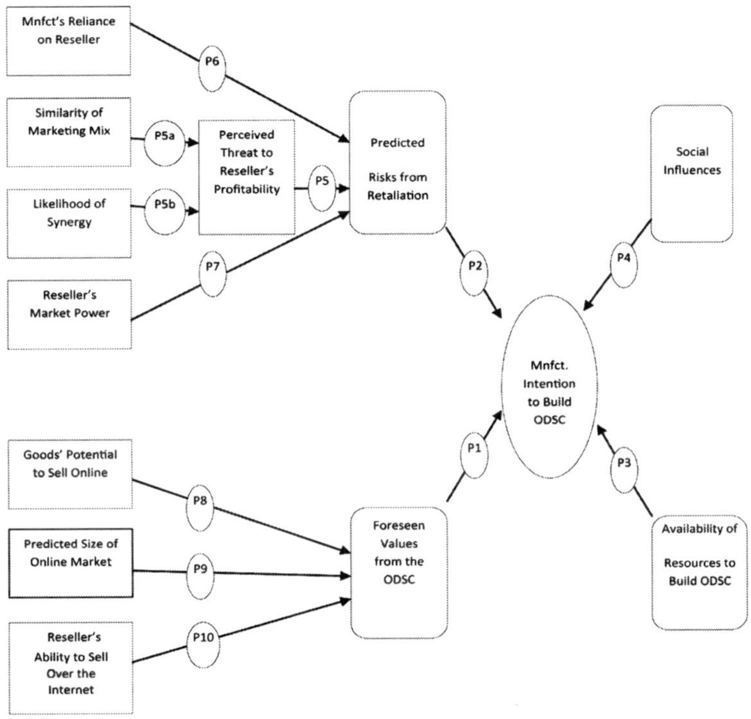

THE STRATEGIC FRAMEWORK

From the Relational Model above, we have developed a Strategic Framework (Figure 5), based on which the manufacturer can make strategic decisions on whether and under what conditions it should initiate or avoid an ODSC, taking into account the potential values of an ODSC and the losses from possible retaliation of traditional intermediaries.

Channel conflict management is crucial for the channel performance and relationship between a manufacturer and its resellers (Gudonavičienė, Alijošienė, & Aukščionis, 2008). Facing channel conflicts and the possibility of retaliation from the reseller, the manufacturer essentially has three options: (1) ignore a possible retaliation and initiate a full scale ODSC, (2) stay away from ODSC in order to save its relationship with the reseller, and (3) establish an ODSC but compromise with the reseller in the scale of the ODSC, segments and products served on the ODSC, or through mutually beneficial programs. Making the appropriate choice may define the success or failure of the manufacturer.

A channel conflict resolution involves three major steps: 1) identifying the potential or existing conflicts (in this paper we focus on potential conflicts), 2) evaluating the possible impact of the conflict, and 3) identifying and implementing a proper channel conflict solution strategy (Gudonavičienė, Alijošienė, & Aukščionis, 2008).

Figure 5. Strategic framework

		Value of ODSC: *High*	Value of ODSC: *Low*
Loss from Channel Conflicts	*High*	Build ODSC and compromise to alleviate channel conflicts	Definitely build no ODSC
	Low	Build ODSC in full possible scale	Build no or low scale ODSC

What strategy to use depends on the estimated values that the manufacturer is expected to gain from the initiation of an ODSC and the perceived risks that it will have to face from channel conflicts. Based on such values and risks, we developed four scenarios: Low value low risk, low value high risk, high value low risk, and high value high risk. Different strategies should be employed under different scenarios.

- *Low Value High Risk Scenario:* Referring to Figure 5, if the estimated value from an ODSC is low but risks resulting from the channel conflicts are high, it won't be worth initiating the ODSC. It simply does not make business sense to take a high risk of losing a reseller that the manufacturer strategically relies on to pursue another sales channel that is not expected to offer the manufacturer substantial values.
- *Low Value Low Risk Scenario:* When the estimated values from a new ODSC are limited and the risks from channel conflicts are low, then the manufacturer should take caution if it chooses to initiate an ODSC. In this case, it is not worth the firm's all-out effort for an ODSC even though the current risk from channel conflicts is low, because such risk may increase when channel power structure changes in the future.
- *High Value Low Risk Scenario:* If the estimated risk from a channel conflicts is low but estimated values from an online direct sales channel is high, then the manufacturer should establish a full-scale ODSC and at the same time try to maintain its relationship with the traditional resellers.
- *Critical Scenario*: when estimated values from the ODSC and risks from channel conflicts are both high, the manufacturer should make efforts to initiate an ODSC and capture the opportunities on the online channel. But at the same time the manufacturer must take measures to maintain a normal relationship with the traditional reseller. The task can be daunting since the likelihood of reseller retaliation is high. Some of the strategies include channel integration and channel offer differentiation (Bauer, Falk, & Hammerschmidt, 2006; Falk et al., 2007; Neslin, 2006). Essentially, the manufacturer can assure the reseller that it will sell different products or sell the same products to different customer segments, those not currently served by the reseller. To ease possible tension, the manufacturer may also seek to create the illusion of differentiation among the customers by using different names and numbers for the same items and introducing minor product modifications (Bucklin et al., 1997). In addition, the manufacturer can establish mutual-promotion or revenue-sharing programs with the reseller. For example, the manufacturer may list its reseller's physical locations on its web site or it can allow a customer to order from the manufacturer's web site and pick up the products from the reseller's physical store. These tactics are likely to help shift the reseller's attention away from the potential threat that the manufacturer's ODSC pose to its profitability and focus on how to build channel synergy and benefit from the manufacturer's new channel.

DISCUSSION

Implications for Research

This paper proposes a Relational Model of a manufacturer's intention to establish an ODSC. The manufacturer's perceived benefits from the ODSC and perceived risks resulting from channel conflicts, social influences, and availability of resources are the major factors contributing to its intention to build an ODSC. The perceived benefits of an ODSC, in turn, are determined by the potential of the manufacturer's products to sell online, the predicted size of online markets, and the intermediary's effectiveness in selling the manufacturer's products. The perceived risk of reseller retaliation is determined by the level of threat to the reseller's profitability, the manufacturer's reliance on the reseller, and the reseller's channel power relative to the manufacturers'. Based on the Relational Model, this paper also develops a Strategic Framework, which manufacturers can use as a guideline to formulate Internet channel strategies as to when and at what level they should initiate an ODSC. Both the relational model and the strategic framework proposed in this study contribute to the theoretical foundation for future research on Internet channel conflicts, particularly in the context of manufacturers.

Implications for Business Practice

The study also contributes substantially to business practice. The online direct sales channel has provided manufacturers with a relatively inexpensive way to reach its customers directly. However, prior to initiating a dual-channel model, manufacturers are faced with the task of integrating its online channel with its existing conventional channels (Rosenbloom, 2007). Otherwise, potential damages from Internet channel conflicts may offset any benefits the manufacturer gain from an ODSC and further, jeopardize its relationships with traditional resellers. The first contribution of this study to business practice is that the Relational Model presented in this paper offers a guideline on what variables and aspects manufacturers should consider prior to initiating an online direct sale channel so as to realize the benefits from the ODSC, and at the same time, minimize the negative impacts of channel conflicts.

Second, the relational framework also has significant practical implications for channel customer management. For instance, when determining whether it should initiate an online direct sales channel, the manufacturer may need to refer to the Relational Framework proposed in this study and estimate potential values of the online direct sales channel by forecasting the market size on the new channel. The manufacturer should determine whether the online direct sales channel will reach sizable new customers, or the customers are expected to come primarily from the firm's prior traditional channels (Rosenbloom, 2007). Also, the manufacturer should assess the possibility that a failure in integrating multiple-channels may result in customer dissatisfaction and loss of customers to competitors.

Third, manufacturers can also use the Strategic Framework proposed in this paper as a guideline to formulate their Internet channel strategies on whether and at what level to launch an ODSC. Essentially, the channel strategy that the manufacturer should employ depends on the intensity of conflicts, which could vary considerably from case to case. When the conflict is expected to be intense, the implementation of a channel alignment program should be planned to resolve the conflicts (Zikmund and Catalanello, 1976). The emphases of such alignment program should be on channel integration, channel offer differentiation (Bauer, Falk, & Hammerschmidt, 2006; Falk et al., 2007; Neslin, 2006), and the reinforcement of channel trust, which is crucial for the channel harmony (Balasubramanian, Konana, & Menon, 2003). For instance, the manufacturer can build a plan to maintain trust of its reseller by making a cooperative gesture and pursuing a network organization structure for repeated, enduring exchange relations with the reseller (Goethals & Vandenbulcke, 2005). It may also consider improving its E-business infrastructure to enhance its relationship coordination with its reseller (Osmonbekov, Bello, & Gilliland, 2009). In addition, the manufacturer may consider employing behavior-based coordination strategy and problem-solving approach to resolve conflicts, enhance collaborative communication and cooperation, and strengthen trust (Sahadev, 2005).

However, when channel coordination is too costly or the costs cannot justify the benefits of managing both online and the traditional offline channels, then a channel alignment program may not work and channel integration thus won't be a feasible strategy (Falk et al., 2007). In that case, manufacturers may have to choose to use a "status quo preserving strategy" and give up the ODSC initiative and focus totally on the traditional offline sales channel (Falk et al., 2007). Or they have to employ a "status quo substitution strategy" and switch completely from the traditional offline channel to a new online direct sales channel (Deleersnyder et al., 2002; Pauwels & Neslin, 2006). "Status quo substitution strategy" is an offensive marketing strategy. The manufacturer must use it with caution. As Neslin et al. (2006) suggest the "status quo substitution strategy" may complicate channel customer management; that is, customers may feel frustrated when being forced to switch to the online direct sales channel and thus may switch to a different vendor or brand. If that constitutes a serious concern, "status quo preservation strategy" may be a preferred strategy.

Finally, if a manufacturer has multiple existing resellers prior to initiating an ODSC, it may use the Relational Model proposed in this paper to assess the estimated values of the ODSC and risks linked to the potential channel conflicts, and then use the strategic framework to evaluate and prioritize different resellers.

Future Research

The study has raised some opportunities for future research. First, the relational model focuses only on the theoretical modeling of the effect of a series of crucial factors on the manufacturer's channel decisions. Future research is need to empirically test parts or the entire model. Second, the Relational Model is formulated strictly from the perspective of a manufacturer. Future studies can also be conducted to explore the ODSC issue from different perspectives, for instance, from the perspective of a reseller (wholesaler, distributor, or retailer). For example, researchers can investigate a reseller's reactions to a manufacturer's decision to establish an ODSC and the factors underlying such reactions. Just as this study contributes to the literature by providing a guideline to manufacturers in their formulation of internet channel strategies, other studies will be useful in providing models guiding resellers in their formulation of Internet channel strategies.

REFERENCES

Abdellaouia, M., Barrios, C., & Wakker, P. P. (2007). Reconciling introspective utility with revealed preference: Experimental arguments based on prospect theory. *Journal of Econometrics*, *138*(1), 356–378. doi:10.1016/j.jeconom.2006.05.025

Abellan-Perpiñan, J. M., Bleichrodt, H., & Pinto-Prades, J. L. (2009). *Journal of Health Economics*, *28*(6), 1039–1047. doi:10.1016/j.jhealeco.2009.09.002

Ajzen, I. (1991). The theory of planned behavior. *Organizational Behavior and Human Decision Processes*, *50*, 179–221. doi:10.1016/0749-5978(91)90020-T

Alberto, S. V. (2005). How potential conflict drives channel structure: Concurrent (direct and indirect) channels. *JMR, Journal of Marketing Research*, *42*(4), 507–515. doi:10.1509/jmkr.2005.42.4.507

Anderson, E., Day, G. S., & Rangan, V. K. (1997). Strategic channel design. *Sloan Management Review*, *38*(4), 59–69.

Anderson, J. C., & Narus, J. A. (1990). Model of distributor firm and manufacturer firm working partnerships. *Journal of Marketing*, *54*, 42–58. doi:10.2307/1252172

Balasubramanian, S., Prabhudev, K., & Nirup, M. M. (2003). Customer satisfaction in virtual environments: A study of online investing. *Management Science, 49*(7), 871–889. doi:10.1287/mnsc.49.7.871.16385

Bauer, H. H., Falk, T., & Hammerschmidt, M. (2006). eTransQual: A transaction process-based approach for capturing service quality in online shopping. *Journal of Business Research, 59*(7), 866–875. doi:10.1016/j.jbusres.2006.01.021

Bucklin, C. B., Thomas-Graham, P. A., & Webster, E. A. (1997). Channel conflict: When is it dangerous? *The McKinsey Quarterly, 3*, 36–43.

Chou, P., Chou, R., & Ko, K. (2009). Prospect theory and the risk-return paradox: some recent evidence. *Review of Quantitative Finance and Accounting, 33*(3), 193–208. doi:10.1007/s11156-009-0109-z

Cragg, P. B., & King, M. (1993). Small-firm computing: motivators and inhibitors. *Management Information Systems Quarterly, 17*(1), 47–60. doi:10.2307/249509

Deleersnyder, B., Geyskens, I., Gielens, K., & Dekimpe, M. G. (2002). How cannibalistic is the Internet channel? *International Journal of Research in Marketing, 19*(4), 337–348. doi:10.1016/S0167-8116(02)00099-X

Eyuboglu, N., & Kahadayi, S. (2005). Dealer-manufacturer alienation in a multiple channel system the moderating effect of structural variables. *Journal of Marketing Channels, 12*(3), 5–27. doi:10.1300/J049v12n03_02

Falk, T., Schepers, J., Hammerschmidt, M., & Bauer, H. H. (2007). Identifying cross-channel dissynergies for multichannel service providers. *Journal of Service Research, 10*(2), 143–160. doi:10.1177/1094670507306683

Fisher, J., Craig, A., & Bentley, J. (2007). Moving from a web presence to e-Commerce: The importance of a business--web strategy for small-business owners. *Electronic Markets, 17*(4), 253–262. doi:10.1080/10196780701635864

Frazier, G. (1999). Organizing and managing channels of distribution. *Journal of the Academy of Marketing Science, 27*(2), 226–240. doi:10.1177/0092070399272007

Goethals, F., & Vandenbulcke, J. (2005). Two basic types of business-to-business integration. *International Journal of E-Business Research, 1*(1), 1–15.

Gudonavičienė, R., Alijošienė, S., & Aukščionis, L. (2008). The peculiarities of conflicts management in marketing channels. *Economics & Management*, 261-271.

Jih, W., Wong, S., & Chang, T. (2005). Effects of Perceived Risks on Adoption of Internet Banking Services: An Empirical Investigation in Taiwan. *International Journal of E-Business Research, 1*(1), 70–88.

Kahneman, D., & Tversky, A. (1979). Prospect theory: An analysis of decision under risk. *Econometrica, 47*(2), 263–292. doi:10.2307/1914185

Klein, L. R. (1998). Evaluating the potential of interactive media through a new lens; search versus experience goods. *Journal of Business Research, 41*(3), 195–203.doi:10.2307/1914185

Kula, V., & Tatoglu, E. (2003). An exploratory study of Internet adoption by SMEs in an emerging market economy. *European Business Review, 15*(5), 324–333. doi:10.1108/09555340310493045

Kusev, P., van Schaik, P., Ayton, P., Dent, J., & Chater, N. (2009). Exaggerated risk: Prospect theory and probability weighting in risky choice. *Journal of Experimental Psychology. Learning, Memory, and Cognition, 35*(6), 1487–1505. doi:10.1037/a0017039

Lee, J. (2004). Discriminant analysis of technology adoption behavior: A case of Internet technologies in small businesses. *Journal of Computer Information Systems, 44*(4), 57–66.

Lee, Y., Lee, Z., & Larsen, K. R. T. (2003). Coping with Internet channel conflict. *Communications of the ACM, 46*(7), 137–142. doi:10.1145/792704.792712

Malone, T. W., Yates, J., & Benjamin, R. I. (1987). Electronic market and electronic hierarchies. *Communications of the ACM, 30*(6), 484–497. doi:10.1145/214762.214766

Modahl, M. (2000). *Now or never: How companies must change Today to Win The Battle for Internet Consumers* (pp. 169–174). New York: HarperCollins.

Mukhopadhyay, S. K., Yao, D., & Yue, X. (2008). Information sharing of value-adding retailer in a mixed channel hi-tech supply chain. *Journal of Business Research, 61*(9), 950–958. doi:10.1016/j.jbusres.2006.10.027

Nair, G., & Pleasance, D. (2005). Mitigating channel conflict. *The McKinsey Quarterly, 3*, 16–17.

Nelson, P. J. (1970). Information and consumer behavior. *The Journal of Political Economy, 78*(2), 311–329. doi:10.1086/259630

Neslin, S. A., Grewal, D., Leghorn, R., Shankar, V., Teerling, M. L., Thomas, J. S., & Verhoef, P. C. (2006). Opportunities and challenges in multichannel customer management. *Journal of Service Research, 9*(2), 95–112. doi:10.1177/1094670506293559

Nielsen, J. F., Høst, V., & Mols, N. P. (2005). Adoption of Internet-based marketing channels by small and medium-sized manufacturers. *International Journal of E-Business Research, 1*(2), 1–23.

Norton, S. W., & Norton, W. Jr. (1988). An Economic perspective on the information content of magazine advertisements. *International Journal of Advertising, 7*(2), 138–148.

NRF/Forrester Research. (2002). *December 2002 NRF/Forrester Online Retail Index*.

Osmonbekov, T., Bello, D. C., & Gilliland, D. I. (2009). The impact of e-business infusion on channel coordination, conflict and reseller performance. *Industrial Marketing Management, 38*(7), 778–784. doi:10.1016/j.indmarman.2008.03.005

Palmer, J. W., & Griffith, D. A. (1998). An emerging role of web site design for marketing. *Communications of the ACM, 41*(3). doi:10.1145/272287.272296

Pauwels, K., & Neslin, S. A. (2006). *Building with bricks and mortar: The revenue impact of opening physical stores in a multichannel environment*. Hanover, NH: Dartmouth College.

Rhee, H., Riggins, F. J., & Kim, C. (2009). The impact of product type and perceived characteristics of the web on multifaceted online shopping behavior. *Journal of Organizational Computing and Electronic Commerce, 19*, 1–29. doi:10.1080/10919390802198907

Rosenberg, L. J. (1974). A new approach to distribution conflict management. *Business Horizons, 17*(5), 67–74. doi:10.1016/0007-6813(74)90042-1

Rosenbloom, B. (2007). Multi-channel strategy in business-to-business markets: Prospects and problems. *Industrial Marketing Management, 36*(1), 4–9. doi:10.1016/j.indmarman.2006.06.010

Sahadev, S. (2005). Exploring the role of expert power in channel management: An empirical study. *Industrial Marketing Management, 34*(5), 487–494. doi:10.1016/j.indmarman.2004.11.002

Schoemaker, P. J. H. (1982). The expected utility model: Its variants, purposes, evidence and limitations. *Journal of Economic Literature*, *20*, 529–563.

Steinfield, C. (2001). Pillars of virtual commerce: Leveraging physical and virtual presence in the new economy. *Info*, *3*(3), 203–213. doi:10.1108/14636690110801923

Steinfield, C., Bouwman, H., & Adelaar, T. (2002). The dynamics of click-and-mortar electronic commerce: Opportunities and management strategies. *International Journal of Electronic Commerce*, *7*(1), 93–119.

Tamura, H. (2005). Behavioral models for complex decision analysis. *European Journal of Operational Research*, *166*, 655–665. doi:10.1016/j.ejor.2004.03.038

Ward, M. R. (2001). Will online shopping compete more with traditional retailing or catalog shopping? *NETNOMICS: Economic Research and Electronic Networking*, *3*(2), 103–117. doi:10.1023/A:1011451228921

Webb, K. L. (2002). Managing channels of distribution in the age of electronic Commerce. *Industrial Marketing Management*, *31*, 95–102. doi:10.1016/S0019-8501(01)00181-X

Wilson, R. (1998, May 15). Manufacturer's dilemma: To sell or not to sell directly.

Yan, R., & Pei, Z. (2009). Retail services and firm profit in a dual-channel market. *Journal of Retailing and Consumer Services*, *16*(4), 306–314. doi:10.1016/j.jretconser.2009.02.006

Zhou, N., Zhuang, G., & Yip, L. S. (2007). Perceptual difference of dependence and its impact on conflict in marketing channels in China: An empirical study with two-sided data. *Industrial Marketing Management*, *36*(3), 309–321. doi:10.1016/j.indmarman.2005.10.001

Zikmund, W. G., & Catalanello, R. F. (1976). Managing channel conflict through channel development. *Journal of the Academy of Marketing Science*, *4*(4), 801–813. doi:10.1007/BF02729839

This work was previously published in International Journal of E-Business Research Volume 6, Issue 4, edited by In Lee, pp. 52-68, copyright 2010 by IGI Publishing (an imprint of IGI Global).

Chapter 14
Governance Mechanisms in Internet-Based Affiliate Marketing Programs in Spain

Paul B. Fox
ESADE - Ramon Llull University, Spain

Jonathan D. Wareham
ESADE - Ramon Llull University, Spain

ABSTRACT

Internet-based affiliate marketing programs have emerged as one of the fastest-growing methods for online retailers to acquire customers and increase sales by tapping into the power of independent web sites to reach a large, diverse audience of potential customers. However, while these programs have proven effective in increasing website traffic and sales, illegal or inappropriate activities on the part of affiliates could negatively impact a retailer's brand in the eyes of customers. This study is an exploratory analysis of governance mechanisms (formal contracts, partner selection, incentives and monitoring) in one-to-many affiliate programs in Spain. Agency theory and transaction cost analysis provide the theoretical background. The conclusion is that there is a significant lack of transparency in the guidance and restrictions communicated to affiliates, and a lack of systematic monitoring of affiliate behavior, which increases the risk of opportunism or misconduct. General recommendations for managers of affiliate programs are considered.

INTRODUCTION

Revenue-sharing affiliate marketing is potentially the most cost-effective method for acquiring new customers on the Internet. Also called pay for performance marketing, an affiliate marketing program consists of an on-line retailer (merchant) who places a link on a third-party website (affiliate). If a visitor to the affiliate site clicks on the link and performs a specified action (e.g. visits the merchant's website, fills out a form, or purchases a product), the affiliate receives a commission. The arrangement has been described as similar to

DOI: 10.4018/978-1-61350-462-8.ch014

having a large, independent sales force working solely for commission and absorbing the total risk associated with marketing a retailer's products (Duffy, 2005).

While Amazon is generally credited with creating the first major affiliate program on the Internet (launched in 1996), Hoffman and Novak first focused the attention of the academic community on this strategy in 2000, concluding that of the various forms of advertising used by online retailer CDnow, their affiliate program was by far the most cost effective (after word-of-mouth, to which they attributed a cost of zero), since it allowed the retailer to "draw a direct line from advertisement to sale" (Hoffman & Novak, 2000, p. 188).

Hoffman and Novak are not the only researchers who have taken an interest in affiliate marketing. Papatla and Bhatnagar (2002) propose a guide to choosing affiliate partners, while Libai Biyalogorsky, and Gerstner (2003) perform a theoretical analysis of affiliate referral fee structures in order to determine the optimal program for different types of affiliates. Other research includes case studies of merchants (e.g. Walthieu, 2000) as well as affiliates (e.g. Moon, 2000).

Despite these initial efforts to bring the topic of affiliate marketing to the fore, recent empirical studies have been few and far between. This might lead one to the conclusion that affiliate marketing was merely a fad which was popular for a time, but has since faded into obscurity. This is not the case. Duffy (2005) recently observed that "affiliate marketing is likely to become the principal mainstream marketing strategy for e-commerce businesses in the future" (p. 161). It is difficult, however, to find exact estimates of market size, as affiliate marketing has yet to be clearly defined, and the entrance of dozens of intermediaries further complicates the situation (Molander, 2005). While ValueClick (2006) estimates the global affiliate marketing sector to be in the range of \$400 - \$500 million, MarketingSherpa places the figure at \$6.5 billion (MarketingSherpa, 2006a). Today, 9-40% of a typical online retailer's sales come from affiliates (MarketingSherpa, 2006b).

In light of the growing practical importance of affiliate marketing, there are critical issues which have not received sufficient attention by researchers thus far. An area of particular interest, and the topic of this paper, is the governance mechanisms used in affiliate programs to control the promotional activities of affiliates. One of the principal advantages of affiliate marketing is its ability to accurately track the behavior of visitors in terms of website visits, lead generation and sales. However, there is a significant risk that while promoting merchants' products and services, affiliates may engage in activities that are fraudulent, unethical, or somehow destructive to the brand value of the merchant. As affiliates are often the first point of contact with a potential customer, negatively-perceived activities on their part could have a disastrous effect on the retailer's business.

The need for further investigation in the area of governance of affiliate marketing programs is corroborated by a recent survey from AffStat (2006), in which nearly 200 affiliate managers were asked their biggest challenge in affiliate marketing. About half indicated some form of governance issue (detecting fraud, properly managing the affiliates, monitoring affiliates for brand risks and monitoring affiliates' use of trademarks in search engines).

Transaction cost analysis (Coase, 1937; Williamson, 1975, 1985, 1991) and agency theory (Eisenhardt, 1989; Jensen & Meckling, 1976; Ross, 1973) provide the theoretical background for the constructs used in this study. Opportunism has appeared in the literature in many different forms (see Wathne & Heide, 2000 for a detailed treatment of the construct), but can be generally described as "some form of cheating or undersupply relative to an implicit or explicit contract" (Wathne & Heide, 2000, p. 48). In the TCA and Agency theory literature, the primary relationship considered is that between a principal and an agent, which in affiliate marketing would be the merchant and its affiliate. However, in the context of affiliate marketing (and in marketing relationships in general), the most important relationship is that between the merchant and the end customer. Therefore, opportunism in this case would include any action on the part of the affiliate which could damage the reputation of the merchant in the eyes of its customers (or potential customers).

The means of limiting channel member opportunism in transaction cost analysis is through the use of governance, and several *mechanisms* are available for managing partner opportunism (Heide, 1994). The authors have identified the following as the most relevant in the affiliate marketing channel: the formal contract, partner selection, incentives, and monitoring. All four governance mechanisms are reviewed in greater detail in the Research Background section.

This paper explores the following research question:

How are governance mechanisms used by merchants in one-to-many, Internet-based affiliate marketing programs in Spain to control the activities of their affiliates?

This is accomplished by combining theoretical analysis with empirical evidence obtained in two phases: first, data was collected from a total of 136 affiliate programs managed by three of the largest affiliate networks in Spain at the time the study was conducted; next, semi-structured interviews were conducted with 11 professionals representing the various roles in the affiliate marketing channel: merchants, advertising agencies, and affiliate networks.

The remainder of this paper is structured as follows. In the following section, background theory is reviewed on types of affiliate programs, the affiliate marketing channel structure, commission pricing, and transaction cost analysis and agency theory applied to the affiliate marketing context. We continue with Data Collection and Analysis, where the research method is described and a summary of the data is provided. The Results section details the findings for each of the four governance mechanisms. Finally, the Discussion section summarizes the findings and describes implications for the management of affiliate programs and is followed by Conclusions.

RESEARCH BACKGROUND

Affiliate Marketing

Amazon.com is widely credited with creating the first affiliate program on the Internet in 1996 (Dysart, 2002; Libai *et al.*, 2003). Since then, Amazon has gone on to develop one of the largest and most successful affiliate programs in the world. According to their website, they currently have over 1,000,000 members worldwide. Many others have followed Amazon's example, making affiliate marketing an important source of customer acquisition (Hoffman & Novak, 2000; Libai *et al.*, 2003).

The benefits from these programs in terms of increased visitor traffic and sales can be significant. However, there are additional risks for retailers when they entrust their marketing efforts to affiliates. According to a recent survey (see Table 1) of affiliate managers (merchants) by AffStat (2006), when asked their biggest challenge in affiliate marketing, a high percentage of the responses were related to governance issues[1]:

The following sections provide further background on the types of affiliate programs in use, the affiliate marketing channel structure and affiliate commission pricing, as well as a discussion of agency theory and transaction cost analysis as applied to governance mechanisms in the affiliate marketing context.

Types of Affiliate Programs

Libai *et al.* (2003) describe two types of affiliate programs, one-to-one and one-to-many. In a one-to-one program, the merchant and affiliate negotiate a unique contract which specifies the

Table 1. Challenges in affiliate marketing

Risk	Relationship Structure	
	Pay per Sale	Pay per Lead
Detecting Fraud	14%	24%
Properly managing the affiliates	18%	22%
Monitoring affiliates for brand risks	6%	4%
Monitoring affiliate use of trademarks in search engines	8%	2%
Total	46%	52%

terms and conditions of the arrangement. In this case, the affiliate site has significant negotiating power with the merchant. The fee arrangement is often long-term and often involves up-front payment of all or a portion of the commissions.

In contrast, Amazon.com has an "open" program, where affiliates link to the Amazon.com site and earn up to an 8.5% referral fee when visitors who click on the links make a purchase. This is an example of a one-to-many program, in which the merchant (Amazon in this case) makes the program available to numerous affiliates and establishes the terms of the agreement including pricing, advertising formats available, and acceptable practices. Affiliates simply decide whether or not to apply to the program. Once the affiliates sign up for the program, the merchant has considerable power in the relationship, and may change the terms of the agreement or cancel the agreement at any time.

One-to-one programs by their very nature are easier for merchants to control: there are fewer affiliates, and the relationship is more direct. However, one-to-many programs are much more difficult, as merchants must keep track of the performance of potentially thousands of affiliates, and often use one or more intermediaries to do so. Due to the acute nature of the governance issues in one-to many programs as compared to one-to-one, the remainder of this research will focus on one-to-many affiliate programs.

Affiliate Marketing Channel Structure

The affiliate marketing channel structure is summarized in Figure 1.

Figure 1. Affiliate marketing value chain

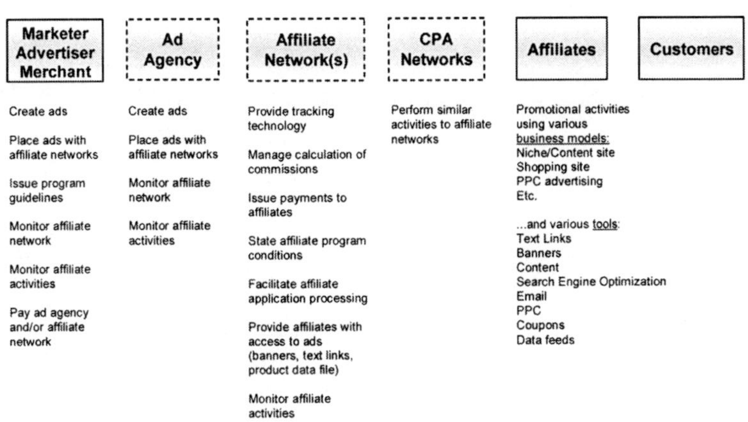

A company can either manage their affiliate program themselves, or use one of the many affiliate marketing networks who perform this function. Proponents of in-house programs claim that they are only cost-effective once they reach a certain size (Ray, 2001, p. 30). Therefore, the vast majority of merchants use one or more affiliate networks to manage their affiliate programs.

The focus of this research is on affiliate programs which use one or more affiliate networks as intermediaries. The proposition is that this indirect relationship between the merchant and the affiliates increases the importance of governance mechanisms. As the affiliate marketing channel gets longer (with additional intermediaries between the advertiser and the consumer) and wider (with additional affiliates added to the channel), governance of the activities of affiliates logically becomes more difficult.

Affiliate Business Models

Affiliates engage in various types of business activities, where affiliate marketing can represent anything from a small source of extra income to the core activity of the business. In a recent survey by PartnerCentric (2006) of 1,041 affiliates, the majority of respondents identified themselves as niche/content sites (44.5%), coupon/discount shopping sites (26.9%) and PPC advertisers (17.5%). The remaining 11.1% included sweepstakes/contest sites, shopping malls, incentive/loyalty sites, personal websites and blog/ezine.

The business models used by affiliates have governance implications, particularly in the areas of partner selection and monitoring. For example, it may be more difficult to monitor an affiliate who promotes merchants using search engine marketing, but does not have a web page.

Affiliates also vary significantly in the size of their operations and contribution to an advertiser's program, but the general rule is that the vast majority of visitors and sales come from a small number of affiliates. Some have claimed that 20% of affiliates account for 80% of transactions (Fox, 2000), while others believe that as much as 95% of sales come from 5% of the affiliates (Ray, 2001).

Affiliate Promotion Tools

In terms of the tools used to promote affiliate programs, the leading responses in the Partnercentric (2006) survey (see Table 2) include:

Much of the governance challenge in affiliate marketing is due to the degree to which affiliates can customize the merchant's message. Therefore, those tools where the affiliate has a significant degree of autonomy in deciding how to represent a merchant tend to add to the monitoring challenge.

Email is one tool which is popular for affiliates since it allows them to send targeted advertis-

Table 2. Promotional tools for affiliate programs

Promotion Tool	Percentage
Text links	19.88%
Banners	18.80%
Content	18.40%
Search engine optimization	11.51%
Email	9.42%
PPC	7.51%
Coupons	6.83%
Data feeds	4.09%

ing to customers at a low cost. However, from a customer perspective, the problem of spam looms large. Consumer concerns regarding spam include "privacy, false email identities, questionable email content, enticement and fraud", and in 2002 the European Parliament dictated that users must opt-in (give prior consent) to receive UCE (unsolicited commercial email) (Sipior, Ward, & Bonner, 2004).

One common strategy used by merchants is to forbid the use of their trademarks in the search engine marketing activities of affiliates. Another approach is to prevent affiliates from using the merchant trademark in the text of advertisements written and placed in search engines. The efficacy of these measures is far from clear, however. While there is a risk of cannibalizing sales from the merchants' other marketing efforts, many have suggested allowing affiliates to bid on trademarks improves reach, effectively "filling the channel" by showing multiple ads to search engine users, which ultimately should result in more sales for the advertiser (Internet Retailer, 2004; Stein, 2004).

Commission Pricing

The Internet as a marketing channel has the unique ability to not only track the amount of advertising delivered, but the amount consumed as well (Hoffman & Novak, 2000). Through the use of a cookie placed on a visitor's computer when they click on an ad, it is possible to track which site the visitor came from (including any affiliate programs associated with the ad) as well as their behavior on the merchant site, up to and including events such as a registration or a purchase, even if that event happens 30-45 days (or in some cases a year, depending on the configuration of the cookie) later. This ability to track visitor behavior has given rise to additional pricing models which emphasize pay-per-performance (PPP), commonly known as cost-per-action or cost-per-acquisition (CPA). The choice of pricing model has important implications for governance. See Table 3 below for a brief description.

Governance Mechanisms: Agency Theory and Transaction Cost Analysis Theory

Agency theory (Eisenhardt, 1989; Jensen & Meckling, 1976; Ross, 1973) addresses the agency relationship in which a principal delegates work to an agent who then performs the work. The two parties are engaged in cooperative behavior, but have different goals and different attitudes towards risk. Agency theory describes this relationship using the metaphor of a contract, and attempts to resolve two problems: a) the desires or goals of the principal and agent conflict; and b) it is difficult or expensive for the principal to verify the agent's activities (Eisenhardt, 1989). The problems in developing effective contracts are a result of certain human assumptions (self-

Table 3. Affiliate commission pricing models

Pricing Model	Description
CPC (Cost-per-Click)	A commission is paid for each unique visitor referred to the merchant site.
CPS (Cost-per-Sale)	A commission is only paid when a visitor makes a purchase.
CPL (Cost-per-Lead)	A commission is only paid when a visitor fills out a form.
Hybrid	A commission is paid for each unique visitor, and an additional commission is paid per sale or lead

interest, bounded rationality and risk aversion) and information asymmetry between the two parties, which give rise to the risks of moral hazard and adverse selection.

Information asymmetry in the case of affiliate marketing occurs due to the fact that one-to-many affiliate marketing programs have potentially thousands of affiliates who are given a significant degree of autonomy as to how they represent the merchants' products and services. Many of these activities are not easily observed due to the highly distributed nature of the Internet. This information asymmetry is exacerbated by the presence of affiliate networks acting as intermediaries, which distances the merchant even further from the activities of its affiliates.

The primary limitation of agency theory is that it focuses exclusively on the formal contract as the unit of analysis. Transaction cost analysis may be a useful tool to extend the analysis to a broader understanding of governance in the merchant-affiliate relationship. Transaction cost analysis (TCA), (Coase, 1937; Williamson, 1975, 1985), has been used extensively in marketing to analyze interfirm relationships (e.g. Anderson & Weitz, 1992; Brown, Chekitan, & Lee, 2000; Dahlstrom McNeilly, & Speh, 1996; Rindfleisch & Heide, 1997). Brown *et al.* (2000) claim that the normative goal of the theory is to minimize the costs of exchange, and that one way to accomplish that objective is to minimize the impact of opportunism on channel members.

The means of limiting channel member opportunism in transaction cost analysis is through the use of governance, which may be broadly defined as "a multidimensional phenomenon, encompassing the initiation, termination and ongoing relationship maintenance between a set of parties" (Heide, 1994, p. 72).

Several *mechanisms* are available for managing partner opportunism. The model provided by Wathne and Heide (2000, p. 44) is used as the basis for the governance mechanisms employed in this research, along with the formal contract as used in agency theory. The governance mechanisms identified as relevant to affiliate marketing include: the formal contract, partner selection, incentives and monitoring.

Governance Mechanisms: Formal Contract

Agency theory describes two basic forms of contracts, outcome-based and behavior-based. The assertion is that outcome-based contracts "coalign the preferences of agents with those of the principal because the rewards for both depend on the same actions and, therefore, the conflicts of self-interest between principal and agent are reduced" (Eisenhardt, 1989, p. 60). Behavior-based contracts, on the other hand, work best in situations where information systems are capable of measuring the agent's behavior, where outcomes are not easily measurable, where agents are risk averse, where the tasks to be performed by the agent are easily programmed in advance, and where the relationship is long-term (Eisenhardt, 1989).

It follows, therefore, that in one-to-many affiliate marketing programs the contracts would tend to be outcome-based. In fact, this is practically the definition of pay-per-action commission pricing. However, it may also be possible to have a hybrid policy towards control, where a mix of outcome-based and behavior-based strategies are used (Oliver & Anderson, 1995). This hybrid model appears to be what is required in affiliate marketing. While the incentives may be outcome-based, many of the conditions stipulated in the contracts between the affiliate network and affiliate address the behavior that the affiliate may or may not engage in while promoting the merchant's products and services. In other words, the contract stipulates not just the outcomes expected from the affiliate, but the expected behavior as well. This will have important implications for monitoring.

Governance Mechanisms: Partner Selection

One method for addressing the problem of opportunism, specifically regarding adverse selection, or "the risk of being locked-in with a supplier who lacks the needed skills" is to "select exchange partners a priori that are not opportunistically inclined or are inherently cooperative with respect to a particular task" (Stump & Heide, 1996, p. 432).

Papatla and Bhatnagar (2002) addressed partner selection in affiliate marketing by proposing guidelines for how to choose affiliate partners, concluding that affiliate partnerships should be established between businesses with related products including: substitutes, strict complements, episodic substitutes and episodic compliments.

In this study, we analyze this governance mechanism through a combination of the criteria for selection set forth in the formal contracts, as well as interviews with professionals involved in affiliate marketing.

Governance Mechanisms: Incentives

The primary benefit of well-structured incentives is a situation where the long-term gains from cooperative behavior exceed the short-term payoffs from opportunism (Wathne & Heide, 2000). An effective incentive structure has also been called a self-enforcing agreement, which "remains in force as long as each party believes himself to be better off by continuing the agreement than he would be by ending it" (Telser, 1980, p. 27).

The main incentives in affiliate marketing programs are outcome-based commissions based on the number of visitors sent to a merchant site, and the conversion of these visits into sales or registrations (see Table 3 for a list of the pricing models in use). The type of incentive used has important implications for partner selection, for monitoring, and for evaluating the success of the program.

In this study, incentives are explored through a categorization of the base commissions included in the affiliate contracts, combined with qualitative data from interviews with professionals involved in affiliate marketing.

Governance Mechanisms: Monitoring

Monitoring may be defined as "an effort made by one party to measure or "meter" the performance of another" (Heide, Wathne, & Rokkan, 2007, p. 426). Monitoring serves to reduce information asymmetry between the two parties, and thereby to reduce opportunism. Agency theory uses the term *information systems* for the same activity. In the case of affiliate marketing, affiliate networks gather real-time statistics on visitors, registrations and sales in the various affiliate marketing programs. In other words, the outcomes of affiliate activities are tracked regularly. However, it is not readily apparent which strategies are used to monitor the affiliate activities in order to ensure they comply with the behavioral restrictions included in the contract.

In this study, monitoring is analyzed through data gathered from the affiliate program guidelines (formal contracts), combined with interviews with professionals associated with affiliate marketing.

DATA COLLECTION AND ANALYSIS

Spain was chosen as the location of the study for two main reasons: 1) it is small enough for the sample to encompass a substantial portion of the affiliate programs in the country; and 2) It contains a much higher proportion of programs with cost per click pricing (as opposed to cost per action) than countries such as the U.S. or the U.K., and therefore has a higher potential for opportunistic behavior on the part of affiliates.

This study explores the first governance mechanism, the formal contract, by reviewing the restrictions placed on affiliate behavior in three areas: a) the program guidelines communicated to affiliates when they are accepted into a program; b) restrictions stipulated in the contract between the affiliate and the affiliate network; and c) restrictions from other intermediaries such as search engines.

Data was collected from a total of 136 programs managed by three of the largest affiliate networks in Spain at the time the study was conducted: Tradedoubler (56 advertisers), Zanox (75 advertisers) and OMG (5 advertisers). These include all of the programs in the three affiliate networks and, as such, represent a significant portion of the overall population in Spain. Not included are programs run on other affiliate networks, or programs run directly by a merchant without using an intermediary. While it is difficult to gauge the total population of affiliate programs in Spain due to a lack of an authoritative list of all programs, the authors estimate that this sample represents upwards of 75% of the population at the time the sample was taken, and perhaps as much as 90%.

Background information was collected from the guidelines for each program including: name of affiliate network, program name and category of business activity. Next, affiliate restrictions were gathered from the program guidelines. The results are summarized below in Table 4:

The remaining constructs (partner selection, incentives and monitoring) are analyzed based on a combination of the program data described above, as well as data gathered in semi-formal interviews conducted with professionals in the various roles in the affiliate marketing channel. 11 interviews were conducted in 2007. Please see Table 5 for an overview of the interviewees.

RESULTS

Formal Contracts

The first question is whether there are restrictions against an affiliate's using the merchant's trademark in the affiliate website URL (Uniform Resource Locator, or the Internet address of the

Table 4. Research results

Question	Number of Affirmative Responses	Percentage of Total
Are there restrictions on the use of the company trademark in the affiliate's URL?	5.00	3.68%
Is search engine marketing prohibited?	20.00	14.71%
Is the affiliate restricted from using the merchant's trademark(s) in advertising copy?	22.00	18.97%*
Is the affiliate restricted from bidding on the merchant's trademark(s)?	34.00	29.31%*
Is the affiliate restricted from bidding on the misspellings of the merchant's trademark?	8.00	6.90%*
Is the affiliate restricted from bidding on the trademarks of the merchant's competitors?	13.00	11.21%*
Is there a restriction on the amount that the affiliate can bid on the merchant's trademark(s)?	1.00	0.86%*
Are visitors from the affiliate site sent to a special URL separate from the merchant's home page?	6.00	5.17%*
Are there restrictions on email marketing?	3.00	2.21%
Are there restrictions related to privacy policies?	2.00	1.47%
Partner Selection: Are there restrictions on the types of affiliates accepted in the program (restrictions on affiliate website content)?	18.00	13.24%

N=136, or * N=116 percentage of total programs which allow search engine marketing

Table 5. Interviewee profiles

	Title	Organization
1	Online Marketing Manager	Large Online Travel Agency
2	Online Advertising and Internet Manager	Online Financial Services Company
3	Online Manager	Internet Dating Site
4	Online Marketing Manager	Hotel Reservation Site
5	Marketing Manager	Event Ticket Sales Online
6	Managing Director	Interactive Marketing Agency
7	Account Director / Consultant	Affiliate Network
8	Account Manager	Affiliate Network
9	Affiliate Manager	Affiliate Network
10	Chief Marketing Director	Large Online Travel Agency
11	Head of Affiliates	Large International Online Travel Agency

affiliate site). While only 3.68% of the merchants expressly forbid this activity, the affiliate network Zanox forbids "the use of domain names registered with copyrights". Tradedoubler states in their affiliate contract that "the affiliate also guarantees that the information and the productions on his website do not infringe the rights of any third party, including intellectual property rights, and that the information and the productions are not offensive, forbidden or objectionable for any reason. *In case of doubt, Tradedoubler reserves the right to finalize its business relationship with the Affiliate* (author's italics)" (Tradedoubler, 2007). OMG makes similar broad statements regarding intellectual property infringements.

The area for which there are the most restrictions placed on affiliate behavior in the formal contracts is in the use of search engine marketing. This is an area where affiliates typically have a great deal of independence, since they write their own ad copy and choose which keywords to bid on. Over 85% of the programs in the study allow search engine marketing in some form. Of these programs, 29.3% prohibit bidding on their trademark(s) in sponsored search and 6.9% forbid bidding on competitors' trademarks. One of the programs reviewed in the current study allows bidding on their trademarks, but limits the maximum amount of the bid. Presumably this is to prevent affiliates from outbidding the ads placed by the merchant directly.

19% of merchants prohibit affiliates from using merchant trademark(s) in the advertising copy used in search engine marketing. This low percentage is somewhat difficult to interpret, however. Search engine policies typically limit third parties' use of trademarks in advertising copy. Merchants may rely on these search engine restrictions rather than making them explicit in the affiliate program guidelines.

An additional control measure is preventing affiliates from sending visitors directly to the merchant's homepage, using instead a special website set up for affiliates. This was the case in 5.17% of the programs.

Email and privacy policies are two related areas. There are three main issues here: whether affiliates clearly communicate their privacy policies to visitors; whether affiliates only send email to individuals who have previously opted-in; and whether merchants restrict and/or monitor the messages communicated via email regarding their products and services. In the affiliate programs we reviewed, a mere 2.2% of merchants (3 programs total) address email marketing in some way. Of these 3 programs, one prohibits its use, and two require visitors to opt-in, or agree to receive promotional email. A mere 1.5% require affiliates to communicate their privacy policies to visitors. These areas, however, are also covered under European Union and Spanish law, which require prior opt-in for unsolicited commercial email (Sipior *et al.*, 2004). In addition, the program guidelines for Zanox expressly forbid spam, which they define as unsolicited email without prior opt-in. However, by not explicitly stating these restrictions in their own program guidelines, merchants may be leaving themselves open for problems later.

More importantly, there are no specific guidelines as to the *content* of the email advertisements. This is surprising, as it leaves the affiliates free to represent the merchants' products and services however they see fit. Furthermore, while merchants can visit affiliate websites in order to monitor the content, it is more difficult to monitor the content of emails sent which contain merchant-related content. One way that many merchants try to address this is by providing pre-approved content for email marketing which the affiliate can paste directly in emails sent to their customers. While this does not address the spam issue, it does help ensure a consistent advertising message on the part of the merchant.

Partner Selection

Only 13.24% of the programs explicitly state restrictions on the types of affiliates who may or may not apply to their program. These restrictions generally include some combination of nudity, sexual material, child pornography, firearms, drug consumption, gambling, illegal activity or violence. The remaining 86.76% make no mention whatsoever of the types of affiliate businesses which are acceptable. A well-known credit card provider had the most stringent policy regarding affiliate website content, adding restrictions against sites with religious material and sites whose content is poorly designed or of poor quality. However, this was the only program which referred to either the design of the site or its quality.

Nevertheless, this data must be evaluated in the context of the general affiliate contracts in the various affiliate networks. Tradedoubler restricts affiliates from placing merchant advertising alongside material which is "pornographic, discriminatory by race, religion or sex, or which infringes the rights of third parties in any way" (Tradedoubler, 2006). OMG's contract states "the Affiliate website does not and shall not display or contain any information or materials or hyper text links to information or materials which are or may be objectively considered to be defamatory, obscene, pornographic, offensive, threatening, blasphemous or liable to incite racial hatred or which promote any illegal activity including (but not limited to) cracking or hacking" (OMG Affiliate Contract, 2007).

While the above restrictions address affiliates who publish illegal or offensive content, they do not consider the types of business models used by affiliates. Furthermore, other criteria used to evaluate affiliate applications based on, for example, their offering related products (Papatla & Bhatnagar, 2002) are not addressed by affiliate program guidelines. Therefore, while merchants may have established internal guidelines in this regard, these are not communicated to the affiliates.

In the interviews, while most merchants mentioned that they pay attention to the business model of the affiliate site, the industry segment, the quality of the design and the ways that ads are shown (even if these criteria are not usually mentioned in the formal program guidelines), most maintain a fairly liberal policy when it comes to accepting affiliates. There is general recognition that one of the primary benefits of affiliate marketing is broad coverage, having their ads shown next to a wide variety of content and to a varied audience, and not necessarily just those visitors who fit the merchant's usual customer segments. One merchant emphasized that "many times it's the ugly sites which bring the most traffic". Another sums up their policy as follows: "the criteria are not strict for accepting sites, but rather it's the tracking (monitoring) that happens later which determines whether the site is a good partner or not."

Another important consideration in partner selection is the commission pricing model used by the merchant. Merchants who use a pure cost per sale model stated that while they try to attract affiliates who will generate a large volume of sales, they are not concerned if the traffic sent by many affiliates does not accomplish this goal. However, if the commission pricing includes cost per click, the merchants may be more restrictive in selection since the merchant is paying a commission for each visitor sent by the affiliate.

Furthermore, if the incentive program uses cost per lead commission pricing, the merchant may not want to accept affiliates who run points and rewards programs which offer visitors incentives for making purchases and filling out forms. One merchant with a CPL program states that "what this model brings us are registrations, but what we really live on as a company is sales. So, it doesn't really help us in the end to have thousands of registrations if, in the end, they're not active (i.e. generate revenue for the merchant)."

The affiliate networks also have an important role when an affiliate applies to a new merchant program, as the network can provide references for affiliates based on their past experience representing other merchants on that affiliate platform. One affiliate network account manager states "we know our affiliates, and we know whether or not a particular affiliate is a good fit for a new program. Also, when a new merchant signs up, we can recommend which of our affiliates can help them to accomplish their goals".

Incentives

The incentives used in affiliate marketing programs are outcome-based commissions based on visitors referred, leads, and sales, or some combination of the three. In Spain, these incentives can be classified into five basic commission pricing models, which are chosen depending on the program objectives of each merchant. Each of the models has its own governance challenges. Table 6 below shows the distribution of the programs in Spain:

The most common model is CPS (cost per sale), where the affiliate is only compensated when a referred visitor makes a purchase. This is the model with the fewest governance challenges, according to the interviewees, since the risk associated with advertising and conversion (getting visitors to make purchases) is transferred to the affiliates. In a pure CPS model, affiliates have an incentive to send "quality" traffic, or visitors who are likely to make a purchase. Otherwise the affiliates receive no compensation for their efforts.

The second-most popular model is CPC + Sale, where the affiliate is paid a high initial cost-per-click commission for each visitor referred, and then an additional commission if a visitor makes a purchase. The objective under this incentive model is similar to a pure CPS model: generating visitor traffic and sales. However, according to one affiliate network representative, this model is more attractive to affiliates "because the merchant absorbs some of the advertising and conversion risk".

One important governance mechanism available to merchants is their ability to change the commission pricing for a given affiliate or for the entire program whenever they want and for any reason. For example, in the case of CPC + Sale,

Table 6. Commission pricing models used in Spain

Pricing Model	Number of programs	Percentage of Total
CPS	60	44.12
CPC + Sale	30	22.06
CPL	22	16.18
CPC	12	8.82
CPC + Lead	12	8.82

if the visitors sent from a particular affiliate do not produce a high conversion rate, the merchant may change that affiliate's commission to a pricing structure focused more on sales.

CPL programs only pay incentives to affiliates when a referred visitor fills out a form on the merchant site. This form can be as simple as a contact form with phone number and email, or as detailed as a credit card or loan application, including banking and employment details. As discussed in the background section, the objective of this compensation structure is to generate leads, but it may be difficult to ensure the quality of the leads. Monitoring is a key governance mechanism for programs using this incentive structure.

CPC programs present the most significant governance issues, as the sole objective is to generate traffic. However, since there is no conversion objective (sales or leads), it can be difficult to gauge the quality of the traffic. According to the affiliate network representatives interviewed, this type of structure is usually used for a short-term program with a traffic quota. Once this quota is reached, either the incentive structure is changed to include a CPS or CPL component, or the program is discontinued.

The risks and governance issues in a CPC + Lead program are similar to those in a CPC + Sale program.

Many merchants operate multiple program levels depending on affiliate performance, and the merchant has the ability to reward an affiliate with promotion to a higher commission level, or to penalize them by demoting them to a lower commission level.

These outcome-based incentives are designed to accomplish program objectives regarding generating traffic, sales and leads. However, they can not address many aspects of affiliate behavior included in the formal contract. In other words, the incentive structures discussed above address the results of the affiliate activities, but not how they achieve those results. Therefore, monitoring is critical for ensuring that both outcomes and behavior comply with a merchant's requirements.

Monitoring

The first area to consider under monitoring is how merchants measure the results of their programs. Some important metrics include EPC, CTR, CR, and AOV. EPC is the average earnings per hundred clicks, calculated as the total commissions divided by the total number of clicks times 100. It represents the ability to turn clicks into commissions, and can be calculated both for an individual affiliate as well as for an entire affiliate program. CTR is the click-through rate, measuring the percentage of impressions (times an ad is shown) which result in a unique visitor clicking on an ad. CR is the conversion rate, or percentage of visitors who take a desired action, such as making a purchase or filling out a form, after clicking on an affiliate link. AOV, or average order value, is calculated by dividing the total sales value by the number of sales.

While the above metrics are useful in evaluating programs and affiliates, the most important metric is return on investment (ROI). There are various ways to calculate this metric: the denominator is usually total marketing cost, but the numerator could be sales, gross profit or net profit, depending on the merchant's specific objectives. While a high ROI is clearly preferable to a low one, one merchant emphasized that they prefer affiliates who bring high volume at a lower ROI to affiliates who have a high ROI but few sales.

While ROI numbers were not disclosed by the interviewees, one media agency representative with experience in the UK claimed that a transactional ROI of less than 120% in the mid- to long- term would be unacceptable, and that this figure would be very difficult to achieve for any program using CPC (which is the case for many affiliate programs in Spain). He claimed that CPC programs tend to benefit the affiliates and the affiliate networks at the expense of the merchants, whose profitability depends on sales, not visitor traffic.

The frequency of monitoring these metrics depends both on the merchant as well as the type of commission pricing used. A program using a CPA model (either CPS or CPL) will usually review the program metrics and individual affiliate metrics bi-weekly or monthly, and changes are made to an individual affiliate's commission structure depending on their performance. However, if a program's commission pricing includes CPC (CPC only, CPC + sale, or CPC + lead), the CR is reviewed daily according to the affiliate networks (after an initial introduction period of a few weeks to a month). If the CR is low for an affiliate, it means that they are sending low quality traffic, where few of the visitors complete a desired action. In this case the commission structure can be changed to reduce the CPC commission, or to change the commission structure to CPC or CPL.

The above relates to the monitoring of outcomes, which is one of affiliate marketing's greatest strengths. However, monitoring affiliate behavior is just as important; the formal contracts include several restrictions on affiliate behavior, especially concerning how an affiliate is allowed to represent merchants in search engine marketing, and to a lesser extent in email marketing.

One of the main ways that merchants tend to recognize fraud or undesirable activity is by monitoring significant changes in the visitors/sales/leads generated by affiliates. For example, an affiliate who usually generates very low volume may all of a sudden generate a large number of clicks, or a very high conversion rate. These could be symptoms of using popups or forced clicks, or even adware, which "steals" commissions from other affiliates. Another example is in CPL programs where affiliates send a large number of registrations which remain inactive, or do not convert into sales.

The affiliate program representatives stated that they try to monitor the way their ads are displayed on affiliate web sites on a monthly or quarterly basis, but that this is done in an unsystematic way, randomly selecting web sites and visiting them one by one.

One of the biggest concerns of merchants is how affiliates use their trademarks (to use the legal term) or brands (to use the marketing term), especially in search engine marketing. One merchant described their approach as follows: "I'm fine with them (affiliates) bringing in traffic with whatever keyword they want, but leave our brand alone."

The monitoring strategies used by merchants in this area include the following: first, they register their trademarks with the search engines, so that only the merchant can bid on its own brand. Then, the merchant monitors the search engines for advertising which violates their program restrictions. If the merchant notices an affiliate misusing their brand, they notify them and perhaps exclude them from the program if the violation is intentional and persistent. Most emphasize that this is not a systematic process, however.

Merchants also use the individuals or departments responsible for their own search engine marketing and search engine optimization to help monitor affiliate behavior. Since it is these individuals' responsibility to keep detailed track of the positioning of the merchant's web site compared to those of competitors for their important keywords, they are often the first to notice inappropriate behavior on the part of affiliates. One affiliate manager also stated that he uses a third-party auditing service to track the merchant's brand, and that he often identified affiliate irregularities in these reports. One merchant summarized it as follows: "while there aren't currently any tools which control this behavior completely, it is still very controllable through coordination with our internal SEO and search marketing efforts."

In terms of email marketing, the affiliate managers interviewed expressed little concern because there are few affiliates engaging in this activity currently. However, this is a concern for the future, as many programs are beginning to recognize its potential, while there are currently few restrictions on affiliate behavior using email marketing included in the formal contracts.

DISCUSSION

Summary of Findings

The purpose of this paper has been to explore the following research question: How are governance mechanisms used by merchants in one-to-many, Internet-based affiliate marketing programs in Spain to control the activities of their affiliates? Four, inter-related governance mechanisms were identified: formal contracts, partner selection, incentives, and monitoring. The overall conclusion is that there is a considerable lack of transparency as to how affiliates are permitted to represent the merchants' products and services. The situation is further complicated by the fact that affiliates must consider not only the specific affiliate program guidelines, but also the network affiliate contract as well as other conditions specified by third party intermediaries such as search engines. This makes the affiliate's task more difficult when it comes to conforming to the expectations of merchants, and it increases the importance of the other three governance mechanisms discussed in this paper. In addition, the merchants have significant power in their management of their affiliate programs: they generally reserve the right to alter or cancel the arrangement with a particular affiliate at any time and for any reason (which need not be communicated to the affiliate). While this may be an effective governance mechanism, greater transparency may help to create more positive merchant-affiliate relationships.

Figure 2 summarizes the conclusions for each governance mechanism.

Implications for Managers of Affiliate Marketing Programs

The following are general recommendations for managers of affiliate marketing programs:

Figure 2. Conclusions for each governance mechanism

Governance Mechanism	**Conclusions**
Formal Contract	There are some limitations placed on the use of search engine marketing by affiliates, primarily related to the use of the merchant's trademarks. The content of the advertisements is not addressed, however, either in search engine or email promotions; this is left to the discretion of the affiliates. Here, the lack of explicit guidelines increases the risk of an affiliate misrepresenting the merchant's business, and thus increases the importance of the other governance mechanisms discussed here.
Partner Selection	The formal contracts clearly prohibit affiliates who use illegal or offensive content or who use the trademark of the merchant in their website address. It is unclear, however, the types of affiliate business models and industry segments which are acceptable. The interviews indicated that while merchants are most interested in affiliates who can generate a high volume of traffic and sales, offer complementary products and have well-designed, professional sites, they still accept the majority of affiliate applications in order to reach the broadest audience of potential customers.
Incentives	Incentives fall into five basic categories of outcome-based commission pricing schemes. The model chosen by a given merchant depends on the objectives for their program and the type of model used has governance implications, particularly regarding the types of affiliates who should be accepted in the program and the level of monitoring required. For example, a CPC program requires much more frequent monitoring to ensure that affiliates are sending high-quality visitors.
Monitoring	**Outcomes:** Outcome-based monitoring activities are focused on tracking the number of visitors, sales and registrations generated by the affiliate, with criteria according to the particular commission pricing scheme in place. Return on investment is the primary metric used to evaluate affiliates and programs. Affiliates who perform well are rewarded by promotion to a better commission level, and those who perform poorly may be penalized by demotion to a lower commission level, or eliminated from the program altogether.
	Behavior: Merchants look for sudden changes in affiliate performance as possible indicators of fraud or irregular activity. Most conduct periodic random checks of affiliate websites, albeit in an unsystematic way, to evaluate the placement of their advertisements. Finally, merchants generally identify affiliates who do not comply with their search engine marketing restrictions on an exception basis through periodic searches and through coordination with their search engine optimization and search engine marketing departments.

- Merchants should provide more explicit guidelines concerning sponsored search restrictions, the types of businesses which are acceptable as affiliates, and the necessity of including privacy policies on affiliate sites that use email marketing.
- Merchants should conduct regular reviews of affiliate web sites, especially the placement and content of merchant-related information.
- Merchants should do regular, systematic keyword searches for their brand and related keywords in order to identify the keywords and advertising messages used to represent the merchant's brand.
- Affiliate managers should coordinate with the merchant's search engine marketing and search engine optimization efforts as a cross-check on affiliate behavior.
- Email should be carefully controlled: either the affiliate should be restricted to using content prepared by the merchant, or they should be required to send the affiliate manager a copy of any promotional emails which include merchant-related content.

CONCLUSION

This has been an exploratory study of the governance mechanisms used by merchants in one-to-many Internet-based affiliate marketing programs in a specific market, Spain, to control the behavior of their affiliate partners. While we cannot ensure that the sample was thoroughly random, we do believe that the relatively small size of the market/population enabled us to capture a large proportion of the overall phenomenon. Of course, the Spanish market may not be representative of the global phenomenon.

Since this was an initial exploratory study, a more predictive approach might be useful in analyzing the relationships between these governance mechanisms and program outcomes, both positive and negative.

In addition, since Spain is a relatively new market in terms of affiliate marketing, future research may expand the study to include more established markets such as the U.K. and the U.S. For example, the U.K. and U.S. may be more restrictive in terms of program guidelines and partner selection, and incentives may be fine-tuned on a more frequent basis. Furthermore, their monitoring efforts may be more extensive than those in Spain, and may shed further light in terms of how best to ensure positive outcomes and behavior, and to identify/avoid opportunism.

ACKNOWLEDGMENT

An earlier version of this paper was submitted to the 20th Bled eConference (June, 2007).

REFERENCES

AffStat. (2006). Affiliate program merchant facts & figures. *Shawn Collins Consulting*. Retrieved from http://www.shawncollinsconsulting.com

Anderson, E., & Weitz, B. (1992). The use of pledges to build and sustain commitment in distribution channels. *JMR, Journal of Marketing Research*, *29*, 18–34. doi:10.2307/3172490

Brown, J. R., Chekitan, S. D., & Lee, D.-J. (2000). Managing marketing channel opportunism: the efficacy of alternative governance mechanisms. *Journal of Marketing*, *64*(2), 51–65. doi:10.1509/jmkg.64.2.51.17995

Coase, R. H. (1937). The nature of the firm. *Economica*, *4*, 386–405. doi:10.1111/j.1468-0335.1937.tb00002.x

Dahlstrom, R., McNeilly, K. M., & Speh, T. W. (1996). Buyer-seller relationships in the procurement of logistical services. *Journal of the Academy of Marketing Science*, *24*(2), 110–124. doi:10.1177/0092070396242002

Duffy, D. L. (2005). Affiliate marketing and its impact on e-commerce. *Journal of Consumer Marketing*, *22*(2/3), 161–163. doi:10.1108/07363760510595986

Eisenhardt, K. M. (1989). Agency theory: An assessment and review. *Academy of Management Review*, *14*(1), 57–74. doi:10.2307/258191

Fox, L. (2000) Affiliate marketing makes headway *Upside*, *12*, 4, 176.

Heide, J. B., Wathne, K. H., & Rokkan, A. I. (2007). Interfirm monitoring, social contracts, and relationship outcomes. *JMR, Journal of Marketing Research*, *XLIV*, 425–433. doi:10.1509/jmkr.44.3.425

Hoffman, D. L., & Novak, T. P. (2000). How to acquire customers on the web. *Harvard Business Review*, (May-June): 2000.

Jensen, M., & Meckling, W. (1976). Theory of the firm: managerial behavior, agency costs, and ownership structure. *Journal of Financial Economics*, *3*, 305–360. doi:10.1016/0304-405X(76)90026-X

Libai, B., Biyalogorsky, E., & Gerstner, E. (2003). Setting referral fees in affiliate marketing. *Journal of Service Research*, *5*(4), 303–315. doi:10.1177/1094670503005004003

MarketingSherpa. (2006a). *Affiliate Summit 2006 wrap-up report: commissions to reach $6.5 billion in 2006*. Retrieved from http://inspirations.marketingsherpa.com/barrier.cfm?contentID=3157

MarketingSherpa. (2006b). 1,041 affiliates reveal how merchants should improve their programs. Retrieved from http://www.marketingsherpa.com/sample.cfm?ident=28526

Molander, J. (2005). The evolution of affiliate marketing. *AffilateTip.com*, Oct 20, 2005. Retrieved from http://affiliatetip.com/news/

Moon, Y. (2000). *BizRate.com* (Case 9-501-024, 1-22). Cambridge, MA: Harvard Business School.

Oliver, R. L., & Anderson, E. (1995). Behavior- and outcome-based sales control systems: Evidence and consequences of pure-form and hybrid governance. *Journal of Personal Selling & Sales Management*, *15*(4).

OMG UK. (2008). *Affiliate contract*. Retrieved February 2008, from https://omgadmin.co.uk/en/clientarea/affiliates/welcome.asp

Papatla, P., & Bhatnagar, A. (2002). Choosing the right mix of on-line affiliates: How do you select the best? *Journal of Advertising*, *31*(3), 69–81.

PartnerCentric. (2006) *Affiliate survey*. Retrieved February 2008, from http://www.marketingsherpa.com/cs/amscr2006/study.html

Pew Internet & American Life Project. (2005). *CAN-SPAM a year later*. Retrieved February 2008, from http://www.pewinternet.org/report_display.asp?r=102

Ray, A. (2001). Affiliate schemes prove their worth. *Marketing London*, *2001*, 29–30.

Rindfleisch, A., & Heide, J. B. (1997). Transaction cost analysis: Past, present and future applications. *Journal of Marketing*, *61*(October), 30–54. doi:10.2307/1252085

Ross, S. (1973). The economic theory of agency: The principal's problem. *The American Economic Review*, *63*, 134–139.

Sipior, J. C., Ward, B. T., & Bonner, P. G. (2004). Should spam be on the menu? *Communications of the ACM*, *47*(6), 59–63. doi:10.1145/990680.990681

Stein, G. (2004). Trademark bidding on Google. *Clickz.com*. Retrieved February 2008, from http://www.clickz.com/showPage.html?page=3341321

Stump, R. L., & Heide, J. B. (1996). Controlling supplier opportunism in industrial relationships. *JMR, Journal of Marketing Research*, *33*, 431–441. doi:10.2307/3152214

Telser, L. G. (1980). A theory of self-enforcing agreements. *The Journal of Business*, *53*(1), 27–44. doi:10.1086/296069

Tradedoubler. (2007). Affiliate contract. Retrieved February 2008, from http://www.tradedoubler.com/pan/aInfoCenter.?textKey=INFORMATION_AFFILIATE_INFOCENTER_AFFILIATE_AGREEMENT

ValueClick. (2006). *Investor relations overview*. Los Angeles: Author.

Walthieu, L. (2000). *Yesmail.com* (9-500-092). Cambridge, MA: Harvard Business School.

Wathne, K. H., & Heide, J. B. (2000). Opportunism in interfirm relationships: Forms, outcomes and solutions. *Journal of Marketing*, *64*(October), 36–51. doi:10.1509/jmkg.64.4.36.18070

Williamson, O. (1975). *Markets and hierarchies*. New York: The Free Press.

Williamson, O. (1985). *The economic institutions of capitalism*. New York: The Free Press.

Williamson, O. (1991). Comparative economic organization: the analysis of discrete structural alternatives. *Administrative Science Quarterly*, *36*(June), 269–296. doi:10.2307/2393356

ENDNOTE

1 The results of the survey are divided by the type of pricing scheme offering. In Pay per Sale programs, the affiliate receives a commission when the customer referred to the merchant site makes a purchase. In Pay per Lead programs, the affiliate receives a commission for each unique visitor sent.

This work was previously published in International Journal of E-Business Research Volume 6, Issue 1, edited by In Lee, pp. 1-18, copyright 2010 by IGI Publishing (an imprint of IGI Global).

Chapter 15
Software Firm Cost Structure and Its Impact on IPOs in the E-Commerce Era

Richard B. Carter
Iowa State University, USA

Troy J. Strader
Drake University, USA

ABSTRACT

The first decade of the e-commerce era saw an increase in activity in the software development industries as new firms were created and existing firms made acquisitions. Many firms pursued a growth strategy and this growth required capital that was often obtained through an initial public offering (IPO) of equity. Software firm cost structures are very different from traditional physical goods firms because their marginal costs are much lower, but what is not known is whether this affects their financing strategies. In this study we compare software firm and traditional firm IPOs using data from 780 IPOs offered during the late dot-com era (1998-2002) to identify differences in firm and offer characteristics, investment risk, initial returns, and underwriting activity. We find that the characteristics and performance of software firm IPOs are significantly different from IPOs offered by traditional firms during this time period providing supporting for our conclusion that firm cost structure should be considered when analyzing IPOs and other strategic issues.

INTRODUCTION

The first decade of the electronic commerce (e-commerce) era, from about 1993 through 2002, was an exciting and tumultuous time. The introduction of the World Wide Web (Web) in the early 1990s, and the ongoing development of related e-commerce applications throughout the 1990s and early 2000s, created demand for new software products and opportunities for new software distribution processes. This led to an increase in activity in the software industries

DOI: 10.4018/978-1-61350-462-8.ch015

including growth in existing companies as well as the creation of new firms.

From 1997 to 2002, software publishing company receipts grew from $61.7 billion to $103.7 billion, annual payrolls grew from $18.4 billion to $34.6 billion, and the number of paid employees rose from 266 thousand to more than 353 thousand (U.S. Census Bureau, 2004). During this same period, the number of firms fell from roughly 12 thousand to 9.9 thousand as competition grew, and firms either failed or were purchased by existing companies. Because many firms chose some form of strategic growth strategy to capitalize on these external opportunities, there was also an increase in the need for capital to finance growth. The choice for many software firms in the late 1990s and early 2000s was an initial public offering (IPO).

In this research we show that the timing and pricing of IPOs is different for software development (SWD) firms than for traditional firms. We argue that because of fundamental differences in cost structures, particularly pre-IPO fixed to variable cost ratios, as well as the "all-or-nothing" nature of SWD prospects, SWD firms needed to acquire capital earlier in their life cycle than traditional firms. SWD firms "load" their operation up front with fixed costs, market the product, and then hope for the best. Traditional firms can make significant adjustments to variable and fixed costs initially and throughout the firm's life.

These basic differences in SWD and traditional firms impact their long-term potential and the nature of the IPO. As a result SWD and traditional firms are viewed differently by the market and their IPO underwriter. If SWD firms wait until the market determines the winners and losers it will be too late for many to recoup their initial investment. Because traditional firms can make necessary changes after the IPO they can often avoid eventual failure. Moreover, the potentially dramatic long-run return of a small percentage of SWD firms, like Microsoft, as compared to the more broadly distributed long-run market values of traditional firms, makes SWD IPOs more like an option than those of traditional firms. Hence, the early after-market value of SWD firms, relative to their anticipated offer price as set by the firm and their IPO underwriter, would be more positively related to risk, like an option, than traditional firms.

IPOs have been studied for decades and, except for occasionally distinguishing technology firms with an indicator variable, these earlier studies grouped all firms together for the purpose of identifying the factors that affect returns, risk, and likelihood of failure (see Ritter & Welch, 2002, for a synopsis of IPO research). To the best of our knowledge no one has separated issuing firms based on their cost structures when analyzing IPO results. In the following we argue that one way to explain the differences in certain offering details is through an examination of a firm's cost structure.

These issues have only been addressed indirectly in previous research. Loughran and Ritter (2004), for example, show that firms that issue seasoned securities have falling operating profit to asset ratios following the offering and it is worse for the small issues. However, nearly 15% of the firms were in industries with extremely low marginal costs similar to software companies. The cost structures of these firms and their product market may have been an important factor in explaining the drop in operating profit. As noted earlier, technology/Internet IPOs, where all firms are considered equal regardless of their cost structure, have been studied to some extent, but there is a dearth of research on the relation between cost structure and IPO outcomes. This is the primary issue addressed by this study.

In the following section we review some of the relevant literature and elaborate on the theoretical differences between SWD and traditional firms and how they impact their IPO characteristics. The section culminates in testable hypotheses. The next section describes the data followed by the methods and results. Finally we provide our conclusions along with implications, and directions for future research.

Software Development vs. Traditional Firms

There are obvious differences between firms with digitizable products and more *traditional* physical product manufacturing, distributing, and retailing firms. While *digitizable* would include non-software products, we are focusing attention on SWD firms to limit extraneous variables that would be introduced with other electronic products and services like wireless communication, online auctions, etc. Traditional firms must manage marginal costs such as materials and labor that are very different from software firms. During the introductory phase of their life cycle they will expend a considerable amount on plant, equipment, and other infrastructure costs. But during the growth phase the proportion of total costs associated with marginal costs can increase significantly. These costs are offset to a great extent by the revenue received from product sales.

Unlike traditional physical product firms, for SWD companies like Microsoft, Adobe, or McAfee, marginal costs are low or nearly non-existent. For example, the cost to provide another copy of an existing antivirus software product is nearly zero. This is especially true for firms that distribute their product directly to consumers through an online channel (Strader & Shaw, 1997). This is a favorable situation for SWD firms as they grow, but the downside is that they typically have a large up-front fixed cost in the introductory phase of their life cycle that is required to offer software to even one user. Consequently, their fixed cost to variable cost (FCVC) ratio is very high at the IPO and higher than for traditional firms.

Many of the up-front fixed costs for SWD firms are associated with the infrastructure activities that support software product development including product management, process management, integration management, operations management, and documentation production (Harter & Slaughter, 2003). As a result we should see a negative relation between FCVC and risk for SWD firms. The infrastructure costs are almost entirely expended prior to the receipt of any sales revenues. Because of these differences in the costs of doing business at varying points in the firm's life cycle, there are inherent differences in the risk of these firms in their early years (Wilbon, 2002).

The fundamental cost structure differences between SWD and traditional firms described above and the need for financing growth in the first decade of the e-commerce era, suggest that we may be able to estimate their importance through their manifestation in the early capital acquisition process. Do cost structure differences affect the characteristics, short term performance, and investment risk related to IPOs across these sectors? If there are differences, then perhaps IPOs should be first categorized and then separately analyzed to predict results. But more importantly do these cost structure differences influence management and strategic decisions with regards to raising equity capital? Do underwriters and the market view software development firm IPOs differently than those of traditional firms?

As the number of IPOs increased dramatically in the 1990s, researchers began to address questions related specifically to IPOs of the unique category of Internet (or e-commerce) firms. These firms were often described as either technology firms, high-tech firms, or Internet firms. The issues studied included investor behavior around the time of the dot-com bubble (Cao & Shi, 2006; Jaggia & Thosar, 2004), the impact of managerial decisions and organizational strategies on firm value and survival after an IPO (Filson, 2004; Rajgopal, Venkatachalam, & Kotha 2002; Wilbon, 2002), prediction of initial returns (Johnston & Madura, 2002; Marshall, Crutchley, & Lending, 2004), and comparisons of high-tech IPOs in Europe and the U.S. (Aaij & Brounen, 2002). The common limitation of these studies is that, when addressing the characteristics of these technology/Internet IPOs, no distinction is made for the cost structure differences in these ventures. For example, an electronics hardware manufacturer, software company, online service provider, and online retailer would be considered to be in the same category. Yet, information products are costly to produce and cheap to reproduce with high fixed costs and low marginal costs (Shapiro & Varian, 1999).

Because SWD firms have substantial fixed costs to recover and a limited product market, only a fixed number will actually be successful. With an *experience good*, like a digitizable product, the consumer must experience the good to value it (Shapiro & Varian, 1999). But once it is experienced and rejected it is too late for SWD firms to recover initial resources through an IPO. On the other hand, for traditional firms many levels of success are likely. This suggests that SWD firms would generally issue an IPO earlier in their life cycle than traditional firms. Moreover, a typical scenario for SWD firms resulting in "winner-take-all" means a very high longer-run IPO return for those investors choosing the right firm. As a result, software firm IPOs appear more like an option than those of traditional firms where the price (or premium) paid for the IPO is positively related to the variance of future returns (see Merton, 1973, for a discussion of the positive relation between option value and the risk or variance of the underlying asset).

Traditionally, IPOs have been viewed from a risk/return standpoint. The riskier the firm going public the lower the offer price to clear the market (Rock, 1986). If we assume the after-market is relatively efficient then we expect the average first day return to be higher for riskier firms because the offer price is discounted more. There is considerable support for this argument (Beatty & Ritter, 1986; Carter & Manaster, 1990; Johnson & Miller, 1988). In the risk/return scenario we expect insider to total offered shares to be lower for riskier firms as owners maintain more ownership in an attempt to allay fears of pending failure (Leland & Pyle, 1977). Less (more) reputable investment banks are relegated the more (less) risky firms (Carter & Manaster, 1990).

If we view the IPOs of SWD firms as an option, where for a relatively low price the underwriters and the market can purchase the opportunity to perhaps own the next Microsoft then the traditional firm IPO relationships described above change. For example, the value of an option increases with the risk of the underlying security such that we expect the offer price to be relatively higher for the riskier SWD offering because of the increase in its option value. In the option scenario, where for a relatively small price both market and underwriter can participate in a potentially dramatic value increasing opportunity, we expect higher fractions of insider to total shares reflecting strong pre-offer demand (Hanley, Kumar, & Seguin, 1993). This should also attract more reputable underwriters to SWD offerings in an attempt to secure long-term business with the next Microsoft (James, 1992).

We summarize the foregoing in eight testable hypotheses:

H1: SWD firms are smaller and younger than traditional firms at their IPO,

H2: SWD firms have higher fixed to variable cost ratios than traditional firms,

H3: SWD firms are riskier than the traditional firms,

H4: SWD firms' insiders are more likely to sell shares at the offer than those of traditional firms,

H5: SWD firms have relatively higher offering prices than traditional firms,

H6: SWD firms' underwriters will be more prestigious than traditional firms,

H7: There is a negative relation between FCVC and risk for SWD firms, and

H8: Risk revealing characteristics of SWD firms' assets are meaningful indicators of their IPO value as an option.

Data

Using the Thomson Financial SDC database we identified all domestic IPOs from January of 1998 through December of 2002. We chose 1998 as a point of inception for data collection because it is several years after the initiation of e-commerce activity which allows for firms to begin operations and develop their products to the point where we can assess their cost structure and its impact on firm performance and IPO activity. The five year period includes a large sample of traditional and software firm IPOs which enables more in-depth analysis. The period also includes several years of activity before and after the dot-com bubble to assess these questions in both a time of increasing and declining IPO activity.

From the SDC sample we removed unit offerings, closed end funds, REITS, limited partnerships, stocks with offering prices of less than $4, and stocks without CRSP returns. The remaining sample consists of 1175 IPOs.

To determine the nature of each IPO firm's business, we read a description of their product and/or service. The description was obtained from the SDC data, from the original prospectus or from the company's website. Software firms included those firms engaged in software development, computer system design and computer programming. Because we are comparing software firms to those with significant marginal costs, for example, materials, labor, etc. we also removed 327 online service firms, where we expect the ratio of fixed costs to variable costs to be substantially greater than traditional firms. Online service firms include those engaging in communications, digital content, cable television, wireless transmission, telephone transmission, information processing, and those described as online service. All other firms were considered traditional.

The other variables from the SDC database include the names of all underwriters managing the offering, the lead underwriter, the date of the offer, the number of shares offered by both the firm and those from the personal holdings of pre-IPO owners, the offer price, and the first reported high and low price in the pre-IPO range.

We supplement the SDC data with a number of additional sources. Individual stock and index returns, as well as outstanding shares for each IPO firm at the time of the offering, are taken from the CRSP database. The after-market prices are used to calculate daily returns, the standard deviation of daily return and the market model beta for each stock. Both the standard deviation and beta are calculated using market day IPO + 20 through day IPO + 120. The initial IPO return (INITRET) is calculated as the last closing bid price on the day of the IPO divided by the offer price minus 1.

Financial statement variables such as sales, net income and total assets in the year prior to the IPO are obtained from Standard and Poor's Compustat database. Wherever partial year data is reported, we annualize it to facilitate comparisons. Because there is often insufficient time to accumulate annual data prior to the IPO and because data from some firms were either purged from or never included in Compustat's database, we do not have financial statement data for all 848 firms in the sample.

The age of the issuing firms at the time of the offering are hand-collected by searching through firm histories included in the Moody's manuals, the original IPO prospectuses, and individual company web sites for the founding date. In cases where the founding dates are not available, we use the incorporation date.

Among the necessary variables is a proxy for underwriter reputation. We use the Carter-Manaster (1990) underwriter reputation ranking (CM) adjusted with the average offer price of the IPOs the underwriter marketed during the period. The product is then divided by 100. The Carter and Manaster reputation variable is a nominal measure from 0 to 9 where a 9 represents the most prestigious underwriter and 0 the least prestigious. As Loughran and Ritter (2004) point out, a potential flaw in the Carter-Manaster methodology is that penny stock underwriters that are never allowed to participate in a syndicate of prestigious underwriters might never be assigned low Carter-Manaster ranks. The final ranking (CMOP) for the late 1990s and early 2000s is from Carter and Dark (2006).

The final data set for the analysis includes 616 traditional firm IPOs and 164 software firms. Table 1 displays the summary statistics for each of the variables.

The data for our sample appears fairly consistent with recent papers, for example, Ljungqvist and Wilhelm (2003, 2005). Their initial return mean is 35.7% whereas ours is 33.7%. They report a ratio of final offer price to expected of 1.06 and ours is also 1.06. The age of their firms is 13.3 and ours is 12.2 whereas their revenues are $232.5 million and ours is $218.2 million. Their Carter and Manaster underwriter rank is 7.3 whereas ours (not reported) is 8.6. The one major difference is in net income where we report 27.2 million and they report -4.5 million. However, this is likely the result of the inclusion of online firms, truncating their data in 2000.

Methods and Results

To test the first six hypotheses, we use a uni-variate analysis, estimating parametric differences in means between the two sub-samples (SWD versus traditional firms). We include the non-parametric difference in medians examination because of the uncertainty in the distribution of the population. A median test was chosen because of its usefulness when the samples are independent and of different sizes.

Our first hypothesis predicts that SWD firms are smaller and younger than traditional firms. This should also translate into smaller gross proceeds for their IPOs. The second hypothesis predicts SWD firms will have higher fixed cost to variable cost ratios (FCVC) than traditional firms. The results are found in Table 2, Panel A.

The software firms are significantly smaller than the traditional firms in terms of assets, revenues, and offer size. Their net income is negative compared to positive mean net income for tradi-

Table 1. Descriptive statistics for 780 IPOs issued from January 1, 1998 through December 31, 2002

Variable	Mean	Standard Deviation	Median
Total assets ($Mill)[a]	1668.00	13012.00	103.84
Revenues ($Mill)	547.83	2207.00	56.36
Net income ($Mill)	27.24	202.91	-1.05
SG&A Expenses/COGS	5.32	26.51	0.63
Beta[b]	1.90	1.59	1.56
Standard deviation (%)	5.85	2.80	5.70
Offer Size ($Mill)	218.15	643.52	88.48
First day return (%)	41.55	79.50	14.52
Age at the time of IPO (Years)	12.19	21.25	5.00
Inside/total offered shares (%)[c]	34.26	43.88	0.00
Offer price/E(Offer price)[d]	1.06	0.31	1.00
Underwriter reputation[e]	1.22	0.30	1.24

[a] Balance sheet and income statement items are from the year prior to the IPO.

[b] Beta and standard deviation of return are estimated using the 100 daily returns 20 days to 120 market days post IPO. Beta was estimated using the market model and the contemporaneous returns from the equally-weighted NASDAQ index.

[c] Inside shares are those offered by the company's pre-IPO owners.

[d] E(Offer price) is estimated as the midpoint of the pre-IPO offer price range.

[e] Underwriter reputation is the product of the Carter and Manaster (1990) ranking and the average offer price for all IPOs underwritten by the subject investment bank. The product is then divided by 100.

Table 2. Difference in means and median tests for 616 traditional firms versus 164 software firms

	Traditional Firms		Software Firms		Differences				
Variables	Mean	Standard Deviation	Mean	Standard Deviation	Diff in Means	t Statistic		F Statistic[f]	
Panel A:									
Total assets ($Mill)[a]	2075.30	14616.00	139.57	408.38	1935.73	3.28	***	6.15	***
Revenues ($Mill)	664.85	2458.70	108.28	481.93	556.57	5.25	***	6.32	***
Net income ($Mill)	39.10	223.00	-17.32	81.26	56.42	5.13	***	6.85	***
Offer Size ($Mill)	247.16	718.17	109.15	134.33	138.01	4.48	***	2.28	**
Age at the time of IPO (Years)	13.66	23.53	6.66	5.41	6.99	6.74	***	2.16	**
SG&A Expenses/COGS	3.46	25.38	12.28	29.43	-8.81	-3.50	***	-11.59	***
Panel B:									
Beta[b]	1.64	1.46	2.91	1.67	-1.27	-8.89	***	-7.39	***
Standard deviation (%)	5.41	2.67	7.50	2.64	-2.09	-8.93	***	-8.38	***
First day return (%)	33.67	70.67	71.11	101.22	-37.44	-4.46	***	-5.44	***
Panel C:									
Inside/total offered shares (%)[c]	29.38	41.44	52.58	47.92	-23.21	-5.66	***	-5.09	***
Offer price/E(Offer price)[d]	1.02	0.26	1.22	0.41	-0.20	-5.95	***	-5.58	***
Panel D									
Underwriter reputation[e]	1.21	0.30	1.25	0.29	-0.04	-1.39		-0.13	

[a] Balance sheet and income statement items are from the year prior to the IPO.

[b] Beta and standard deviation of return are estimated using the 100 daily returns 20 days to 120 market days post IPO. Beta was estimated using the market model and the contemporaneous returns from the equally-weighted NASDAQ index.

[c] Inside shares are those offered by the company's pre-IPO owners.

[d] E(Offer price) is estimated as the midpoint of the pre-IPO offer price range.

[e] Underwriter reputation is the product of the Carter and Manaster (1990) ranking and the average offer price for all IPOs underwritten by the subject investment bank. The product is then divided by 100.

[f] The t test is a difference in means test and the F statistic represents a median test. Significance at the 10, 5 and 1% levels is indicated by one, two and three asterisks, respectively.

tional firms and their fixed cost to variable cost ratio is nearly four times that of the traditional firms. The results support both H1 and H2.

The third hypothesis predicts that SWD firms are riskier than traditional firms. Again we estimate differences in means and medians for measures of risk. The results are in Panel B of Table 2. We chose both the standard deviation of early returns to the IPOs as well as the market beta as estimates of the inherent risk of the firms. Both are estimated using the 100 daily returns from 20 to 120 market days post IPO and the equally-weighted NASDAQ return is used to estimate beta. The results for both variables support hypothesis H3.

We also looked at the difference in the run-up in first day return. We expect that the market found the SWD firms more attractive during the period of study, driving up the after-market price. The result, displayed in Table 2 below the risk measures, shows that the return for SWD firms is indeed more than twice that for the traditional firms.

The fourth hypothesis predicts that more insiders sell with the offering for SWD firms than for traditional firms, while H5 predicts that the offer price is relatively higher for SWD firms than for traditional firms. The variable we use to test H4 is the ratio of shares being sold by the firm's pre-

IPO owners to the total shares offered. For H5 we use the ratio of the final offer price to the mean of the pre-IPO offer price range as denoted in the original offering application (S1). The results are in Panel C of Table 2. In both cases the variables are significantly greater for the SWD firms than that for traditional firms and support H4 and H5.

Finally, H6 predicts that more reputable underwriters market the IPOs of the SWD firms than the traditional firms. As a reputation proxy we use CMOP the underwriter reputation variable. The hypothesis predicts the SWD underwriters are more prestigious than the traditional firms' underwriters. The results, found in Panel D of Table 2 are not conclusive. While the reputation measure for SWD is greater than that for the traditional firms, the difference is not significant.

To test H7 we use an ordinary least squares multi-variate regression model. The hypothesis predicts that SWD firms become less risky with increasing FCVC. This is more likely for SWD firms than traditional firms simply because of the importance of variable costs for traditional firms—as explained previously. Using a dummy variable to separate the software firms from all other firms in the sample (SWDummy) we can test differences in the relation between dependent and independent variables for each firm type while incorporating all of the data. We include control variables to reduce the possibility that FCVC is a proxy for other risk factors. The model is displayed in equation (1).

$$STDRet = a + b1*SWDummy + b2*FCVC + b3*SWDummy*FCVC + BX + e, \quad (1)$$

where STDRet is the standard deviation of after-market return, BX is a vector coefficients and control variables and e is an error term. The results are found in Table 3.

The relation between FCVC and STDRet for traditional firms is represented by coefficient estimate b2 and the relation for software firms is the sum of b2 and b3. In the table we have already summed these coefficients. The control variables include one plus the logarithm of the age of the firm at the time of offer (LNAGE), the logarithm of total assets (LNTA) and the logarithm of 100 + the net income of the firm (LNNI). Both total assets and net income are in millions of dollars.

The regression is significant at the 1% level ($F = 38.56$) and the adjusted R^2 (0.224) suggests the model fits the data reasonably well. The coefficient of SWDummy (b1) is positive and significant implying that the software firms are riskier in general than the traditional firms and the coefficient b2, also positive and significant, indicates that for traditional firms risk increases with FCVC – consistent with traditional theory for operating risk. For the software firms, not only is the adjusted coefficient (b2 + b3) negative and marginally significant (at better than the 10% level) but the change from the traditional firms is negative and significant at the 1% level indicating that the market perceives higher FCVC ratios as less risky for this type of firm – supporting hypothesis H7. All of the control variable coefficients are consistent with the notion that older and larger firms are less risky.

In a second regression we test the relation between the relative offering value and the risk measure (STDRet). We have hypothesized (H8) that risk revealing characteristics of the firms' assets are meaningful indicators of a SWD firm's value as an option. We also expect it to be more meaningful than it would be for traditional firms because of the all-or-nothing nature of the SWD firm.

We use the ratio of the offer price to the expected price (OP/EOP) as a proxy for the relative value of the offering firm. We expect a significant positive coefficient for STDRet for the SWD firms and we expect that it is significantly greater than the coefficient for the traditional firms. The model is displayed in equation (2).

$$OP/EOP = a + b1*SWDummy + b2*STDRet + b3*SWDummy*STDRet + BX + e, \quad (2)$$

Table 3. Multiple least squares regression (risk)

Independent Variable	All Firms Coefficient[c]	
Intercept	18.51	
SWdummy[a]	1.65	
	5.33	***
Fixed/variable costs (FCVC) Traditional IPOs[b]	0.01	
	2.76	***
SWDummy*FCVC	-0.01	
Δ: Traditional to software firms *(-2.73***)*	[2.85]	*
Log total assets (LNTA)	-0.77	
	-8.33	***
Log (1+age) (LNAGE)	-0.18	
	-2.19	**
Log net income (LNNI)	-1.19	
	-1.55	
Adjusted R^2	0.22	
F Statistic	*38.56*	***
IPOs with complete data	*780*	

NOTE: The dependent variable is the standard deviation of daily returns (STDRet) as measured from offering day plus 20 market days through offering day plus 120 market days. Chi square statistics are in brackets ([ ]) and t and F statistics are in italics.

[a] SWdummy is a binary variable where a zero represents firm engaged in software development and zero represents traditional firms.

[b] Fixed costs are sales, administrative and general expenses divided by costs of goods sold. All income and balance sheet items are from pre-offering reports.

[c] Significance at the 10, 5 and 1% levels is indicated by one, two and three asterisks, respectively.

where BX is a vector of coefficients and control variables (LNOFFSIZE, LNAGE, and CMOP). LNOFFSIZE is the natural logarithm of the gross proceeds to the firm. The results are found in Table 4. The coefficient estimate of STDRet for SWD firms (b2 plus b3) has already been summed for presentation in the table.

The regression is significant at the 1% level (F = 33.18) and the adjusted R^2 (0.234) suggests the model fits the data reasonably well. The coefficient b1 is not significant indicating that after controlling for other offering particulars there is no difference between the relative offer price for SWD firms and traditional firms. The coefficient for STDRet for the SWD firms is positive and significant and, consistent with our hypothesis H8, implies that risk is a meaningful indicator of value for SWD firms. While the STDRet coefficient for traditional firms is also positive it is significantly smaller (nearly a third smaller) than that for the SWD firms. This is supportive of hypothesis H8 and also supportive of the notion that the IPOs of SWD firms can be viewed more as an option than those of traditional firms. The positive coefficient for LNOFFSIZE and negative coefficient for LNAGE indicate that firms with higher relative offering prices are larger and younger than those with lower offering prices.

CONCLUSION AND FUTURE RESEARCH

The primary conclusion of this study is that firm cost structures (the degree to which marginal costs are associated with creation of a product unit) do

Table 4. Multiple least squares regression (relative IPO value)

Independent Variable	All Firms Coefficient[c]	
Intercept	-1.30	
SWdummy[b]	-0.19	
	-1.25	
STDRet (Traditional IPOs)	2.58	
	3.33	***
SWDummy*STDRet (SWD IPOs)	7.30	
Δ: Traditional to software production firms *(2.15**)*	[12.92]	***
Log offering size (LNOFFSIZE)	0.10	
	4.31	***
Log of firm age (LNAGE)	-0.01	
	-2.49	**
Underwriter Reputation Measure (CMOP)	0.04	
	0.70	
Adjusted R^2	0.23	
F Statistic	*33.18*	***
IPOs with complete data	*780*	

NOTE: The dependent variable, the actual offer price divided by the expected offer price (OP/EOP), is regressed on LNOFFSIZE, LNAGE and CMOP. Chi square statistics are in brackets ([]) and t and F statistics are in italics.

[a] LNOFFSIZE is the natural logarithm of the gross proceeds from the IPO, LNAGE is natural logarithm of the age of the firm at the IPO and CMOP is the adjusted Carter and Manaster (1990) reputation measure.

[b] SWdummy is a binary variable where a zero represents firm engaged in software development and zero represents traditional firms.

[c] Significance at the 10, 5 and 1% levels is indicated by one, two and three asterisks, respectively.

matter to firms and to the market, affecting both the timing and pricing of capital acquisition. There are significant differences in most of the characteristics and outcomes when comparing software development (SWD) firms with traditional firms during the 1998 through 2002 period. These cost structure differences affect risk, initial returns, and reputation of the investment bank chosen to underwrite the issue. SWD firm characteristics at the time of their IPO indicate that they are younger and smaller. Yet the relation between operating risk and the riskiness of after-market return is relatively lower for software firms than for traditional firms. This is consistent with the notion that software firms, those with high fixed to variable cost ratios, are viewed more like an option than traditional firms, where fixed to variable cost ratios are indicators of pre-IPO risk and are positively related to the IPO price.

These results have implications for IPO stakeholders. Investors, both underwriters and the market, must realize that with a limited number of software firm winners, there are multiple losers. While overall, taking many positions in software firms may be relatively less costly than that for traditional firms, many will result in significant if not total loss.

There are several directions available for future research in this area. A limitation of the current study is that it analyzes IPOs during a unique period of activity (1998-2002). We chose this period because of the rapidly increasing number of software firms that went public. However, this time frame also includes the *dot com bubble,* a period of falling stock prices and a general decline in the number of IPOs in the latter years. The methodology applied to this study could be utilized to verify the results using a broader time period perhaps starting with the introduction of e-commerce activity, IPOs offered in 1994 through 1997, or the period following when there was a decline in IPO activity, IPOs offered in 2003 through 2007. Either of these time periods could be used to attempt to confirm the results of this present study.

The current study primarily focuses on short term outcomes such as characteristics at the time of the IPO and initial returns. A future study could address the factors that affect longer term outcomes such as returns after one or more years as well as likelihood of firm failure. A third extension would be to compare online service firm versus traditional firm IPOs. Online services and software firms have similar cost structures which may produce similar findings to this current study when comparing their characteristics and short term outcomes.

REFERENCES

Aaij, S., & Brounen, D. (2002). High-tech IPOs: A tale of two continents. *Journal of Applied Corporate Finance*, *15*(1), 87–92. doi:10.1111/j.1745-6622.2002.tb00343.x

Beatty, R. P., & Ritter, J. R. (1986). Investment banking, reputation and the underpricing of initial public offerings. *Journal of Financial Economics*, *15*, 213–232. doi:10.1016/0304-405X(86)90055-3

Cao, M., & Shi, S. (2006). Signaling in the Internet craze of initial public offerings. *Journal of Corporate Finance*, *12*(4), 818–833. doi:10.1016/j.jcorpfin.2005.11.001

Carter, R. B., & Dark, F. H. (2006). *Underwriter reputation and IPO issuer alignment 1981-2000* (working paper). Ames, IA: Iowa State University.

Carter, R. B., & Manaster, S. (1990). Initial public offerings and underwriter reputation. *The Journal of Finance*, *45*(4), 1045–1068. doi:10.2307/2328714

Filson, D. (2004). The impact of e-commerce strategies on firm value: lessons from Amazon.com and its early competitors. *The Journal of Business*, *77*, S135–S154. doi:10.1086/381640

Hanley, K. W., Kumar, A. A., & Seguin, P. J. (1993). Price stabilization in the market for new issues. *Journal of Financial Economics*, *34*(2), 177–197. doi:10.1016/0304-405X(93)90017-6

Harter, D. E., & Slaughter, S. A. (2003). Quality improvement and infrastructure activity costs in software development: a longitudinal analysis. *Management Science*, *49*(6), 784–800. doi:10.1287/mnsc.49.6.784.16023

Jaggia, S., & Thosar, S. (2004). The medium-term aftermarket in high-tech IPOs: patterns and implications. *Journal of Banking & Finance*, *28*(5), 931–950. doi:10.1016/S0378-4266(03)00040-2

James, C. (1992). Relationship specific assets and the pricing of underwriter services. *The Journal of Finance*, *47*(5), 1865–1885. doi:10.2307/2328999

Johnson, J. M., & Miller, R. E. (1988). Investment banker prestige and the underpricing of initial public offerings. *Financial Management*, *17*(2), 19–29. doi:10.2307/3665523

Johnston, J., & Madura, J. (2002). The performance of Internet firms following their initial public offering. *Financial Review*, *37*(4), 525–550. doi:10.1111/1540-6288.00027

Leland, H. E., & Pyle, D. H. (1977). Information asymmetries, financial structure, and financial intermediation. *The Journal of Finance*, *32*(2), 371–387. doi:10.2307/2326770

Ljungqvist, A., & Wilhelm, W. J. (2003). IPO pricing in the dot-com bubble. *The Journal of Finance*, *58*(2), 723–752. doi:10.1111/1540-6261.00543

Ljungqvist, A., & Wilhelm, W. J. (2005). Does prospect theory explain IPO market behavior? *The Journal of Finance*, *60*(4), 1759–1790. doi:10.1111/j.1540-6261.2005.00779.x

Loughran, T., & Ritter, J. R. (2004). Why has IPO under-pricing changed over time? *Financial Management*, *33*(3), 5–37.

Marshall, B. B., Crutchley, C. E., & Lending, D. (2004). Early Internet IPOs versus subsequent entrants. *Journal of Economics & Finance*, *28*(1), 104–116. doi:10.1007/BF02761458

Merton, R. C. (1973). Theory of rational option pricing. *The Bell Journal of Economics and Management Science*, *4*(1), 141–183. doi:10.2307/3003143

Rajgopal, S., Venkatachalam, M., & Kotha, S. (2002). Managerial actions, stock returns, and earnings: the case of business-to-business Internet firms. *Journal of Accounting Research*, *40*(2), 529–556. doi:10.1111/1475-679X.00060

Ritter, J. R., & Welch, I. (2002). A review of IPO activity, pricing, and allocations. *The Journal of Finance*, *57*(4), 1795–1828. doi:10.1111/1540-6261.00478

Rock, K. (1986). Why new issues are underpriced. *Journal of Financial Economics*, *15*, 187–212. doi:10.1016/0304-405X(86)90054-1

Shapiro, C., & Varian, H. R. (1999). *Information rules: A strategic guide to the network economy*. Boston: Harvard Business School Press.

Strader, T. J., & Shaw, M. J. (1997). Characteristics of electronic markets. *Decision Support Systems*, *21*(3), 185–198. doi:10.1016/S0167-9236(97)00028-6

U. S. Census Bureau. (2004). *2002 economic census: Software publishers: 2002*. Retrieved from http://www.census.gov/prod/ec02/ec0251i06t.pdf

Wilbon, A. D. (2002). Predicting survival of high-technology initial public offering firms. *The Journal of High Technology Management Research*, *13*(1), 127–141. doi:10.1016/S1047-8310(01)00052-9

This work was previously published in International Journal of E-Business Research Volume 6, Issue 1, edited by In Lee, pp. 19-31, copyright 2010 by IGI Publishing (an imprint of IGI Global).

Chapter 16
An Empirical Study of the Impact of Brand Name on Personal Customers' Adoption of Internet Banking in Hong Kong

T. C. E. Cheng
The Hong Kong Polytechnic University, Hong Kong SAR, China

W. H. Yeung
The Hong Kong Polytechnic University, Hong Kong SAR, China

ABSTRACT

This study investigates whether brand name has an impact on personal customers' adoption of Internet banking (IB) in Hong Kong. Based on the Technology Acceptance Model (TAM) with proper adaptation to the online environment, we propose a model comprising an external factor, namely brand name, and three constructs of technology adoption, namely perceived credibility, perceived usefulness and perceived ease of use to assess customers' intention to use IB. In addition, we investigate the difference in the brand name's effect on customers' adoption of IB between banks with larger market shares and banks with smaller market shares. We collected data through a questionnaire survey and analyzed the data using structural equation modelling (SEM) to investigate the inter-relationships between the constructs and assess the overall explanatory power of the model. The results provided support for the model and confirmed its robustness in predicting the impact of brand name on personal customers' adoption of IB in Hong Kong.

DOI: 10.4018/978-1-61350-462-8.ch016

INTRODUCTION

Internet banking (IB) in Hong Kong has become increasingly important because it is one of the popular delivery channels for banking services accepted by customers in the cyber age. With IB customers can perform their banking transactions anywhere and at any time. Banks can benefit from much lower operating costs by offering IB, which requires fewer staff and physical branches. Although many banks in Hong Kong have developed and provided IB services since 1999, the adoption rate has not taken off and still lags behind other western countries. In 2006, there were about 3.77 million Internet users in Hong Kong while there were 111 million Internet users in China. *The Computer Industry Almanac* projected that the worldwide Internet population would reach 1.46 billion by 2007. However, Internet banking services have not taken off in Hong Kong as expected. An AC Nielsen Study found that Hong Kong's online banking population was 926,000 in 2004 and the penetration rate was only at 26.6% (out of 3.48 million Internet users in 2004). Therefore, there are considerable business opportunities for Internet banking services in Hong Kong given the increasing number of Internet users.

The Bank of East Asia and Dah Sing Bank pioneered IB services in Hong Kong in late 1999. However, their customers' adoption rates are relatively lower than that of the HSBC Group, i.e., HSBC and Hang Seng Bank. The IB services offered by different banks are more or less the same because of keen competition, which results in quick replication of IB functions and security measures among rival banks. According to the figures of the Hong Kong Monetary Authority (2008), there were 44 banks in Hong Kong offering IB services, of which 31 provided personal Internet bank accounts. Nonetheless, according to ACNielsen (2004), the IB market in Hong Kong was dominated by the HSBC group, with a market share of over 50%, while non-HSBC banks, e.g., the Bank of East Asia, Dah Sing Bank, Standard Chartered Bank, the Bank of China (Hong Kong) and Citibank, etc., held the rest of the market share. In view of this phenomenon, there is a need to identify factors influencing personal customers' intention to use IB, which may include the impact of brand name (which is related to market share), and the perceived credibility, perceived usefulness and perceived ease of use of IB.

There is an abundance of research on the impact of brand name on consumers' intention to purchase physical goods. Nonetheless, there are few studies on the impact of brand name on consumers' intention to acquire services, in particular IB services. While the studies of Berry (2000), de Chernatony and Dall'Olmo (1998), and Keller (1998) formed the basis for understanding the effects of brand name on customers' acquisition of services, their notions lack empirical testing and validation. In this regard, an understanding of the impact of brand name on personal customers' intention to use IB is desirable. We believe that brand name's impact on service adoption can be empirically tested by using the traditional Technology Acceptance Model (TAM) with proper adaptation to the online environment. The recent study by Cheng, Lam, and Yeung (2006) demonstrated the robustness of TAM in predicting customers' intention to use IB with a focus on perceived usefulness, perceived ease of use and web security to meet their banking needs. Considering the potential impact of brand name on IB adoption and seeing that security and privacy issues have attracted increasing research attention in the web literature, we seek to fill the research gap by using brand name as an external factor and adding a new construct of perceived credibility that includes both the security and privacy factors in TAM for empirical testing.

In summary, we aim to first identify factors that affect personal customers' intention to use IB, which include brand name, perceived credibility, perceived usefulness and perceived ease of use of IB. Second, we construct a theoretical model to account for personal customers' intention to use IB based on TAM, in which brand name is treated as an external factor and perceived credibility, perceived usefulness and perceived ease of use as three constructs of IB adoption. Third, we study whether the brand names of IB providers with larger market shares have a different degree of effect on customers' adoption of IB from the brand names of IB providers with smaller market shares. This paper is organized as follows. In Section 2 we review the relevant literature on brand name, perceived credibility and TAM, and formulate our model. In Section 3 we discuss the research methodology and in Section 4 we present and discuss the results. In Section 5 we discuss the results and their theoretical and managerial implications and conclude the research findings, followed by some suggestions for future research.

Literature Review and Model Formulation

Much research has shown that brand name has a direct impact on customers' intention of purchasing a product. For IB services, brand name as an external factor is believed to have a significant direct relationship with the three constructs of perceived credibility, perceived usefulness and perceived ease of use embedded in TAM. The theoretical background, which purports to link the brand name of the IB provider to personal customers' intention to use the IB services provided, is discussed in the following.

Brand Name

The American Marketing Association defines brand as "name, term, sign, symbol, or design, or a combination of them, intended to identify the goods and services of one seller or group of sellers and to differentiate them from those of competition". Shrimp (2003) defined brand as "a convenient and appropriate label for describing an object of concerted marketing efforts". In many research studies, "brand name" is used interchangeably with "brand" as "brand" is used as a noun. The word "brand", when used as a noun, refers to a company name, a product name or a unique identifier such as a logo or trademark. In this study, "brand name" is used interchangeably with "brand". Dodds, Monroe, and Grewal (1991) and Olson (1977) endorsed the fact that brand name is an external information cue affecting customers' evaluation and choice decisions. When consumers experience a branded good or service over time, the power of brand will reside in their minds. Brand name can build up consumers' trust in the goods and services of the producer or the provider (Aaker, 1996). According to Dave (2003), "a brand stands for the immediate image, emotions, or perceptions people experience when they think of a company or product". The brand of a good or service represents all the tangible and intangible quality, image, lifestyle and status associated with possessing and using the good or service. Consumers' intention to purchase a good or service is based on their perceptions and predictions of the quality and perceived credibility of the good or service (Robert & Hugh, 2004). Brand name can enhance the perceived quality and perceived credibility of goods and services and can reduce the perceived risk of owning and using the goods and services, which together will increase the customers' purchase intention (Van Osselaer and Alba, 2000).

Perceived Credibility

According to Wang, Wang, Lin, and Tang (2003), perceived credibility is an independent variable that reflects user security and privacy concerns in the acceptance of IB and therefore it affects customers' intention to use IB services. Consumers are concerned about the level of security and the protection of data privacy when providing sensitive information online (Warrington, Abgrab, & Caldwell, 2000). The banking sector is strongly associated with high levels of trust related to security and privacy issues in the physical environment, which is to be expected to extend to IB.

Security evidence presented on a bank's web site can help ease customers' concerns and increase their confidence in using IB services. Perceived credibility consists of two important dimensions, namely security and privacy. While credibility is broadly defined as the believability of an entity, perceived credibility is defined as "the degree to which a person believes that using a particular system would be free of security and privacy threats. Security and privacy features have been heavily emphasized in the e-commerce context" (Belanger, Hiller, & Smith, 2002; Friedman, Kahn, & Howe, 2000; Shankar, Urban, & Sultan, 2002). Lee and Turban (2001) remarked that consumers are reluctant to adopt IB because of security and privacy concerns and risk related issues. Security refers to the protection of information or systems from unsanctioned intrusions or outflows. Security has been widely recognized as one of the main obstacles to the adoption of IB (Aladwani, 2001; Bestavros, 2000; Furnell & Karweni, 1999). Security concerns arise from the use of an open network via which customers are afraid that their personal and financial information will become available to others and can be used for fraudulent purposes. Privacy has been identified as a major impediment to e-commerce (U.S. Public Interest Research Group, 2000). Privacy represents the control of transactions between persons and others, the ultimate aim of which is to enhance autonomy and minimize vulnerability (Margulis, 1977). Although recent research studies have revealed that perceived credibility has important influences on users' willingness to engage in online shopping, banking and the exchange of money and sensitive personal information (Friedman et al., 2000; Gefen, Karahanna, & Straub, 2003; Hoffman, Novak, & Peralta, 1999; Wang et al., 2003), the relationships between brand name, perceived credibility and intention to use IB are under-explored and have not been empirically tested.

Technology Acceptance Model

The Technology Acceptance Model (TAM) is an adaptation of the Theory of Reasoned Action (TRA) for modelling user acceptance of information systems. It attempts to provide an explanation for the determinants of computer acceptance and is capable of explaining user behaviour across a broad range of end-user computing technologies and user acceptance. Davis (1989) developed TAM with the understanding that "users' adoption of a computer system" depends on their "behaviour intention to use" the system, which in turn depends on their attitude. Attitude consists of two beliefs, namely "perceived ease of use" and "perceived usefulness". According to Davis (1989), perceived ease of use is defined as "the degree to which a person believes that using a particular system would be free of effort". Perceived usefulness is defined as "the degree to which a person believes that using a particular system would enhance his or her job performance".

The literature on social influence processes, cognitive instrumental processes and communication processes, i.e., Theory of Planned Behaviour (TPB), TAM, Innovations Diffusion Theory (IDT) and Extended TAM (TAM2), has provided the theoretical framework to account for the adoption of IB. In this study we choose TAM as the baseline model because it is a well-known and well-tested model for predicting user acceptance of a technology. Also, as IB is concerned with individual adoption of technology, the well-tested TAM is better than IDT and TAM2. The underpinning theory of IDT is the communication process and IDT is exploratory in nature. There are four key constructs underpinning IDT, including innovation, communication channels, social system and time. For TAM2, Venkatesh and Davis (2000) added a number of determinants to the TAM to develop, test and extend it (TAM2). TAM2 explains perceived usefulness and usage intention in terms of social influence processes, i.e., subjective norm, voluntariness, and image and cognitive instrumental processes, and perceived ease of use. Longitudinal data are required to apply TAM2. Therefore, on comparing TAM with IDT and TAM2, we adopted the former as it is the most suitable model for our study.

We set out to test the robustness of TAM with regard to its applicability in predicting personal customers' intention to use IB services in Hong Kong. Because the traditional TAM may not be able to fully address the pertinent issues surrounding the adoption of IB, we extend TAM by adding an independent variable (construct) and an external factor, which are the "perceived credibility" and the "brand name" of the IB provider, respectively. In the original TAM, attitude is a mediating construct between beliefs and intention. Nonetheless, Davis, Bagozzi, and Warshaw (1989) excluded attitude as a mediating construct between beliefs and intentions based on the observation that attitude does not generally intervene between beliefs and intentions. Several recent studies based on TAM did not take attitude into account in their models either, as empirical evidence has shown that attitude does not fully mediate the effects of beliefs on behavioural intention (Moore & Benbasat, 1991; Rogers, 1995, Venkatesh & Davis, 2000). We also adopt this approach in this study.

Perceived Usefulness and Perceived Ease of Use

Extensive research on the adoption of information technology and information systems has provided evidence of the significant effect of perceived usefulness and perceived ease of use on users' intention to use a technology (Davis et al., 1989; Venkatesh & Davis, 1996, 2000; Venkatesh & Morris, 2000). In the context of IB, perceived usefulness and perceived ease of use are significant factors that affect customers' intention to use IB while perceived ease of use is a significant antecedent to perceived usefulness (Cheng et al., 2006).

Brand Name's effect on Perceived Credibility, Perceived Usefulness, Perceived Ease of Use and Intention to Use IB

We discussed in Section 2.1 that brand name can strengthen the perceived quality and perceived credibility of goods and services, which in turn can increase consumers' purchase intentions (Saunders, 1993; Van Osselaer & Alba, 2000). Rao and Monroe (1989) demonstrated in their meta-analyses that the relationship between brand name and perceived quality is positive and statistically significant. On the other hand, Matthew, Ostrom, Roundtree, and Bitner (2000) interpreted perceived quality of IB services as perceived usefulness and perceived ease of use of IB services. Perceived quality is interpreted as perceived usefulness and perceived ease of use in a self-service technology environment and shares the same meaning, including (i) meeting an increasing need for quality self-service technology, (ii) better than alternatives, which include easy to use, time savings and value for money, and (iii) performing its function. Therefore, brand name's effects on perceived credibility, perceived usefulness and perceived ease of use (i.e., perceived quality) and intention to use IB are potentially interrelated and can be formulated as hypotheses for empirical testing.

Model and Hypothesis Development

Based on the brand name literature, and prior TAM and IB studies, we construct a theoretical model based on TAM in Figure 1, and formulate the nine research hypotheses (Figure 2) from the model. Specifically, we seek to test the hypotheses in the context of personal customers' adoption of IB in Hong Kong.

In addition, we wish to further investigate brand name's effect on IB adoption by comparing the effect of the brand names of a group of banks with larger market shares, i.e., the HSBC group,

Figure 1. The hypothesized model

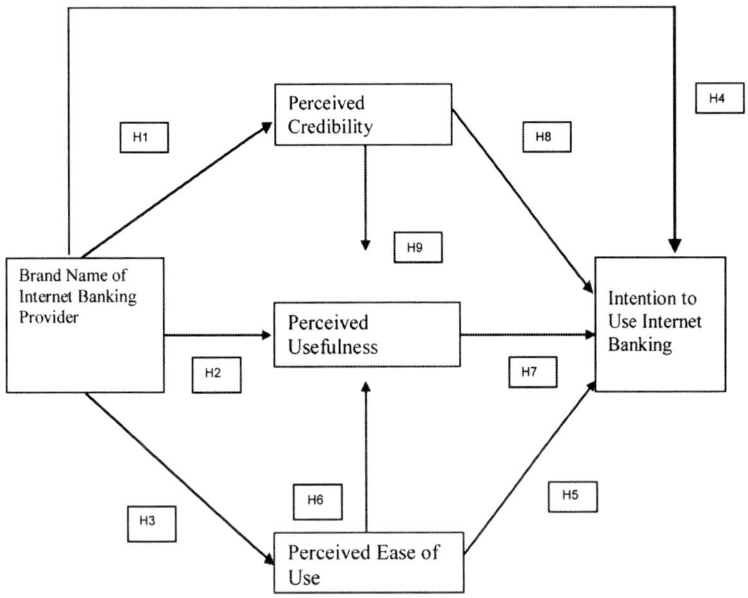

Figure 2.

Hypothesis	Details
H1	Brand name has a positive effect on perceived credibility of IB.
H2	Brand name has a positive effect on perceived usefulness of IB.
H3	Brand name has a positive effect on perceived ease of use of IB.
H4	Brand name has a positive effect on intention to use IB.
H5	Perceived ease of use has a positive effect on intention to use IB.
H6	Perceived ease of use has a positive effect on perceived usefulness of IB.
H7	Perceived usefulness has a positive effect on intention to use IB.
H8	Perceived credibility has a positive effect on intention to use IB.
H9	Perceived credibility has a positive effect on perceived usefulness of IB.

with those of another group of banks with smaller market shares, i.e., non-HSBC banks. Hong Kong's IB market is dominated by the HSBC group, controlling over 50% of the IB market. We wish to test the perception that users who consider using IB services tend to use the services of well-established banks. Therefore, we formulate the null hypothesis in Figure 3.

Figure 3.

Hypothesis	Details
HA	Brand name effects of different banks on personal customers' adoption of IB are the same.

RESEARCH METHODOLOGY

Research Strategy

We conducted an empirical study guided by a theoretical framework built on the original TAM with proper adaptations and extensions. To collect data, we designed a questionnaire by adapting the instruments and scales developed for TAM and from the brand name literature.

Sample Size

Because usage and experience of IB is yet to become a common phenomenon in Hong Kong, a simple random sample of the population is likely to result in a sample of respondents with limited IB experience. In view of this, a convenience sample of Internet users who are presumed to have knowledge of using IB services was taken. We collected 269 valid responses through a questionnaire survey for data analysis. There is hence a limitation that our survey was about the perceptions of Internet users. The data were collected by two trained research assistants, who interviewed personal customers outside bank branches, university campuses, large securities trading firms, various government offices and coffee chops, etc.

Measurement and Operationalization of the Constructs

In devising a useful measurement instrument for our study, instruments from the following sources and scales were adapted:

- The questionnaire developed by Grace and O'Cass (2005) on brand name.
- The questionnaire developed by Warrington et al. (2000) on perceived credibility.
- The questionnaire developed by Davis (1989) for TAM on perceived usefulness, perceived ease of use and intention to use.

Survey Questionnaire and Pilot Test

Based on the theoretical model (Figure 1), we devised a survey questionnaire with 26 items (Appendix 1). These 26 items relate to brand name and user acceptance behaviour. To gather more personal information, there are other seven demographic and general questions. We developed the survey questionnaire in English and did not translate it into Chinese because we believed that the respondents, i.e., Internet users who are potential IB users, generally have the ability to understand English. We used a 7-point Likert scale (1 = strongly disagree, 2 = disagree, 3 = slightly disagree, 4 = neutral, 5 = slightly agree, 6 = agree and 7 = strongly agree). We conducted a pilot test and made minor changes to the questionnaire after the pilot test so as to improve its accuracy and acceptance by future potential respondents.

RESULTS AND DISCUSSIONS

Demographic Characteristics of the Respondents

The descriptive statistics of the respondents' demographic characteristics were analyzed and presented in Table 1. Of the 269 respondents, 42% were male; 38% were 21 - 30 in age; 43% were 31 - 40 in age and 18% were 41 - 50 in age. Few were under 20 or over 50. There were 62% respondents who intended to use the HSBC group for IB. Other demographic details are also shown in Table 1.

Confirmatory Factor Analysis

We used the theoretical model to test model fitness by performing a Confirmatory Factor Analysis (CFA) on the data. We fed the data into the AMOS 7.0 programme for testing the hypothesized model. Table 2 shows the fit statistics of the original model.

Table 1. Descriptive statistics of respondents' demographic characteristics

Demographic profile	Details	Frequency	Per cent (%)
Gender	Male	112	41.6
	Female	157	58.4
Age	16 to 20	2	0.7
	21 to 25	42	15.6
	26 to 30	60	22.3
	31 to 35	88	32.7
	36 to 40	29	10.8
	41 to 45	33	12.3
	46 to 50	14	5.2
	Above 50	1	0.4
Education level	Secondary	83	30.9
	Diploma	19	7.1
	Associate degree	2	0.7
	Bachelors degree	108	40.1
	Masters degree	56	20.8
	Doctoral degree	1	0.4
Occupation	Executive/ Managerial	44	16.4
	Professor/ Teacher	7	2.6
	Professional	13	4.8
	Sales/ Marketing	62	23.0
	Customer Services	69	25.7
	Clerical	31	11.5
	Student	11	4.1
	Others	32	11.9
Category of banks respondents intended to use for IB	HSBC & Hang Seng Bank	166	61.7
	Standard Chartered, Citibank, the Bank of East Asia, the Bank of China (HK), Dah Sing Bank and other banks	103	38.3
IB services respondents intended to use	Enquiry service	55	20.4
	Deposit service	18	6.7
	Loan service	1	0.4
	Securities trading service	8	3.0
	Investment fund service	2	0.7
	Insurance service	1	0.4
	Multiple services	126	46.8
	Not specified	58	21.6

Table 2. Fit statistics of the original model

Fit Statistics	Suggested	Obtained Values
Chi-square	N/A	703.88
Degree of Freedom (DF)	N/A	290
Chi-square significance	P < or = 0.05	0.00
Chi-square/DF (Wheaton et al., 1977)	<5.0	2.43
GFI (Joreskog and Sorbom, 1988)	>0.90	0.85
NFI (Bentler and Bonett, 1980)	>0.90	0.94
CFI (Bentler, 1990)	>0.90	0.96
RMSEA (Steiger and Lind, 1980)	<0.05	0.07
PCLOSE	>0.50	0.00

MODEL ASSESSMENT

Conformance to Expected Parameter Estimates

In the original model, an RMSEA of 0.07 showed a marginal fit. The P value for closeness of fit was 0.000, i.e., not > 0.50 and hence further investigation was warranted. The model was re-estimated by freeing covariances as suggested by AMOS. After modification, all the key fit statistics showed significant improvement in Table 3. Overall, these fit indices indicated a good model fit.

Reliability of Parameter Estimates

Any parameter of a theoretical model having a critical ratio (C.R.) not exceeding ±1.96 can be regarded as unimportant to the model. Table 4 shows that all C.R. values were larger than 1.96 with the exception of Perceived Usefulness ← Brand Name (C.R. = -0.124, p = 0.901) and Intention ← Brand Name (C.R. = -2.122, p = 0.034).

Additional Measurements and Potential Bias

In order to demonstrate that the measurement model has a satisfactory level of validity and reliability, we examined and assessed construct reliability and variance extracted (or average variance extracted) (Fornell & Larcker, 1981).

Indicator Loadings

Table 5 shows that each indicator of the five latent variables (constructs) exceeded the critical values for the 0.05 significance level. Thus, all the variables were significantly related to their specified constructs, verifying the posited relationships among the indicators and constructs.

Construct Reliability and Construct Variance Extracted

Reliability refers to the degree to which a set of latent construct indicators are consistent in their measurements. Variance extracted refers to the amount of shared or common variance among the indicators or manifest variables for a construct. Table 5 shows that, for our hypothesized model, both values for construct reliability and variance extracted exceeded the recommended level of 0.70 and 0.50, respectively.

Non-Response Bias

In order to assess the possible existence of non-response bias, we divided the respondents into two groups. The first group of respondents (N

Table 3. Fit statistics of the re-specified model

Fit Statistics	Suggested	Obtained Values
Chi-square	N/A	405.01
Degree of Freedom (DF)	N/A	281
Chi-square significance	P < or = 0.05	0.00
Chi-square/DF (Wheaton et al., 1977)	<5.0	1.44
GFI (Joreskog and Sorbom, 1988)	>0.90	0.90
NFI (Bentler and Bonett, 1980)	>0.90	0.97
CFI (Bentler, 1990)	>0.90	0.99
RMSEA (Steiger and Lind, 1980)	<0.05	0.04
PCLOSE	>0.50	0.97

Table 4. Parameter estimates of the re-specified model: Regression weights

	Estimate	S.E.	C.R.	P
Perceived Credibility < --- Brand Name	0.390	0.046	8.437	***
Perceived Ease of Use < --- Brand Name	0.410	0.044	9.350	***
Perceived Usefulness < --- Brand Name	-0.003	0.027	-0.124	0.901
Perceived Usefulness < --- Perceived Ease of Use	0.717	0.058	12.457	***
Perceived Usefulness < --- Perceived Credibility	0.291	0.047	6.155	***
Intention < --- Brand Name	-0.062	0.029	-2.122	0.034
Intention < --- Perceived Usefulness	0.430	0.088	4.879	***
Intention < --- Perceived Credibility	0.320	0.057	5.629	***
Intention < --- Perceived Ease of Use	0.327	0.086	3.784	***
Brand Name 1 < --- Brand Name	1.000			
Brand Name 2 < --- Brand Name	1.003	0.024	41.075	***
Brand Name 3 < --- Brand Name	0.986	0.029	34.123	***
Brand Name 4 < --- Brand Name	0.992	0.031	31.663	***
Brand Name 5 < --- Brand Name	0.912	0.032	28.830	***
Perceived Credibility 1 < --- Perceived Credibility	1.000			***
Perceived Credibility 2 < --- Perceived Credibility	0.993	0.029	34.605	***
Perceived Credibility 3 < --- Perceived Credibility	0.950	0.032	29.510	***
Perceived Credibility 4 < --- Perceived Credibility	0.953	0.033	29.217	***
Perceived Usefulness 1 < --- Perceived Usefulness	1.000			
Perceived Usefulness 2 < --- Perceived Usefulness	0.995	0.033	30.390	***
Perceived Usefulness 3 < --- Perceived Usefulness	0.989	0.040	24.649	***
Perceived Usefulness 4 < --- Perceived Usefulness	0.993	0.037	26.677	***
Perceived Usefulness 5 < --- Perceived Usefulness	1.038	0.038	27.583	***
Perceived Usefulness 6 < --- Perceived Usefulness	1.009	0.038	26.713	***
Perceived Usefulness 7 < --- Perceived Usefulness	0.982	0.038	25.815	***
Intention 1 < --- Intention	1.000			
Intention 2 < --- Intention	0.997	0.018	54.393	***
Intention 3 < --- Intention	0.970	0.020	49.036	***
Perceived Ease of Use 7 < --- Perceived Ease of Use	1.000			
Perceived Ease of Use 6 < --- Perceived Ease of Use	1.057	0.042	25.170	***
Perceived Ease of Use 5 < --- Perceived Ease of Use	1.053	0.041	25.730	***
Perceived Ease of Use 4 < --- Perceived Ease of Use	1.061	0.044	24.100	***
Perceived Ease of Use 3 < --- Perceived Ease of Use	1.086	0.045	23.952	***
Perceived Ease of Use 2 < --- Perceived Ease of Use	1.071	0.046	23.541	***
Perceived Ease of Use 1 < --- Perceived Ease of Use	1.080	0.045	24.009	***

Table 5. Construct reliability and variance extracted values

	Loading	Square loading	Indicator error	Construct reliability	Variance extracted	> Standard level
BN1	0.947	0.897	0.141			
BN2	0.944	0.891	0.150			
BN3	0.955	0.912	0.115			
BN4	0.940	0.884	0.158			
BN5	0.920	0.846	0.183			
Sum BN	**4.706**	**22.146**	**0.747**	**0.967**	**0.856**	**Yes**
PC1	0.951	0.904	0.083			
PC2	0.948	0.899	0.089			
PC3	0.939	0.882	0.096			
PC4	0.936	0.876	0.102			
Sum PC	**3.774**	**14.243**	**0.370**	**0.975**	**0.906**	**Yes**
PU1	0.917	0.841	0.160			
PU2	0.929	0.863	0.184			
PU3	0.937	0.878	0.176			
PU4	0.928	0.861	0.125			
PU5	0.903	0.815	0.119			
PU6	0.900	0.810	0.128			
PU7	0.912	0.832	0.144			
Sum PU	**6.426**	**41.293**	**1.036**	**0.976**	**0.851**	**Yes**
PEOU1	0.923	0.852	0.196			
PEOU2	0.919	0.845	0.193			
PEOU3	0.925	0.856	0.158			
PEOU4	0.926	0.857	0.134			
PEOU5	0.913	0.834	0.142			
PEOU6	0.898	0.806	0.151			
PEOU7	0.886	0.785	0.144			
Sum PEOU	**6.390**	**40.832**	**1.118**	**0.973**	**0.839**	**Yes**
INT1	0.980	0.960	0.037			
INT2	0.978	0.956	0.040			
INT3	0.968	0.937	0.056			
Sum INT	**2.926**	**8.561**	**0.133**	**0.985**	**0.955**	**Yes**

= 133) consisted of the questionnaires collected by the two research assistants within the first 15 days. The second group of respondents (N = 136) consisted of the questionnaires collected after 15 days. A t-test was performed on the responses of the two groups to see if they were different. The results showed no differences in the responses of the two groups, suggesting that there was no evidence of non-response bias.

MULTI-GROUP ANALYSIS

HSBC Group and Non-HSBC Banks

In order to further test the impact of brand name on personal customers' adoption of IB, we split the sampled data into two groups, namely the HSBC group and non-HSBC banks. The sample size of the HSBC group was 166 customers while the sample size of the non-HSBC banks was 103 customers. A respondent was classified as an HSBC group customer when his/her response to the question "Which category of banks' IB services in Hong Kong are you likely to use in the coming months?" was HSBC or Hang Seng Bank. Otherwise, he/she was classified as a non-HSBC banks customer, which means he/she responded by indicating that he/she was a customer of the Standard Chartered Bank, Citibank, the Bank of Asia, the Bank of China (HK), Dah Sing Bank or other banks. We applied SEM using AMOS 7.0 to analyze the multi-group model. We used the re-specified model as a starting point for multi-group analysis for the HSBC group and non-HSBC banks. Please refer to Table 6 for details.

The same path diagram was applied to the two groups for the HSBC group and non-HSBC banks, i.e., CMIN/DF was < 3.0. Although the HSBC group and non-HSBC banks had the same path diagram, there was no requirement that the parameters have the same values in the two groups. The weightings of the items of the constructs among multi-groups were not restricted and the weightings of the paths of the model between the two groups were not restricted either. When it was hypothesized that the unconstrained model for multi-group analysis was correct for both the HSBC group and non-HSBC banks, AMOS 7.0 was used to test for measurement invariance and structural invariance for the HSBC group and the non-HSBC banks and the unconstrained models for the HSBC group and non-HSBC banks.

Measurement Invariance and Structural Invariance

Measurement invariance means the latent variables are consistently measured by their underlying indicators across different groups. This is checked by the chi-square difference and the p value between the unconstrained model and the constrained model where all the factor weightings of the indicators on a latent construct are restricted to be equal among all the groups. When the p value comparing the unconstrained model with the measurement invariance model is greater than 0.05, then the measurement invariance model is valid. The results are given in Table 7, which suggest measurement invariance.

Table 6. Multi-group analysis parameters

	Model with MI	Unconstrained model for multi-group analysis
Number of distinct sample moments	351	702
Number of distinct parameters to be estimated	70	140
Degrees of freedom	281	562
Chi-square	405.01	876.90
Degrees of freedom	281	562
CMIN/DF	1.44	1.56
Probability level	0.00	0.00

In the calculation of degrees of freedom for the unconstrained model, all of the numbers for sample moments, parameters to be estimated and degrees of freedom are exactly doubled.

Table 7. Measurement invariance: Assuming the unconstrained model is correct

Model	DF	CMIN	P	NFI Delta-1	IFI Delta-2	RFI rho-1	TLI rho2
Measurement invariance	21	9.585	.984	.001	.001	-.002	-.002

Measurement invariance was supported by the fact that the p value was >0.05

Table 8. Measurement invariance: Assuming the unconstrained model is correct

Model	DF	CMIN	P	NFI Delta-1	IFI Delta-2	RFI rho-1	TLI rho2
Measurement invariance	21	9.585	.984	.001	.001	-.002	-.002
Structural invariance	30	24.652	.742	.002	.002	-.002	-.002

Structural invariance means constraints are imposed on (1) the weightings of the items of the constructs between both groups, and (2) the weightings of the paths of the model between both groups. Structural invariance is present across the two groups when the p value is > 0.05 and the results are given in Tables 8 and 9.

The results on structural invariance show that there was no difference between the HSBC group and non-HSBC banks in brand name's effect on personal customer's adoption of IB. A summary of AMOS outputs for the unconstrained models, the measurement invariance models and the structural invariance models are given in Appendix 2. The results in the structural invariance columns are the focus of this analysis. Reading from the structural invariance columns in the regression weights output, all the C.R. values were larger than 1.96 with the exceptions of Brand Name <--> Perceived Usefulness (C.R. = -0.126, p = 0.900) and Brand Name <--> Intention (C.R. = -2.448, p = 0.014). This implies that Hypotheses H2 and H4 were not supported. The results applied to both the HSBC group and non-HSBC banks.

Results of Hypothesis Testing

Based on the SEM analysis of the empirical data, Hypotheses H1, H3, H5, H6, H7, H8 and H9 were supported, whereas Hypotheses H2 and H4 were not supported. The findings are summarized in Table 10.

CONCLUSION AND IMPLICATIONS

Conclusion

The primary objective of this research is to study personal customers' intention to use IB in Hong Kong. We built a theoretical model based on TAM extended with brand name as a new external factor and perceived credibility as an additional new construct. We collected 269 usable data and used SEM to test the model. The results supported

Table 9. Structural invariance: Assuming the measurement invariance model is correct

Model	DF	CMIN	P	NFI Delta-1	IFI Delta-2	RFI rho-1	TLI rho2
Structural Invariance	9	15.067	.089	.001	.001	.000	.000

Structural invariance was supported by the fact that the p value is >0.05

Table 10. Research findings

Hypotheses	Effects	Testing result		
		Integrated model	HSBC group	Non-HSBC banks
H1	Brand name has a positive effect on perceived credibility	Supported	Supported	Supported
H2	Brand name has a positive effect on perceived usefulness	*Not Supported*	*Not Supported*	*Not Supported*
H3	Brand name has a positive effect on perceived ease of use	Supported	Supported	Supported
H4	Brand name has a positive effect on intention	*Not Supported*	*Not Supported*	*Not Supported*
H5	Perceived ease of use has a positive effect on intention	Supported	Supported	Supported
H6	Perceived ease of use has a positive effect on perceived usefulness	Supported	Supported	Supported
H7	Perceived usefulness has a positive effect on intention	Supported	Supported	Supported
H8	Perceived credibility has a positive effect on intention	Supported	Supported	Supported
H9	Perceived credibility has a positive effect on perceived usefulness	Supported	Supported	Supported

that perceived credibility, perceived usefulness and perceived ease of use have a positive impact on intention to use IB. While brand name as an external factor has positive relationships with perceived credibility and perceived ease of use, it has no direct relationships with perceived usefulness and intention to use IB. The findings also showed that there is no difference between a well-known brand, i.e., the HSBC group and a less well-known brand, i.e., non-HSBC banks, in brand name's effect on personal customers' intention to use IB. All in all, this empirical study measured and evaluated the impact of brand name on personal customers' intention to use IB in Hong Kong based on TAM. The impact of brand name through the three constructs of perceived credibility, perceived usefulness and perceived ease of use on the intention of adopting IB services was empirically tested and verified. This study contributes to both academic research and bank management regarding developing effective promotion strategies and marketing activities for IB services in Hong Kong. Effective promotion strategies and marketing activities are important to strengthening brand name, which in turn can enhance perceived credibility, perceived usefulness and perceived ease of use of IB.

Implications for Bank Management

Our research findings support that brand name has an impact on personal customers' adoption of IB in Hong Kong with its relationships with the three constructs of perceived credibility, perceived usefulness and perceived ease of use. As a result, IB providers should concentrate on marketing their IB services with information regarding web security and privacy (i.e., perceived credibility), perceived usefulness and perceived ease of use. These are implications of the results that perceived credibility, perceived usefulness and perceived ease of use are positively related to intention to use IB while brand name has no direct relationship with intention to use IB. Security concerns arise from the use of an open network where customers are afraid that their personal and financial information will become available to third parties through the Internet and can be used for fraudulent purposes. In order to create customer confidence in IB services, security evidence should be presented on the bank's web site. Banks should consider presenting a third-party guaranteeing security and the appearance of a privacy statement that promises to minimize risk and privacy breaches on their web sites. Security systems used for IB services should also be presented to customers in both technical and non-technical terms. Moreover, banks can outline the procedures and information on how to deal with problems if they occur and can provide instructions on their web sites on how to use IB services safely. The more users' risk and privacy concerns are reduced, the more likely they will adopt IB services.

To cater for customer needs for ease of use IB services, banks should provide high quality online system services. These services should be compatible with customers' application software and hardware and be user friendly. For example, there should be an easy login procedure and a high speed response to customers' input, while functions are easy navigate. Personalized services for different customer groups with identified and preferred service dimensions should also be provided. To promote the usefulness of IB services, providers should create awareness of IB benefits, which include allowing customers to control their accounts from anywhere, at a convenient time and at a low cost. Banks can use all forms of media advertising such as leaflets, brochures, web pages, etc. Bank tellers and customer service officers in branch networks can provide information about IB to potential adopters. Information should include reference to time savings, the convenience of access anywhere and at any time, low cost and information availability. In addition, banks should design their web sites as effective delivery channels and offer information beyond banking services, i.e., various types of financial and non-financial information.

REFERENCES

Aaker, D. A. (1996). *Building strong brands*. NY: Free Press.

ACNielsen. (2004). *Hong Kong online banking study*. Retrieved from http://acnielsen.com/news/20050714_ap.shtml

Aladwani, A. (2001). Online banking: A field study of drivers, development challenges, and expectations. *International Journal of Information Management*, *21*, 213–225. doi:10.1016/S0268-4012(01)00011-1

Anderson, E., & Weitz, B. (1989). The use of pledges to build and sustain commitment in distribution channels. *JMR, Journal of Marketing Research*, *29*, 18–34. doi:10.2307/3172490

Bagozzi, R. P., & Yi, Y. (1991). Multi-trait multi-method matrices in consumer research. *The Journal of Consumer Research*, *17*, 426–439. doi:10.1086/208568

Belanger, F., Hiller, J. S., & Smith, W. J. (2002). Trustworthiness in electronic commerce: The role of privacy, security, and site attributes. *The Journal of Strategic Information Systems*, *11*, 245–270. doi:10.1016/S0963-8687(02)00018-5

Berry, L. L. (2000). Cultivating service brand equity. *Journal of the Academy of Marketing Science*, *28*, 128–137. doi:10.1177/0092070300281012

Bestavros, A. (2000). Banking industry walks "tightrope" in personalization of Web services. *Bank Systems and Technology*, *37*, 54–56.

Cabral, L. M. B. (2000). Stretching firm and brand reputation. *The Rand Journal of Economics*, *31*, 658–673. doi:10.2307/2696353

Chan, S. C., & Lu, M. T. (2004). Understanding Internet banking adoption and use behavior: A Hong Kong perspective. *Journal of Global Information Management*, *12*, 2–43.

Cheng, T. C. E., Lam, D. Y. C., & Yeung, A. C. L. (2006). Adoption of Internet banking: An empirical study in Hong Kong. *Decision Support Systems*, *42*(3), 1558–1572. doi:10.1016/j.dss.2006.01.002

Dave, D. (2003). *How to build and maintain a powerhouse brand: Branding basics*. Retrieved from http://www.DaveDolak.com

Davis, F. D. (1989). Perceived usefulness, perceived ease of use, and user acceptance of information technology. *MIS Quarterly*, *13*, 318–339. doi:10.2307/249008

Davis, F. D., Bagozzi, R. P., & Warshaw, P. R. (1989). User acceptance of computer technology: a comparison of two theoretical models. *Management Science*, *35*, 982–1002. doi:10.1287/mnsc.35.8.982

de Chernatony, L., & Dall'Olmo, R. F. (1998). Defining a "brand": Beyond the literature with experts' interpretations. *Journal of Marketing Management*, *14*, 417–443. doi:10.1362/026725798784867798

Dodds, W., Monroe, K., & Grewal, D. (1991). Effects of price, brand and store information on buyers' product evaluation. *JMR, Journal of Marketing Research*, *28*, 307–319. doi:10.2307/3172866

Doney, P., & Cannon, J. (1997). An examination of the nature of trust in buyer-seller relationships. *Journal of Marketing*, *61*, 35–51. doi:10.2307/1251829

Featherman, M. S., & Pavlou, P. A. (2002). *Predicting e-services adoption: A perceived risk facets perspective*. Paper presented at the Eighth American's Conference on Information Systems, Dallas, Texas.

Fornell, C., & Larcker, D. F. (1981). Evaluating structural equation models with unobservable variables and measurement error. *JMR, Journal of Marketing Research*, *18*, 39–50. doi:10.2307/3151312

Friedman, B., Kahn, P. H. Jr, & Howe, D. C. (2000). Trust online. *Communications of the ACM*, *43*, 34–40. doi:10.1145/355112.355120

Furnell, S., & Karweni, T. (1999). Security implications of electronic commerce: A survey of consumers and business. *Internet Research. Electronic Networking Applications and Policy*, *9*, 372–382. doi:10.1108/10662249910297778

Gefen, D., Karahanna, E., & Straub, D. W. (2003). Trust and TAM in online shopping: An integrated model. *MIS Quarterly*, *27*, 51–90.

Grace, D., & O'Cass, A. (2005). Service branding: Consumer verdicts on service brands. *Journal of Retailing and Consumer Services*, *12*, 125–139. doi:10.1016/j.jretconser.2004.05.002

Hoffman, D. L., Novak, T. P., & Peralta, M. (1999). Building consumer trust online. *Communications of the ACM*, *42*, 80–85. doi:10.1145/299157.299175

Hong Kong Monetary Authority. (2008). *Banking policy and supervision*. Retrieved from http://www.info.gov.hk/hkma/eng/bank/index.htm

Keller, K. L. (1998). *Strategic brand management: Building, measuring, and managing brand equity*. Upper Saddle River, NJ: Prentice Hall.

Larzelere, R. E., & Huston, T. L. (1980). The dyadic trust scale: toward understanding interpersonal trust in close relationships. *Journal of Marriage and the Family*, *42*, 595–604. doi:10.2307/351903

Lee, M., & Turban, E. (2001). A trust model for consumer Internet shopping. *International Journal of Electronic Commerce*, *6*, 75–91.

Margulis, S. (1977). Conceptions of privacy: Current status and next steps. *The Journal of Social Issues*, *33*, 5–12.

Matthew, L. M., Ostrom, A. L., Roundtree, R. I., & Bitner, M. J. (2000). Self-service technologies: Understanding customer satisfaction with technology based service customers. *Journal of Marketing*, *64*, 50–65.

Moore, G. C., & Benbasat, I. (1991). Development of an instrument to measure the perceptions of adopting an information technology innovation. *Information Systems Research, 2*, 192–222. doi:10.1287/isre.2.3.192

Nunnally, J. C. (1978). *Psychometric theory* (2nd ed.). NY: McGraw-Hill.

Olson, J. C. (1977). Price as an informational cue: Effects on product evaluations. In A. G. Woodside et al. (Eds.), *Consumer and industrial buying behavior* (pp. 267-286). NY: North Holland.

Oxoby, R. J., & Finnigan, H. (2007). Developing heuristic-based quality judgments: Blocking in consumer choice. *Psychology and Marketing, 24*, 295–393. doi:10.1002/mar.20162

Podsakoff, P. M., Mackenzie, S. B., Lee, J. Y., & Podsakoff, N. P. (2003). Common method biases in behavioral research: A critical review of the literature and recommended remedies. *The Journal of Applied Psychology, 88*, 879–903. doi:10.1037/0021-9010.88.5.879

Rao, A. R., & Monroe, K. B. (1989). The effect of price, brand name, and store name on buyers' perception of product quality: an integrative review. *JMR, Journal of Marketing Research, 26*, 351–357. doi:10.2307/3172907

Rogers, E. M. (1995). *Diffusion of innovations* (4th ed.). New York: The Free Press.

Saunders, J. (1993). Branding financial services. *International Journal of Bank Marketing, 11*(6), 32–38. doi:10.1108/02652329310045747

Shankar, V., Urban, G. L., & Sultan, F. (2002). Online trust: A stakeholder perspective, concepts, implications, and future directions. *The Journal of Strategic Information Systems, 11*, 325–344. doi:10.1016/S0963-8687(02)00022-7

Shrimp, T. A. (2003). *Advertising promotion: Supplementary aspects of integrated marketing communications* (5th ed.). Orlando, FL: Dryden Press.

Suh, B., & Han, I. (2002). Effect of trust on customer acceptance of Internet banking. *Electronic Commerce Research and Applications, 1*, 247–263. doi:10.1016/S1567-4223(02)00017-0

U.S. Public Interest Research Group. (2000, April 25). *Public comment on barriers to electronic commerce, response to call by U.S. Department of Commerce* (65 []. Boston: USDC.]. *Federal Register*, 15898.

Van Osselaer, S. M. J., & Alba, J. W. (2000). Consumer learning and brand equity. *The Journal of Consumer Research, 27*, 1–16. doi:10.1086/314305

Venkatesh, V., & Davis, F. D. (1996). A model of the antecedents of perceived ease of use: development and test. *Decision Sciences, 27*, 451–481. doi:10.1111/j.1540-5915.1996.tb01822.x

Venkatesh, V., & Davis, F. D. (2000). A theoretical extension of the technology acceptance model: four longitudinal field studies. *Management Science, 46*, 186–204. doi:10.1287/mnsc.46.2.186.11926

Venkatesh, V., & Morris, M. G. (2000). Why don't men ever stop to ask for directions? Gender, social influence, and their role in technology acceptance and usage behavior. *MIS Quarterly, 24*, 115–139. doi:10.2307/3250981

Wang, Y. S., Wang, M. Y., Lin, H. H., & Tang, T. I. (2003). Determinants of user acceptance of Internet banking: An empirical study. *International Journal of Service Industry Management, 14*(5), 501–518. doi:10.1108/09564230310500192

Warrington, T. B., Abgrab, N. J., & Caldwell, H. M. (2000). Building trust to develop competitive advantage in e-business relationships. *Competitiveness Review, 10*, 160–168.

APPENDIX 1. 26 ITEMS IN THE SURVEY QUESTIONNAIRE

1. The brand name tells me a lot about what to expect from the Internet banking services of this bank.

2. The brand name tells me a lot about the Internet banking services of this bank.

3. The brand name of this bank means something about the Internet banking services to me.

4. The brand name of this bank sends a message to me about the Internet banking services.

5. The brand name of this bank tells me everything I need to know about the Internet banking services of this bank.

6. Internet banking services are easy to use.

7. Internet banking is an easy way to conduct banking transactions.

8. Learning to operate Internet banking is easy for me.

9. It is easy for me to remember how to perform tasks using Internet banking.

10. I believe it is easy to get Internet banking to do what I want it to do.

11. Using Internet banking does not require a lot of effort.

12. Internet banking provides a clearer interface than other channels.

13. Internet banking makes it easier for me to conduct banking transactions.

14. Internet banking provides me greater control over my finances.

15. Internet banking allows me to manage my finances more efficiently.

16. Internet banking is a convenient way to manage my finances.

17. Internet banking is more user-friendly than other banking channels such as phone banking, ATM, branches, etc.

18. Internet banking eliminates time constraints and can be used at any time.

19. Internet banking eliminates geographic limitation and can be used anywhere with an Internet connection.

20. I will use Internet banking for my personal needs in the coming 6 months.

21. Using Internet banking services for banking transactions is something I will do in the coming 6 months.

22. I am interested in using Internet banking for different banking services such as securities trading services, insurance services, investment funds services, and MPF services, etc.

23. I would feel secure sending sensitive information during the use of Internet banking services.

24. Internet banking is a secure means through which to perform banking transactions.

25. I would feel totally safe when using Internet banking services.

26. Overall, Internet banking is a safe environment to perform banking transactions.

APPENDIX 2. REGRESSION WEIGHTS OF CONSTRAINED, MEASUREMENT INVARIANCE AND STRUCTURAL INVARIANCE MODELS (TABLE 11)

Table 11.

Regression weights		Unconstrained				Measurement invariance				Structural invariance			
		HSBC		Non HSBC		HSBC		Non HSBC		HSBC		Non HSBC	
		C.R.	P	C.R.	P	C.R.	P	C.R.	P	C.R.	P	C.R.	P
	BN < -- > PC	7.196	***	4.683	***	7.247	***	4.767	***	8.380	***	8.380	***
	BN < -- > PEOU	7.122	***	6.290	***	7.245	***	6.635	***	9.364	***	9.364	***
	BN < -- > PU	0.036	0.971	-0.233	0.816	0.029	0.977	-0.258	.796	-0.126	0.900	-0.126	0.900
	PEOU < -- > PU	9.323	***	8.163	***	9.680	***	8.987	***	12.444	***	12.444	***
	PC < -- > PU	4.065	***	5.146	***	4.092	***	5.349	***	6.414	***	6.414	***
	BN < -- > INT	-0.307	0.759	-3.201	0.001	-0.298	0.766	-3.217	0.001	-2.448	0.014	-2.448	0.014
	PU < -- > INT	3.470	***	3.756	***	3.487	***	3.776	***	4.872	***	4.872	***
	PC < -- > INT	4.505	***	2.439	0.015	4.493	***	2.453	0.014	5.673	***	5.673	***
	PEOU < -- > INT	3.644	***	1.384	0.166	3.674	***	1.354	0.176	3.874	***	3.874	***

This work was previously published in International Journal of E-Business Research Volume 6, Issue 1, edited by In Lee, pp. 32-51, copyright 2010 by IGI Publishing (an imprint of IGI Global).

Chapter 17
A Study of the Impact of Individual Differences on Online Shopping

Jianfeng Wang
Indiana University of Pennsylvania, USA

Linwu Gu
Indiana University of Pennsylvania, USA

Milam Aiken
University of Mississippi, USA

ABSTRACT

Previous studies explored the effects of individual cognitive and affective reactions on online shopping and those of individual differences on usage of information systems. However, few studies evaluated how individual differences affect online shopping. This paper draws on the theory of reasoned action (TRA) and the technology acceptance model (TAM) as it attempts to gain understanding of individual difference impacts on online shopping attitudes. The model was tested using data collected from a Web survey, and 171 questionnaires were collected and analyzed. Results show that online shopping experience, level of Internet usage, cognitive absorption, personal innovativeness, and computer self-efficacy positively influence online shopping attitudes, and that perceived usefulness is a significant mediator, but only between online shopping attitudes and two individual differences variables: the level of Internet usage and cognitive absorption.

INTRODUCTION

In 2008, about 1.4 billion people (22% of the world's population) used the Internet (Internet World Stats, 2009), and in the United States, almost 60% of the 220 million Internet users spent $138 billion online, an increase of 7.6% from the prior year (U.S. Census Bureau, 2009). Yet, there are many questions remaining about why consumers buy online. According to the Nielson Company (2009), about 40% of the world's online population has made a purchase online (an increase of 10% from two years ago), and 60% of global online

DOI: 10.4018/978-1-61350-462-8.ch017

consumers used their credit card for a recent online purchase. Given the rapid growth of electronic commerce, continued research on the factors that influence online shopping is vital.

Within IS usage research, individual differences can be classified into four categories: cognitive style, personality traits, demographics, and situational factors (Alavi & Joachimsthaler, 1992; Karahanna, Ahuja, Srite, & Galvin, 2002). Many variables have been discussed in studies of online shopping such as product perceptions, shopping experience, perceived characteristics of the Web, consumer characteristics, and consumer risks (Chang, Cheung, & Lai, 2005), and demographic information has been collected by der Heijden, Verhagen, & Creemers (2003), Shih (2004), and Torkzadeh and Dhillon (2002). However, no single study has explored all four categories of individual differences, and this paper attempts to address this shortfall by incorporating TRA (Fishbein & Ajzen, 1975) and TAM (Davis, 1989) into the study of online shopping attitudes.

Theoretical Foundation

Davis (1989) found that an individual's attitude towards using a computer system is directly affected by two beliefs: perceived usefulness and perceived ease of use. The goal of TAM (Figure 1) is to explain the general determinants of computer acceptance, and this model has been used in many studies to explore user perceptions of system use and the probability of adopting an online system.

Further, TAM has been used to predict user acceptance (Lederer, Maupin, Sena, & Zhuang, 2000). The model helps researchers understand the factors of technology adoption that affect user acceptance and why people resist using computers (Chung & Tan, 2004). A few studies have augmented TAM with other theories in order to improve its specificity and explanatory power (e.g., Hu, Chau, Sheng, & Tam, 1999; Legris, & Collerette, 2003), and the model has also been extended to explore the user acceptance of an online system (Chung & Tan, 2004; Shang, Chen, & Shen, 2005; Shih, 2004).

Attitudes toward behavior and subjective norm are two independent variables used in TRA (Fishbein & Ajzen, 1975). The first variable in the model (Figure 2) is defined as the individual's positive or negative feelings about performing a behavior, and the second is defined as an individual's perception of whether people important to the individual think the behavior should be performed. TRA explicitly describes the mechanisms through which individual differences influence behavior, and it proposes that attitudes are impacted by people's beliefs about whether others think they should perform the action. TRA has been used for predicting cognitive and affective behavior using the belief–attitude relationship in social psychology (Shih, 2004), and Ajzen and Fishbein (1980) have shown how individual differences such as personality, cogni-

Figure 1. Technology acceptance model (TAM) (From Davis, 1989)

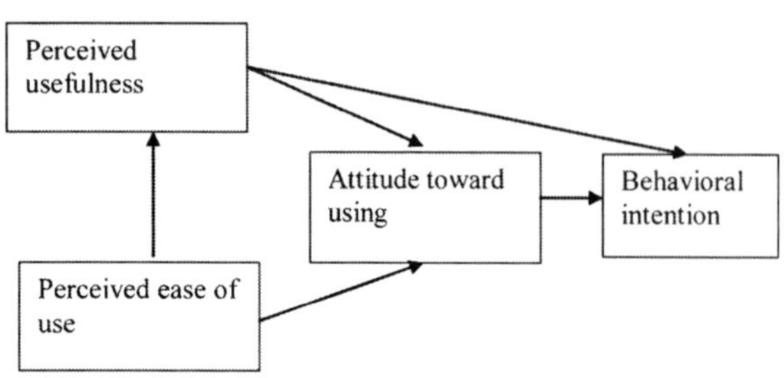

Figure 2. Theory of reasoned action (TRA) model (From Fishbein and Ajzen, 1975)

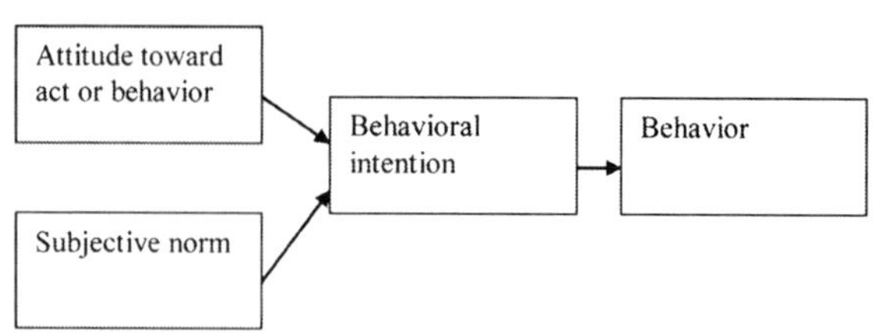

tive style, demographic traits, situational factors, and behavior, reciprocally influence each other.

Integrating TAM with other theories could improve its specificity and explanatory power (Legris et al., 2003; Shang et Al., 2005). For example, normative beliefs and self efficacy were added to the TAM in the research of online shopping performed by Vijayasarathy (2004). By incorporating factors from both TRA and TAM, a new model can be used to predict and explain how individual differences affect online attitudes. In this model (Figure 3), perceived risk is treated as one of the two mediating variables between individual differences and online shopping attitudes. Online consumers are concerned with risks about the security of transmitting credit card information over the Internet, and they can also be worried about buying something without touching it (Bhatnagar et al., 2000). Obviously, when consumers perceive lower risks associated with online shopping, their online shopping attitudes become more positive.

RESEARCH AND HYPOTHESIS

Perceived Usefulness

Davis (1989) defines perceived usefulness as "the degree to which a person believes that using a particular system would enhance his or her job performance," and the effects of perceived usefulness on online shopping have been tested in previous studies (e.g., Shang et al. 2005; Vijayasarathy, 2004). Perceived usefulness is operationalized as perceived value (Shih, 2004). In this paper, we define perceived usefulness as the degree to which an individual believes that online shopping will enhance his/her benefits. Therefore, perceived usefulness concerns cost reduction, product value, and time-saving in our study. Cost saving might increase customer satisfaction and loyalty (Phillips et al., 1997). Time-saving is an important perceived consequence of online shopping, and is a vital predictor of online buying (Limayem, Khalifa, & Frini, 2000). In addition, product value

Figure 3. Research model of individual differences on online shopping attitudes

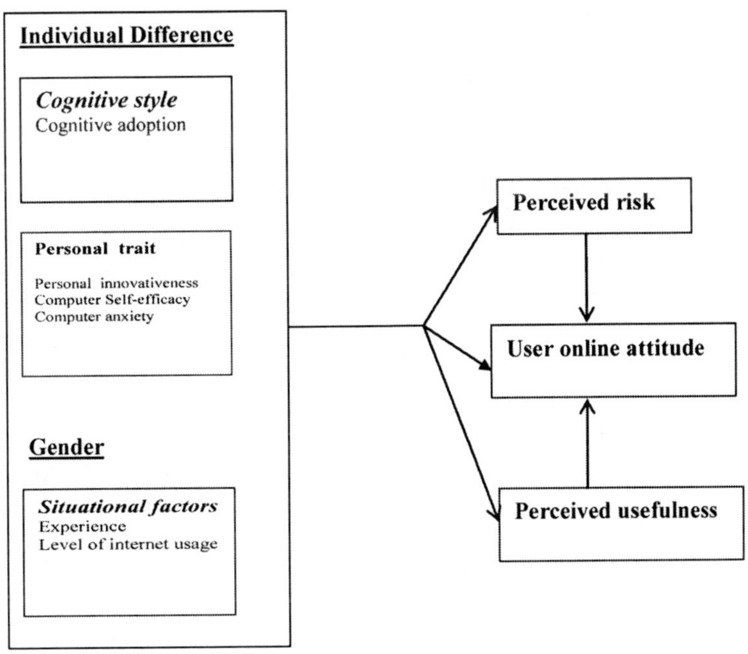

is an important factor that influences the success of Internet commerce (Torkzadeh & Dhillon, 2002).

Perceived Risks

Perceived risks are commonly thought of as the uncertainty felt in a situation. It is defined as "a combination of uncertainty plus seriousness of outcome involved" (Bauer, 1967). Further, perceived risks have significant and negative influences on online shopping attitudes (Jarvenpaa, Tractinsky, & Vitale, 2000; Kimery & McCord, 2002). In at least one study (Chang et al., 2005), these risks have been measured by asking respondents to assess whether buying goods online was risky and whether they were concerned with system security, credit card fault, and product risk.

Perceived risks are prominent barriers to the completion of online purchases (Hoffmann, Novak, & Peralita, 1999; Jarvenpaa & Tractinsky, 1999). Consumers who think they are risk-averse have more online purchases (den Poel & Leunis, 1999), and perceived risks are antecedents of attitudes towards online purchasing (der Heijden et al., 2003). Further, consumers might have a low intention of purchasing due to the inconvenience of online shopping (Hoffman, Novak, & Chatterjee, 1995), and they might hesitate to use their credit card numbers for online shopping because they have concerns for security (Salkin, 1999). Like perceived usefulness, which relates to personal traits (Karahanna et al., 2002), individuals can have different perceived risks when evaluating products and services for purchase (Featherman & Pavlou, 2003).

Individual Differences

Individual differences influence behavior via their effects on individual beliefs (Karahanna et al., 2002), and cognitive style, personality traits, demographic features, and situational factors are four groups of differences often used in social psychology research.

Cognitive Style

Cognitive style refers to experience and common sense and an individual's characteristic way of thinking, using information, solving problems, and making decisions (Alavi & Joachimsthalter, 1992; Karahanna et al., 2002). Cognitive style is also correlated with job function and job level. In this paper, cognitive style focuses on cognitive absorption or personal cognitive state, which was defined by Agarwal and Karahanna (2000) as "a state of deep involvement with software." It impacts the intrinsic motivations of online shopping and can also influence online shopping indirectly through its effects on perceived usefulness. Higher cognitive absorption is associated with a satisfied shopping experience, and consumers with high cognitive absorption online will be more likely to have a positive online shopping attitude and be more likely to shop online (Shang et al., 2005). Therefore, we propose:

H1a: Cognitive absorption positively influences perceived usefulness.
H1b: Cognitive absorption negatively influences perceived risks.
H1c: Cognitive absorption positively influences online shopping attitudes.

Personality

Personality is defined as "the cognitive and affective structures maintained by individuals to facilitate judgment to the events, people, and situation" (Alavi & Joachimsthalter, 1992).

Online shopping self–efficacy is a consumer's self–assessment of his/her capabilities to shop online. Computer self-efficacy refers to an individual assessment of one's capability to use a computer and has been found to affect computer usage (Compeau & Higgins, 1995). Computer anxiety is the tendency of an individual to be uneasy, apprehensive, or fearful about the current or future use of computers (Parasuraman & Igbaria, 1990).

While computer self-efficacy relates to individual judgments about their ability, computer anxiety is a negative affective reaction toward computer use.

Innovativeness refers to "the degree to which an individual is receptive to new ideas" (Chang et al., 2005) and has a significant positive relationship with online shopping (Blake et al., 2003; Goldsmith, 2002).

The perceptions about the innovation and perceived behavioral control were significant predictors of the intentions to adopt online services (Tan & Teo, 2000).

We propose that:

H2a: Personal innovativeness and computer self-efficacy have positive but computer anxiety has negative influences on perceived usefulness.

H2b: Personal innovativeness and computer self-efficacy have negative but computer anxiety has positive influences on perceived risks.

H2c: Personal innovativeness and computer self-efficacy have positive but computer anxiety has negative influences on online shopping attitudes.

Demographic Features

Demographic information of individuals such as gender, age, race, income level, and education are commonly used as independent variables. Differences between males and females were found to affect decision-making and information preferences (Powell & Johnson, 1995). Males were found to purchase more online in some studies, but there were no differences in others (Burroughs & Sabherwal, 2001; Chang, Torkzadeh, & Dhillon, 2004; Donthu & Garcia, 1999; Li et al.,1999), and there are mixed reports on whether consumers' income and age have significant positive effects on online shopping (Bhatnagar et al., 2000; Goldsmith & Goldsmith, 2002). To simplify our analysis, we use gender as a moderating dummy variable in the model, where male is assigned a value of 1 and female a value of 0.

We propose that:

H3a: Gender positively influences perceived usefulness.

H3b: Gender negatively influences perceived risks.

H3c: Gender positively influences online shopping attitudes.

Situational Factors

User situational factors include training, experience, and user involvement. Experience using a Web site is found to be one of the important factors in e-commerce (Jarvenpaa & Todd 1996). Online shopping experience, which is typically used as a variable of the situational factor construct, is concerned with user-friendliness and the fun of shopping online (Shang et al., 2005), and some studies suggest that online shopping experience relates to Internet acceptance (Lederer et al., 2000; Pitknow & Kehoe, 1996) and evaluated effort, compatibility, and playfulness (Chang et al., 2004). The second variable for the situational factors is the level of Internet usage which influences perceived consequences (Chang et al., 2005).

We propose that:

H4a: Level of Internet usage and shopping experience positively influence perceived usefulness.

H4b: Level of Internet usage and shopping experience negatively influence perceived risks.

H4c: Level of Internet usage and shopping experience positively influence online shopping attitudes.

Attitudes

Belief is associated with the individual subjective assessment present with some behavior, whereas attitudes are associated with individual affective feelings about presenting the behavior (Hubona & Blanton, 1996). Shih (2004) defined attitudes as "the positive or negative feelings of an individual toward a specific behavior." Attitude has been characterized as a person's preference of a certain response to a concept or object and it is defined as "the extent to which a consumer likes on-line shopping and considers it to be a good idea" (Vijayasarathy, 2004).

Previous studies' results show that perceived usefulness and perceived risk are highly relevant in assessing attitude towards on-line shopping. For example, Kimery and McCord (2002) found perceived risk reduction had a positive influence on the attitude towards online shopping, and Torkzadeh and Dhillon (2002) found shipping errors had negative effects on online shopping attitude.

We propose that:

H5a: Perceived usefulness positively influences online shopping attitudes.
H5b: Perceived risks negatively influence online shopping attitudes.

RESEARCH AND METHODOLOGY

Variables

The dependent variable in our study is the online shopping attitude of an individual consumer. The independent variables are grouped into four categories: personality traits, cognitive style, demographic features, and situational factors, and there are two mediate variables: perceived risks and perceived usefulness. All variables except gender, online shopping experience, and level of Internet usage, are measured by asking users their perceptions, and all instruments use a 7-point Likert-type scale, where 1= strongly disagree and 7= strongly agree.

Independent Variables

This paper uses measures of computer anxiety developed by Parasuraman and Igbaria (1990), personal innovativeness by Karahanna et al. (2002), cognitive absorption by Shang, et al. (2005), and computer self–efficacy by Compeau and Higgins (1995). There are 10 items for computer self-efficacy, 3 items for personal innovativeness, 14 items for cognitive absorption, and 5 items for computer anxiety. The online shopping experience is measured by the frequency of online shopping purchases, and the level of Internet usage is measured in terms of hours per week.

Mediator Variables

The measure of perceived usefulness consists of Internet shopping convenience, product value, time-saving, and cost reduction. The perceived risks are measured by shipping errors and uncertainty. Internet shopping convenience, product value, and shipping errors are all measured using the scale developed by Torkzadeh and Dhillon (2002). Time saving was measured following the scale developed by Vijayasarathy (2004), cost reduction using Kuk (2004), and uncertainty using Devaraj, Fan, and Kohli (2002).

Dependent Variables

Online shopping attitude is measured using the scale provided by Limayem et al. (2000). We used the 4 items of attitudes with a 7-point Likert-type scale, where 1= "strongly disagree" and 7= "strongly agree."

Subjects and Procedure

A survey instrument containing 60 questions for 11 variables (see Appendix 1) was administered to students in multiple sections of an undergraduate MIS course and students in an MBA course at a

Table 1. Weekly hours of Internet use

participants	No. of hours online weekly
30	More than 15 hours
64	10-15 hours
60	5-10 hours
16	fewer than 5 hours

large state university in the northeast region of the United States.

DATA ANALYSIS AND RESULTS

Data Summary

The mean age of the 171students (99 males, 72 females) who participated in the survey was 21.53 years, and 81.87% of the students were between 18 and 22. There were no significant differences among the age groups, and 137 were White, 16 were Black, 12 were Asian, and 6 were categorized as "others." As shown in Table 1, only about 10% of those surveyed spent fewer than 5 hours online, and only 4% never made a purchase online (see Table 2).

TESTING THE RESEARCH MODEL

Construct Reliability

The reliability (internal consistency) of the items in each scale was examined using Cronbach's alpha to confirm the adequacy of the measures for testing hypotheses (see Table 3). The coefficients confirmed the results of earlier studies and provided confidence in testing the hypotheses.

Table 2. Number of online purchases

participants	Frequency of online purchases
10	More than 10 times
29	5 to 10 times
54	3-5 times
60	1-3 times
7	0 time

Testing the Effects of Individual Differences on Online Shopping Attitude.

Multiple regression analysis was used to test the individual difference effects on the online shopping attitude. Table 4 shows that the effects of all the individual difference variables except gender are significant. Computer anxiety negatively influences online shopping attitudes. Experience, level of Internet usage, cognitive absorption, personal innovativeness, and computer self-efficacy positively affect online shopping attitudes. So H1c, H2c, and H4c are accepted, but H3c is rejected.

Testing the Effects of Perceived Usefulness and Perceived Risks on Online Shopping Attitude

Perceived usefulness and perceived risks both are shown to have significant effects on online shopping attitude (Table 5), although in different directions. Therefore, H5a and H5b are accepted. Perceived risks negatively affect online shopping attitudes, just as Kimery and McCord (2002) predicted. Perceived usefulness has a positive influence on online shopping attitudes.

Table 3. Construct reliability test

Constructs	Number of items	Construct reliability
Cognitive absorption	14	.847
Computer Self-Efficacy	10	.896
Computer Anxiety	5	.747
Personal innovativeness	3	.779
Perceived usefulness	18	.895
Perceived risks	6	.832
Online shopping Attitude	4	.857

Table 4. The effects of individual differences on online shopping attitudes

Dependent variable	Independent variable	Beta	t	Sig.
Online shopping attitude	Gender	-.041	-.977	.330
	Experience	.112	1.998	.047
	Level of Internet usage	.536	8.911	<.001
	Cognitive absorption	.162	3.335	.001
	Personal innovativeness	.128	2.869	.005
	Computer Self-efficacy	.146	2.795	.006
	Computer anxiety	-.087	-2.024	.045

Table 5. The effects of perceived usefulness and perceived risks on online shopping attitudes

Dependent variable	Independent variable	Beta	t	Sig.
Online shopping attitude	Perceived Usefulness	.582	9.732	<.001
	Perceived Risks	-.210	-3.504	.001

Mediation Test

Three serial regressions are required in testing mediation as suggested by Baron and Kenny (1986). The regressions of perceived usefulness and risks on the individual difference variables were done in the first step to test whether the independent variables affect the mediate variables. The independent variables must have significant effects on the mediators. Otherwise, there is no mediation. Tables 6 and 7 show that cognitive absorption, computer self–efficacy, and level of Internet usage significantly influence perceived usefulness, but they have no significant effects on perceived risks. In addition, online-shopping experience significantly influences perceived risks, but it has no significant effects on perceived usefulness. Therefore, H1a is accepted and H1b is rejected. H2a, H4a, and H4b are partially accepted. Among the independent variables, only personal innovativeness significantly affects both perceived usefulness and perceived risks at the 0.2% and 2% level of significance respectively. Only computer anxiety is not shown to have any effects on perceived usefulness or perceived risks. Gender has an insignificant effect on perceived risks.

Table 6. The effects of individual differences on perceived usefulness

Dependent variable	Independent variable	Beta	t	Sig.
Perceived usefulness	Gender	-.019	-.317	.752
	Experience	-.003	-.034	.973
	Level of Internet usage	.265	3.146	.002
	Cognitive absorption	.195	2.871	.005
	Personal innovativeness	.196	3.130	.002
	Computer Self-efficacy	.300	4.106	<.001
	Computer anxiety	.058	.960	.339

Table 7. The effects of individual differences on perceived risks

Dependent variable	Independent variable	Beta	t	Sig.
Perceived risks	Gender	.146	1.927	.056
	Experience	-.218	-2.160	.032
	Level of Internet usage	.044	.407	.684
	Cognitive absorption	.036	.412	.681
	Personal innovativeness	-.188	-2.343	.020
	Computer self-efficacy	-.073	-.773	.440
	Computer anxiety	.101	1.311	.192

The regression of individual difference variables on online shopping attitude is the second step to check how independent variables affect the dependent variables (Table 4). The regression results indicate that all the individual difference variables except gender significantly affect online shopping attitudes.

The third step is a pooled regression of the dependent variables on both independent variables and mediate variables (Table 8). During this third step, perceived usefulness and perceived risks are shown to have significant effects on online shopping attitude (Table 8).

If there is mediation, and the above conditions all hold, the coefficients (the absolute values) of the independent variables must be smaller in the third regression than those in the second regression. In Table 8, level of Internet usage, cognitive absorption, and computer anxiety are significant individual differences, but they are insignificant with respect to perceived risks (Table 7). As a result, perceived risks cannot be considered as a significant mediator. In Table 6, computer anxiety is an insignificant individual difference with respect to perceived usefulness, but the level of Internet usage and cognitive absorption are significant. The coefficients of the two individual differences variables are smaller in Table 8 than in Table 4, and thus, the perceived usefulness is a significant mediator, but only between online shopping attitudes and two individual differences variables: the level of Internet usage and cognitive absorption.

Table 8. Coefficients of pooled regressions of online attitude on individual differences effects

Dependent variable	Independent variable	Beta	t	Sig.
Online attitude	Gender	-.019	-.468	.640
	Experience	.084	1.579	.116
	Level of Internet usage	.495	8.523	<.001
	Cognitive absorption	.132	2.830	.005
	Personal innovativeness	.069	1.579	.116
	Computer Self-efficacy	.083	1.618	.108
	Computer anxiety	-.084	-2.072	.040
	Perceived usefulness	.178	3.340	.001
	Perceived risks	-.132	-3.163	.002

Discussions and Business Implications

All the tested individual differences variables except gender had significant effects on online-shopping attitude, as expected. Online shopping experience, level of Internet usage, cognitive absorption, personal innovativeness, and computer self-efficacy positively influence online shopping attitudes. The results imply online shopping attitudes can be changed when consumers gain more knowledge about the Internet, when they use the network more, and when they gain more confidence in their capabilities with it and related applications.

Improved online shopping attitudes can boost online sales, and long-term investment in e-commerce is worthwhile. Though there are many people still not willing to disclose their credit card information or banking information online or are afraid of shopping online, the number will most likely decrease year by year.

The findings in this survey are also confirmed by recent developments in the business world such as increasing sales on Amazon and EBay. Our survey subjects are college undergraduates and graduates very active in using online services and entertainment. They spend a large amount of time online, and when these students get older, we expect that a larger percentage of the overall population consequently will be shopping over the Internet.

The coefficient of the level of Internet usage in the regression is.536 (Table 4). With the decreasing cost of using broadband Internet service and decreasing cost in computing overall, more people will try to be online and spend more time online. In addition, online businesses should make their online services or shopping systems easier, more secure, and more enjoyable as the coefficient of the cognitive absorption has a positive influence on both online shopping attitudes and perceived usefulness (Tables 4 and 5).

Cognitive absorption, level of Internet usage, personal innovativeness, and computer self-efficacy all affect perceived usefulness, but gender, experience, and computer anxiety had no significant effects on perceived usefulness. This implies that an easy and enjoyable online system is very important to attract users, even those who are not comfortable using computers or surfing the Internet. As long as consumers are willing to or can be motivated to use online systems, they will slowly discover the usefulness of the technology and their computer anxiety we be reduced.

Personal innovativeness and online shopping experience both significantly affected perceived risks negatively. That is, more innovative people perceive e-commerce to be less risky, and more online shopping experience reduces perceived risks.

A better understanding of the effects of individual differences upon online shopping attitudes will help e-businesses develop better strategies. Researchers might find that the TRA model is a very good predictor of consumer behavior, and practitioners can use the results to better design marketing and promotion programs and provide a more personalized service.

CONCLUSION AND LIMITATIONS

This paper explores the effects of four individual difference categories on perceived usefulness and risks and online shopping attitudes, based on the TRA and TAM models. The online shopping experience, the level of Internet usage, cognitive absorption, personal innovativeness, and computer self-efficacy all affect online shopping attitudes, which in turn can affect e-commerce sales, but this latter relationship was not explored.

There are some limitations in the research, however. First, the paper focuses on online shopping attitudes only; not upon how these attitudes could influence e-commerce sales. In addition, a better conceptualization of perceived risks might be needed. That is, the instrument for measuring perceived risks included shipping errors and uncertainty, but it did not include product risks and transaction risks. Finally, subjects were not sampled randomly but came from a specific population, i.e., college students. This population uses the Internet, but their shopping behavior might be different from older Web users with higher incomes, for example.

REFERENCES

Agarwal, R., & Karahanna, E. (2000). Time flies when you're having fun: Cognitive absorption and beliefs about information technology usage. *MIS Quarterly*, *24*(4), 665–694. doi:10.2307/3250951

Ajzen, I., & Fishbein, M. (1980). *Understanding attitudes and predicting social behavior*. Englewood Cliffs, NJ: Prentice-Hall.

Alavi, M., & Joachimsthalter, E. (1992). Revisiting DSS implementation research: A meta-analysis of the literature and suggestions for researchers. *MIS Quarterly*, *16*(1), 95–116. doi:10.2307/249703

Baron, R., & Kenny, D. (1986). The moderator-mediator variable distinction in social psychological research: Conceptual, strategic, and statistical considerations. *Journal of Personality and Social Psychology*, *51*(6), 1173–1182. doi:10.1037/0022-3514.51.6.1173

Bauer, R. (1967). Consumer behavior as risk taking. In D. Cox, *Risk taking and information handling in consumer behavior*. Cambridge, MA: Harvard University Press.

Chang, J., Torkzadeh, G., & Dhillon, G. (2004). Re-examining the measurement models of success for Internet commerce. *Information & Management*, *41*, 577–584. doi:10.1016/S0378-7206(03)00091-0

Chang, M., Cheung, W., & Lai, V. (2005). Literature derived reference models for the adoption of online shopping. *Information & Management*, *42*(4), 543–559. doi:10.1016/S0378-7206(04)00051-5

Compeau, D., & Higgins. (1995). Computer self-efficacy: Development of a measure and initial test. *MIS Quarterly*, *19*(2), 189–211. doi:10.2307/249688

Davis, F. (1989). Perceived usefulness, perceived ease of use, and user acceptance of information technology. *MIS Quarterly*, *13*(3), 318–339. doi:10.2307/249008

den Poel, D., & Leunis, J. (1999). Consumer acceptance of the Internet as a channel of distribution. *Journal of Business Research*, *45*(3), 249–256. doi:10.1016/S0148-2963(97)00236-1

der Heijden, H., Verhagen, T., & Creemers, M. (2003). Understanding online purchase intentions: Contributions from technology and trust perspectives. *European Journal of Information Systems*, *12*(1), 41–48. doi:10.1057/palgrave.ejis.3000445

Devaraj, S., Fan, M., & Kohli, R. (2002). Antecedents of B2C channel satisfaction and preference: Validating e-commerce metrics. *Information Systems Research*, *13*(3), 316–333. doi:10.1287/isre.13.3.316.77

Featherman, M., & Pavlou, P. (2003). Predicting e-services adoption: A perceived risk facets perspective. *International Journal of Human-Computer Studies*, *59*(4), 451–474. doi:10.1016/S1071-5819(03)00111-3

Fishbein, M., & Ajzen, I. (1975). *Belief, attitude, intention, and behavior: An introduction to theory and research*. Reading, MA: Addison-Wesley.

Hoffman, D., Novak, T., & Chatterjee, P. (1995). Commercial scenarios for the Web: Opportunities and challenges. *Journal of Computer-Mediated Communication*, *1*(3).

Hoffman, D., Novak, T., & Peralita, M. (1999). Building consumer trust online. *Communications of the ACM, 42*(4), 80–85. doi:10.1145/299157.299175

Hu, P., Chau, P., Sheng, O., & Tam, K. (1999). Examining the technology acceptance model using physician acceptance of telemedicine technology. *Journal of Management Information Systems, 16*(2), 91–112.

Hubona, G., & Blanton, J. (1996). Evaluating system design features. *International Journal of Human-Computer Studies, 44*(1), 93–118. doi:10.1006/ijhc.1996.0005

Internet World Stats. (2009). *Internet world stats: Usage and population statistics*. Retrieved January 24, 2009, from http://www.internetworldstats.com/stats.htm

Jarvenpaa, S., & Todd, P. (1996). Consumer reactions to electronic shopping on the World Wide Web. *International Journal of Electronic Commerce, 1*(2), 59–88.

Jarvenpaa, S., & Tractinsky, N. (1999). Consumer trust in an Internet store: A cross–cultural validation. *Journal of Computer-Mediated Communication, 5*(2), 1–35.

Jarvenpaa, S., Tractinsky, N., & Vitale, M. (2000). Consumer trust in an Internet store. *Information Technology and Management, 1*(1-2), 45–71. doi:10.1023/A:1019104520776

Karahanna, E., Ahuja, M., Srite, M., & Galvin, J. (2002). Individual differences and relative advantage: The case of GSS. *Decision Support Systems, 32*(4), 327–341. doi:10.1016/S0167-9236(01)00124-5

Kimery, K., & McCord, M. (2002). Third party assurances: Mapping the road to trust in e-retailing. *Journal of Information Technology Theory and Application, 4*(2), 63–81.

Kuk, G. (2004). Effectiveness of vendor-managed inventory in the electronics industry: Determinants and outcomes. *Information & Management, 41*(5), 645–654. doi:10.1016/j.im.2003.08.002

Lederer, A., Maupin, D., Sena, M., & Zhuang, Y. (2000). The technology acceptance model and the World Wide Web. *Decision Support Systems, 29*(3), 269–282. doi:10.1016/S0167-9236(00)00076-2

Legris, P., Ingham, J., & Collerette, P. (2003). Why do people use information technology? A critical review of the technology acceptance model. *Information & Management, 40*(3), 191–204. doi:10.1016/S0378-7206(01)00143-4

Limayem, M., Khalifa, M., & Frini, A. (2000). What makes consumers buy from Internet? A longitudinal study of online shopping. *IEEE Transactions on Systems, Man, and Cybernetics, 30*(4), 421–432. doi:10.1109/3468.852436

Nielson Company. (2009). *Global online shopping report, February 2008*. Retrieved January 24, 2009 from http://www.nielsen.com/solutions/GlobalOnlineShoppingReportFeb08.pdf

Parasuraman, S., & Igbaria, M. (1990). An examination of gender differences in the determinants of computer anxiety and attitudes toward microcomputers among managers. *International Journal of Man-Machine Studies, 32*(3), 327–340. doi:10.1016/S0020-7373(08)80006-5

Pitknow, J., & Kehoe, C. (1996). Emerging trends in the WWW user population. *Communications of the ACM, 39*(6), 106–108. doi:10.1145/228503.228525

Salkin, S. (1999). Fear of buying. *Logistics Management and Distribution Report, 8*(5), 101–104.

Shang, R., Chen, Y., & Shen, L. (2005). Extrinsic versus intrinsic motivations for consumers to shop online. *Information & Management, 42*(3), 401–413. doi:10.1016/j.im.2004.01.009

Shih, H. (2004). An empirical study on predicting user acceptance of e-shopping on the Web. *Information & Management*, *41*(3), 351–368. doi:10.1016/S0378-7206(03)00079-X

Tan, M., & Teo, T. (2000). Factors influencing the adoption of Internet banking. *Journal of the Association for Information Systems*, *1*(1).

Torkzadeh, G., & Dhillon, G. (2002). Measuring factors that influence the success of Internet commerce. *Information Systems Research*, *13*(2), 187–204. doi:10.1287/isre.13.2.187.87

U.S. Census Bureau. (2009). *Service sector statistics quarterly retail e-commerce sales*. Retrieved January 24, 2009, from http://www.census.gov/mrts/www/ecomm.html

Vijayasarathy, L. (2004). Predicting consumer intentions to use on-line shopping: The case for an augmented technology acceptance model. *Information & Management*, *41*(6), 747–762. doi:10.1016/j.im.2003.08.011

APPENDIX 1.

Cognitive absorption (Agarwal & Karahanna, 2000)

1. Time appears to go by very quickly when I am using the Web
2. I often spend more time on the Web than I had intended.
3. While using the Web, I am able to block out most other distractions.
4. While on the Web, I am immersed in the task I am performing.
5. When on the Web, I get distracted by other attentions very easily.
6. I have fun interacting with the Web.
7. Using the Web provides me with a lot of enjoyment.
8. I enjoy using the Web.
9. Using the Web bores me.
10. I feel that I have control over my control with the Web.
11. The Web allows me to control my computer interaction.
12. Using the Web excites my curiosity.
13. Interacting with the Web makes me curious.
14. Using the Web arouses my imagination.

Computer Self-Efficacy (Compeau & Higgins, 1995)

I could complete the job using a software package...

1. ... if there was no one around to tell me what to do as I go.
2. ... if I had never used a package like it before.
3. ... if I had only the software manuals for reference.
4. ... if I had seen someone else using it before trying it myself.
5. ... if I could call someone for help if I got stuck.
6. ... if someone else had helped me get started.
7. ... if I had a lot of time to complete the job for which the software was provided.
8. ... if I had just the built-in help facility for assistance.
9. ... if someone showed me how to do it first.
10. ... if I had used similar packages before this one to do the same job.

Computer Anxiety (Parasuraman & Igbaria, 1990)

1. Computers are a real threat to privacy in this country.
2. Sometimes I am afraid the data processing department will lose my data.
3. I am anxious and concerned about the pace of automation in the world.
4. I am easily frustrated by computerized bills.
5. I am sometimes frustrated by increasing automation in my home.

Personal innovativeness (Karahanna et al., 2002)

1. I use only computer tools that have a proven track record.
2. Among my colleagues and coworkers, I tend to be among the first to try new computer tools.
3. I leave it to others to work out the "bugs" in the new computer tools before I will consider them.

Perceived usefulness

Time saving (Vijayasarathy, 2004)

1. It is important to minimize payment time.
2. It is important to minimize time pressure when shipping.

Internet shopping convenience (Torkzadeh & Dhillon, 2002)

3. It is important to make shopping easy.
4. It is important to minimize effort of shopping.
5. It is important to minimize queuing time.
6. It is important to minimize time to select a product.
7. It is important to minimize personal hassle.
8. It is important to minimize payment time.
9. It is important to minimize time pressure when shopping.

Internet product value (Torkzadeh & Dhillon 2002)

10. It is important to minimize product cost.
11. It is import to minimize tax cost.
12. It is important to ensure quality of product.
13. It is important to maximize

Cost reduction (Kuk, 2004)

14. Inventory holdings/levels should be reduced.
15. Lower transportation costs due to more efficient planning.
16. Faster inventory turns.
17. Less wastage as work in progress is affected when quality issues are encountered.
18. Fewer product returns due to low quality or obsolescence.

Perceived risk

Internet shipping errors (Torkzadeh & Dhillon, 2002)

1. I worry about receiving wrong products.
2. I am concerned about shipping errors.

Uncertainty (Devaraj et al., 2002)

3. It was easy for me to get relevant quantitative (price, taxes, etc.) information.
4. I believe that it was possible for me to evaluate the various alternatives.
5. The store's Web site provided adequate information.
6. The online site provided sufficient information for the product.

Attitude (Limayem et al., 2000)

1. Online shopping is a good idea.
2. I like to shop through the Web.
3. Purchasing through the Web is enjoyable.
4. Online shopping is exciting.

This work was previously published in International Journal of E-Business Research Volume 6, Issue 1, edited by In Lee, pp. 52-67, copyright 2010 by IGI Publishing (an imprint of IGI Global).

Chapter 18
The Influence of Quality on E-Commerce Success:
An Empirical Application of the Delone and Mclean IS Success Model

Ultan Sharkey
National University of Ireland Galway, Ireland

Murray Scott
National University of Ireland Galway, Ireland

Thomas Acton
National University of Ireland Galway, Ireland

ABSTRACT

This research addresses difficulties in measuring e-commerce success by implementing the DeLone and McLean (D&M) model of IS success (1992, 2003) in an e-commerce environment. This research considers the influence of quality on e-commerce success by measuring the information quality and system quality attributes of an e-commerce system and the intention to use, user satisfaction and intention to transact from a sample of respondents. This research provides an empirical e-commerce application of the updated IS success model proposed by DeLone and McLean (2003). This paper found significant relationships between Information Quality and System Quality and three success dimensions: intention to use, user satisfaction and intention to transact. It found the following constructs to be most important in predicting success: ease of understanding, personalisation and reliability. In particular, that reliability is more important than usability where transactions are concerned and security though important, is not the most important factor.

DOI: 10.4018/978-1-61350-462-8.ch018

INTRODUCTION

Measuring the success of information systems is critical to the process of evaluating IS strategies and investments. However, a comprehensive evaluation framework has been difficult to define and the identification of a dependant variable has remained elusive. Companies experience particular difficulty in evaluating the success of e-commerce initiatives resulting in a significant barrier to entry for many businesses, particularly small to medium sized enterprises. The updated success model proposed by DeLone and McLean (2003) (D&M IS success model) provides guidance for development in this area, especially as a model for developing comprehensive e-commerce success measures. This paper attempts to respond to the call made by DeLone and McLean (2003) for empirical studies to measure the interaction between success dimensions in order to isolate the effect of particular independent variables. Specifically, this paper provides one of the first e-commerce applications of the updated IS success model and explores the relationships that exist between factors of information and system quality and e-commerce success metrics.

THEORETICAL BACKGROUND

The difficulties posed in assessing the success of an online presence as espoused by Hasan and Tibbits (2000) in that the system implemented must make a measurable contribution, implies that the measurement of the success of the system is fundamental to the ability of management to determine the contribution to business. Zhang, Keeling, and Pavur (2000) proposed that validated evaluation criteria would be critical to future decision making for online systems design. However the definition of "success" is controversial as the term itself is multi-dimensional and can be assessed at various levels of the organisation using many differing criteria.

DeLone and McLean (1992) sought to better define the dependant variable "success" in information systems by categorising six interrelated dimensions of IS success—System Quality, Information Quality, Use, User Satisfaction, Individual Impact and Organisational Impact—and encouraged future studies to acknowledge the multidimensional, interdependent and contextual nature of IS success. This model informed a large number of studies and has been updated as the D&M IS success model (2003) to include the variables "service quality" and "net benefits".

DeLone and McLean (2003) also demonstrated the possibility of adapting their IS success model to the e-commerce environment. It is argued that, as the DeLone and McLean Model (D&MM) is based on communications theory, it is highly suited to measuring the IS and communications phenomenon that is the Internet (DeLone & McLean, 2003, p. 34). This process model, (as shown in Figure 1) is derived from a well accepted and tested model of information systems' success (Molla & Licker, 2001; Seddon & Kiew, 1994) and is argued to be appropriate to the communications and commerce process common to e-commerce systems.

Measures of IS Quality

DeLone and McLean (2003) encourage the selection of success dimensions and measures contingent on the objectives and context of the system under study. In terms of e-commerce systems, information quality and system quality have been identified as critical features of determining web site success (Bell & Tang, 1998; Zhang et al., 2000).

Information quality aspects of a system are a measure of the information system outputs, the primary example of such being the production of reports (DeLone & McLean, 1992). E-commerce system users may seek transactional, customer service and marketing services as well as information such that account must be taken of the features distinguishing e-commerce systems

Figure 1. Updated D&M IS success model (2003)

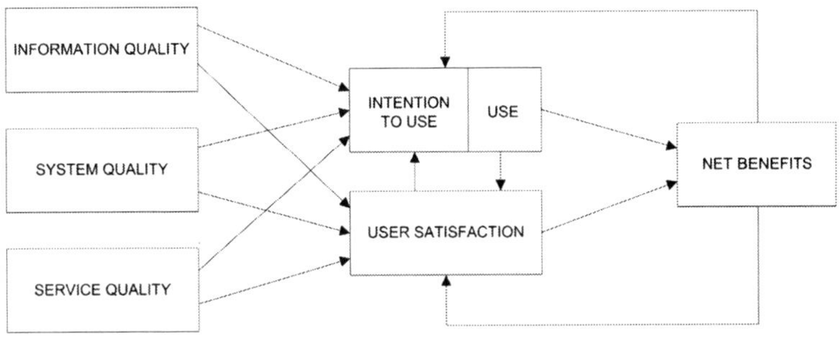

from conventional information systems (Molla & Licker, 2001).

A distinct measurement of information quality remains elusive (Lee, Strong, Kahn, & Wang, 2002 p. 143; Wand et al., 1996, p. 87) yet the concept is accepted as a critical concern of management (Lee et al., 2002). The concept is especially important in the context of e-commerce as the provision of information is the basic goal of any web site (Huizingh, 2000). It has been identified as a factor in information system success (DeLone & McLean, 1992), in web site home page performance (Zang et al., 2000) and in web site success in the context of electronic commerce (Delone & McLean, 2003; Liu & Arenett, 2000). It is sometimes referred to as data quality (Wand et al., 1996) or information content (Ranganathan & Ganapathy, 2002).

Many attributes of the Information Quality concept have been proposed and measured in various types of studies. (Accuracy: Zhang et al., 2000; Timeliness: McKinney, Yoon, & Zahedi, 2002; Reliability: McKinney et al., 2002; DeLone & McLean, 1992; Relevance: McKinney et al., 2002; Scope: McKinney et al., 2002; Lohse & Spiller, 1998; Huizingh, 2000). It has been described as a multi-dimensional concept, (Lee et al., 2002; Lin & Lee, 2006; Wand et al., 1996) and in identifying it as a dimension of web quality, Aladwani and Palvia (2002) suggest its measurement should be multi-dimensional in nature also. The updated D&M IS success model provides five information quality constructs of e-commerce success that will be considered for use in this study: Completeness, Ease of Understanding, Personalisation, Relevance, and Security.

The second feature of online quality is the quality of the system serving that information. System quality refers to the measurement of the information processing system (DeLone & McLean, 1992; Negasha, Ryan, & Igbaria, 2003). In this case the information processing system is the company's online presence. Some researchers refer to system quality as the ease of finding the information on the system (Keevil, 1998). The failure of an online system will cause a user to "mouse-click away" resulting in non-use (Molla & Licker, 2001). Thus, online system quality contains traditional system quality attributes such as reliability, accuracy, flexibility, response time and ease of use (DeLone & McLean, 1992) as well as the more encompassing concept of seamless site performance (Molla & Licker, 2001). The updated D&M IS success model provides for five system quality constructs of e-commerce success that will be considered for use in this study: Usability, Availability, Reliability, Adaptability and Response Time.

Measures of IS Success

In the context of Internet e-commerce, the success of the system must necessarily be based on the goals of the management of that system (Hasan

Figure 2. Research framework

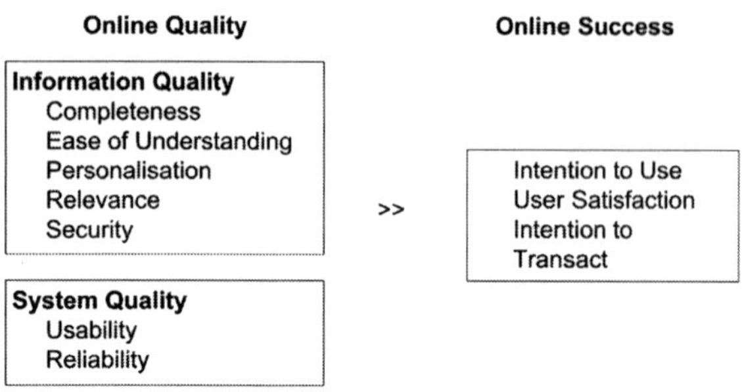

& Tibbits, 2000). The role of context has been stressed as important in the explanation of system success (DeLone & McLean, 1992, 2003). Seddon (1997) argued that use may be a goal in itself or a means to an end, thus use may be equated with success if use is the goal as set by the stakeholder. In the context of online commerce, the success of the system must necessarily be based on the goals of the management of that system.

The choice of success constructs is contingent on the objective of the study (management or stakeholder objectives) or the context of the organisation (DeLone & McLean, 2003). Past measures have included number of hits, sales leads generated, and revenue (Golden, Hughes, & Gallagher, 2003), number of queries received, number of hits and constructive feedback (Webb & Sayer, 1998). This suggests that queries received, hits and feedback were what held value for the stakeholders. A hit may become a query that generates a sales lead, which develops into revenue. Thus, a 'hit' may not be sufficient to cause revenue but remains necessary. Without the original 'hit', no revenue could be generated. DeLone and McLean (2003) posited that success, in this context, should be determined with reference to the nature of the systems' use, the patterns of user navigation, number of site visits and the number of transactions executed.

RESEARCH FRAMEWORK

The research framework below (Figure 2), adapted from the D&M IS success model, is based on the many studies that have validated the model (Seddon & Kiew, 1994; Seddon, 1997; Taylor & Todd, 1995) and the extension of this model to the e-commerce context (Molla & Licker, 2001). The constructs representing Information Quality, System Quality and Online Success have been selected from the literature, contingent on the context and objectives of the investigation and are presented below.

In the context of the above framework, the following hypotheses (Table 1) are posited.

Research Instrument

In order to measure the factors of online quality and system success, a research instrument, consisting of two questionnaire surveys, were compiled from existing, tested and verified instruments. The pre-task questionnaire was administered to the sample before they were directed to begin with the subject system. Certain tasks were then provided to the

Table 1. Hypotheses - The influence of quality on success

Hypotheses
H1: Information Quality has a positive influence on Intention to Use
H2: Information Quality has a positive influence on User Satisfaction
H3: Information Quality has a positive influence on Intention to Transact
H4: System Quality has a positive influence on Intention to Use
H5: System Quality has a positive influence on User Satisfaction
H6: System Quality has a positive influence on Intention to Transact

sample to attempt and the post-task questionnaire was then provided to the sample for completion.

The pre-task questionnaire, using adapted components of the Questionnaire for User Interface Satisfaction (QUIS), collected information about the respondents' past experience on the Internet, their level of confidence in various aspects of Internet usage and the importance of various aspects of computing technologies. The post-task questionnaire was developed from a pool of 206 items collated from an extensive review of the literature. In order to determine the most rigorous of these items it was determined whether each study had tested the instrument for validity, reliability, whether a pilot had been conducted and whether the system under study was Internet-based. The matrix produced by this analysis is available in the appendices. A final instrument of 62 items was thus created. The main sources for the constructs implemented are available in the appendices. Both of these questionnaires were measured using a seven-point Likert scale.

Research Process

The research instrument was tested using a sample of 27 graduate students from different graduate programmes in an Irish university (13 female, 11 male, mean age = 25.67, SD = 5.033). Many other studies in commerce and information systems have utilised students as representative of the Internet using population (Ahuja, Gupta, & Raman, 2003; Garrity, Glassberg, Kim, Sanders, & Shin, 2005; Ives, Olson, & Baroudi, 1983; Lee & Kozar, 2006; Negasha et al., 2003; van Iwaarden, van der Wiele, Ball, & Millen, 2004;). Aladwani and Palvia (2001) administered a 55-item instrument to 104 students between 18 and 21 years old. Palmer (2002) conducted his survey with 35 undergraduate and MBA students. The sample size used in this study is in line with these previous studies and is also in accordance with guidelines proposed by Lewis (1994). In this study the sample of 27 were not fragmented into smaller cells, but remained as a single unit for all analyses. Further, students are suitable for these studies as they are wide users of the Internet (Lala, Arnold, Sutton, & Guand, 2002, p. 245). Garrity et al. (2005) argued for the use of students as an appropriate sample, stating their place as present and future consumers of web technology. This argument supports the population validity of this research in that the purposive sample may be representative of the wider population of e-commerce consumers.

Data collection was carried out by distributing the questionnaires to the sample before and after completion of predetermined tasks in a computer suite. The research process consisted of three aspects. A pre-task questionnaire was distributed to help inform the findings from the post-task questionnaire. To determine the quality of a system or its success the participants must have an awareness of the workings of that system. Thus, four tasks were designed to a) create a user account, b) upload pictures into an album, c) share that album, and d) attempt a purchase of prints until payment details were requested. After completion of these tasks the participants were then asked to complete the post-task questionnaire. A photographic e-commerce system was chosen for the basis of this study. It was chosen for the potential to allow e-commerce oriented tasks and the suitability in measuring the constructs of online quality as defined by this study.

RESULTS

Results of the pre-task questionnaire showed 90% of respondents were aged between 20.6 and 30.7 years. Two respondents were aged more than 30 years. Almost 80% of the respondents indicated they were either Confident or Very Confident in finding information on the Internet; in contrast just 37.5% indicated a similar level of confidence in purchasing using the Internet. The most important attributes of computing technology to the respondents were usefulness of the system (83%); rate information is displayed at (75%) and system reliability (71%). The least important aspects were enjoyment (33%) and overall attractiveness (25%).

Composite constructs were created using the SPSS 'compute' command to produce un-weighted means of each of the constructs items. These scores were then recoded into numeric variables in SPSS. Reliability analysis using Cronbach's Alpha for the composite constructs (Figure 3) are presented below.

Using SPSS linear regression analyses were carried out on the individual constructs with respect to the success variables as an exploration of the relationships with the dependent variables. Then, to determine the inter-relationships and interplay between the constructs and to verify the logic of the research framework, MANOVA analyses were carried out. Regression was completed for all thirty-one combinations of the five independent variables of information quality. The Adjusted R^2 values and significance levels computed for each of the 31 combinations of the composite variables, with Intention to Use as the dependent variable, were entered into a spreadsheet. The data were ordered by the level of significance, then by the Adjusted R^2 value. This produced a table with the most significant relationships at the top with the corresponding greatest Adjusted R^2 values. The most significant of the combinations, with Intention to Use as the dependent variable, was {Completeness, Personalisation} (Adj. $R^2 = 0.212$). The combination of five composite independent variables was not significant. The individual composites that were the strongest predicators were Personalisation (Adj. $R^2 = 0.206$ at the 0.05 level); Security (Adj. $R^2 = 0.168$ at the 0.05 level); and Completeness (Adj. $R^2 = 0.135$ at the 0.05 level) respectively.

Intention to Use

Multiple regression taking Intention to Use as dependent variable (Table 3) indicate that, of the information quality (IQ) constructs, Personalisation is the most influential (Adj. $R^2 = 0.206$). The elimination of Completeness resulted in a higher F value ($F = 6.969$, $p < 0.05$). The final model indicated Intention to Use is most dominantly influenced by Personalisation ($\beta = 0.49$, $p < 0.05$).

Similarly, multiple regression analysis was carried out with the constructs of system quality (SQ) as independent variables. This process was completed for the three possible combinations of the two independent variables representing system quality. The Adjusted R^2 values and significance levels computed for each of the 3 combinations of the composite variables with Intention to Use as the dependent variable were entered into a spreadsheet. The data were ordered by the level of significance, then by the Adjusted R^2 value. The combination of Usability and Reliability was the strongest predicator with an Adj. R^2 value of

Figure 3. Reliability analyses for constructs and composites

Construct (No. of items)		α (Individual)	Composite (No. of items)		α (Composite)
Completeness	(5)	0.88			
Ease of understanding	(5)	0.89			
Personalisation	(5)	0.88			
Relevance	(5)	0.84			
Security	(4)	0.94	Information Quality	(24)	0.95
Usability	(6)	0.89			
Reliability	(5)	0.93	System Quality	(11)	0.90
Intention to Use	(8)	0.94			
User Satisfaction	(5)	0.92			
Intention to Transact	(5)	0.88	System Success	(18)	0.95

Table 3. Ranking of quality factors which influence Intention to Use

Information Quality		**System Quality**	
1	Personalisation	1	Usability
2	Completeness	2	Reliability
3	Security		
4	Relevance		
5	Ease of Understanding		

Table 4. Ranking of quality factors which influence User Satisfaction

Information Quality		**System Quality**	
1	Ease of Understanding	1	Usability
2	Personalisation	2	Realiability
3	Completeness		
4	Relevance		
5	Security		

0.212 ($p < 0.05$). The individual Independent variables relationships with Intention to Use were both significant, Usability Adj. $R^2 = 0.191$ ($p < 0.05$); and Reliability Adj. $R^2 = 0.111$ ($p < 0.1$) respectively.

Multiple regression analyses taking Intention to Use as dependent variable indicate that Usability is the most influential construct with Adj. $R^2 = 0.191$. The final model indicated Intention to Use (Figure 4) is most dominantly influenced by Usability ($\beta = 0.476$, $p < 0.05$). From an analysis of the beta coefficients a ranking of the influencing factors may be determined.

User Satisfaction

A table of relationships (Table 4) sorted by Adjusted R^2 and significance was created with User Satisfaction as the dependent variable. All 31 IQ combinations were found to be significant at the 0.05 confidence level. The combination of all five factors of information quality resulted in a high Adj. R^2 value of 0.342 ($p < 0.05$). The highest Adj. R^2 value found was {Ease of Understanding, Personalisation} ($p \leq 0.001$).

Multiple regression analyses taking User Satisfaction as the dependent variable indicate that Ease of Understanding and Personalisation are the most influential constructs. The elimination of Completeness from the model resulted in a higher F value ($F = 9.261$, $p < 0.05$). The final model indicated User Satisfaction is most dominantly influenced by Ease of Understanding ($\beta = 0.412$, $p < 0.05$) and less so by Personalisation ($\beta = 0.391$, $p < 0.05$).

Figure 4. The influence of information quality on Intention to Use

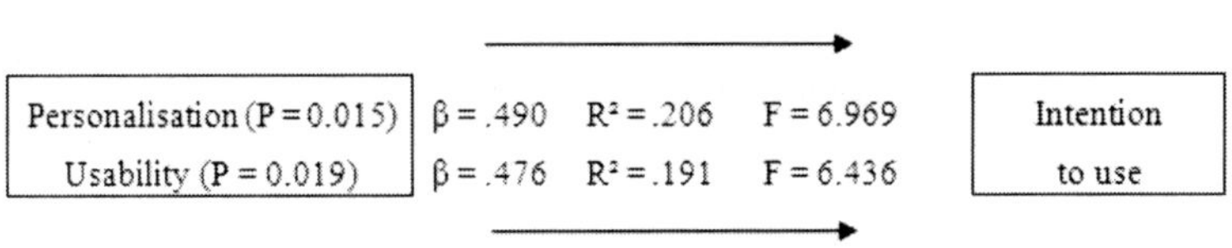

Figure 5. The influence of quality on User Satisfaction

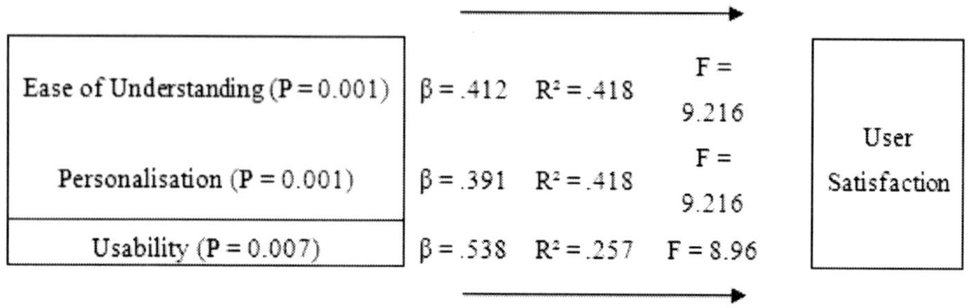

The combined SQ composite variable {Usability, Reliability} was found to be the strongest predictor of User Satisfaction (Adj. $R^2 = 0.291$ at $p < 0.05$). The relationships of the individual variables were less significant in isolation, with {Usability} Adj. $R^2 = 0.257$ ($p < 0.05$), and {Reliability} Adj. $R^2 = 0.187$ ($p < 0.05$), and although usability was the strongest individual contributor, when combined with reliability the effect provides a stronger predictor.

Multiple regression analyses taking User Satisfaction as dependent variable with the constructs of System Quality as independent variables indicate that Usability is the most influential construct with Adj. $R^2 = 0.257$. The final model indicated User Satisfaction (Figure 5) is most dominantly influenced by Usability ($\beta = 0.538$, $p < 0.05$).

Intention to Transact

The combined composite IQ variable {Ease of understanding, Personalisation} was found to be the strongest predicator of Intention to Transact (Adj. $R^2 = 0.533$ at $p < 0.05$). The combination of the five composites with Intention to Transact (Table 5) as the dependent variable resulted in a strong Adj. R^2 value of 0.485 ($p < 0.05$).

Multiple regression analyses taking Intention to Transact as the dependent variable indicate that Ease of Understanding and Personalisation are the most influential IQ constructs. A higher Adj. R^2 value resulted when Security was eliminated from the analysis ($F = 14.106$, $p < 0.05$). The final model indicated Intention to Transact is most dominantly influenced by Ease of Understanding ($\beta = 0.486$, $p < 0.05$) and less so by Personalisation ($\beta = 0.400$, $p < 0.05$).

Table 5. Ranking of quality factors which influence Intention to Transact

Information Quality		System Quality	
1	Ease of Understanding	1	Reliability
2	Personalisation	2	Usability
3	Security		
4	Relevance		
5	Completeness		

The combined composite SQ variable {Usability, Reliability} was found to be the strongest predicator of Intention to Transact (Adj. $R^2 = 0.431$ at $p < 0.05$). The relationships of the individual variables were also strong, {Usability} Adj. $R^2 = 0.323$ ($p < 0.05$), and {Reliability} Adj. $R^2 = 0.320$ ($p < 0.05$).

Multiple regression analyses taking Intention to Transact as dependent variable with the constructs of System Quality as independent variables indicate that Usability and Reliability are the most influential. The retention of both constructs results in a significant F value ($F = 17.223$, $p < 0.05$). The final model indicated Intention to Transact (Figure 6) is most dominantly influenced by Reliability

Figure 6. The influence of information quality on Intention to Transact

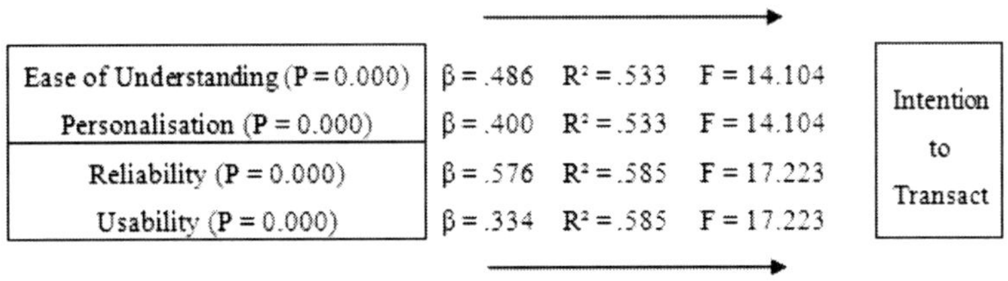

(β= 0.576, $p < 0.05$) and less so by Usability (β= 0.334, $p < 0.05$).

DISCUSSION

The DeLone and McLean Model of Success in an E-Commerce Context

This paper provides one of the first e-commerce applications of the updated IS success model proposed by DeLone and McLean (2003) and clarifies the existence of relationships between factors of quality proposed in the revised D&M model and further establishes which factors are most influential in success as defined through the model.

This study supports the contention that there is no single success variable (DeLone & McLean, 1992 p. 80), and shows there is a complex interplay of variables to be understood when seeking to measure online success. DeLone and McLean (1992) suggested that use and user satisfaction were singularly and jointly affected by system quality (SQ) and information quality (IQ). This statement is supported by this research. This paper reports the factors of SQ and IQ which influence Intention to Use, User Satisfaction and Intention to Transact and also their relative influence.

The 1992 study by DeLone and McLean highlighted the interdependency that exists among the constructs of IS success. This was further relayed in the 2003 updated model suggesting careful attention be paid to the "multidimensional and interdependent nature of IS success" (DeLone & McLean, 2003, p. 11) and in other studies (Lin & Lee, 2006). This study supports this theory by providing empirical evidence of the interdependency of constructs. Specifically this study highlights the fact that some variables obtain extra influence when combined with other influential variables. Indeed the influence of some independent variables is accentuated only in the presence of others. This influence is stronger than either of the two variables individually. This presents difficulty in interpretation in isolating dependant variables. The strong Cronbach alpha measures however suggest that the multidimensional nature of this

Table 6. Hypotheses testing - The influence of quality on success

Hypothesis	Result
H1: Information Quality has a positive influence on Intention to Use	Accepted
H2: Information Quality has a positive influence on User Satisfaction	Accepted
H3: Information Quality has a positive influence on Intention to Transact	Accepted
H4: System Quality has a positive influence on Intention to Use	Accepted
H5: System Quality has a positive influence on User Satisfaction	Accepted
H6: System Quality has a positive influence on Intention to Transact	Accepted

study was the correct approach and indicates that the measurement of e-commerce success should accommodate this multidimensional nature.

Intention to Use

The research found that of the Information Quality factors, Personalisation has the strongest influence on Intention to Use. The respondents who ranked the system highly for Personalisation indicated an intention to return to the site. This implies that increased awareness by the target market of Personalisation features in an online system will increase the number of visitors returning to the system or customer loyalty.

The research found that of the System Quality factors, Usability has the strongest influence on Intention to Use, implying that increased Usability of an online system will increase the numbers of visitors returning to the system or customer loyalty. This research confirms the importance of usability in the success of e-commerce systems. It also confirms Seddon and Kiew's (1994) argument that ease of use is a central part of system quality.

User Satisfaction

The factors that were found to be most influential on User Satisfaction were also the factors that are most influential on Intention to Transact {Ease of Understanding, Personalisation}. The implication is that a satisfied user will have a higher Intention to Transact. As the conversion of a site visitor into a purchaser is a goal of e-commerce systems (Berthon, & Prendergast, 1997) this research confirms the importance of User Satisfaction's inter-relatedness with Intention to Transact.

The requirement of a system to be personalised usually entails the creation of accounts (Huizingh, 2000). This research positively supports the creation of user accounts as an indicator of e-commerce success. The creation of user accounts allows for greater information capturing capabilities (Hagel & Armstrong, 1997, p. 143) which, in turn, allows for the information served to be further personalised and easier for the registered user to understand. This logic is supported by the synergistic relationship seen between Ease of understanding and Personalisation in this research. The more personalised the information served, the easier it is understood and the further satisfied the customer will be.

Intention to Transact

The strongest relationships found by this research were the independent variables' relationships with Intention to Transact. Considering the relatively low-level of confidence in transacting using the Internet shown by the sample in the pre-task questionnaire, this would suggest that Intention to Transact should be the most contentious of the three success variables. The emergence of Ease of understanding as the most influential factor in determining Intention to Transact is therefore not surprising. Where the information describing the offering is easy to understand and personalised, the higher the Intention to Transact.

The security construct was found to be significant in determining Intention to Transact but was removed from regression analysis models. This suggests that security is of lesser importance in determining Intention to Transact than the other factors used in this research. This would imply that security is necessary but not sufficient for an Intention to Transact to be conducted. This result is interesting given the low-level of confidence in transacting exhibited by the sample. One would have expected that security would have been of prime importance to such a sample; however it appears that security is more of a contributory factor to Intention to Transact rather than a dependent one. A precise explanation of this result would require further data collection, however it may be speculated that security was regarded as an *a priori* feature of e-commerce transactions by the sample and hence taken for granted. Alternatively, the strong association of the sample for high levels of personalisation may have connoted a certain level of comfort in the process of transacting and as a result encouraged the development of a sense of security in the process. The authors encourage further research to focus on the influence of personalisation and transaction confidence on security as a dependent or contributory factor to success.

This research does confirm that inadequate security adversely affects the development of e-commerce (Molla & Licker, 2001). This logic may be extended to the other factors of Information Quality which produced high adjusted R^2 figures but did not remain in the final model developed through regression analyses. Thus while content intended to boost transactions must, of necessity, be relevant and complete it must also be personalised and easy to understand to have an acute influence. In this case the interplay of variables creates an interesting level of complexity in affecting success measures.

Ranganathan and Ganapathy's (2002) research found that security was the best predicator of transactional intent. In their exploratory study, the provision of modes of transmitting transactional data and the provision for individual accounts with password and log-in for consumers was thought to influence security's importance in attracting more custom. In this research however, security was found to be of significant importance but was not found to be the most influential. The strong influence of Personalisation manifested as user accounts is confirmed by this research but further study may be required to understand the interdependencies between Personalisation and Security.

The factors that were found to be most influential from a System Quality perspective were Reliability and Usability respectively. The relative strength of Reliability over Usability has implications for developers especially in the balance between a reliable system and one that is not over-burdensome on the user. Trading the functionalities of a system against each other has been accepted as necessary in practice (van Iwaarden et al., 2004). This research indicates that where there is a decision to be made between the reliability of the system and usability, the features that increase Reliability at the expense of Usability may be preferred. This research confirms the importance of speedy recovery from errors (Liu & Arnett, 2000) in online transactional systems. This is of particular interest to the design of transactional pages online. Reliability has importance to the contract undertaken, order fulfilment and billing as well as a professional reputation online. However, ensuring the order is accurate and reliable, may reduce usability in ensuring the user has selected what they wish or ensuring fraudulent details are refused. This study supports such practices as using order confirmation screens seeking a final approval at the expense of adding an extra few clicks to the process.

Limitations

While the original IS success model carries the relationships first to individual and organisation impacts (DeLone & McLean, 1992), and the updated model to net benefits from success (DeLone & McLean, 2003), this study does not include these final constructs and seeks only to reveal the relationships between identified factors of quality and factors of success in the context of Internet based systems.

The sample population for this study who participated in the experiment and completed the questionnaires were graduate students of the commerce faculty at the National University of Ireland, Galway. Characteristics of the sample not recorded through the questionnaires, not least the size of the sample, may indicate that the results may not be sufficiently indicative of the larger population of Internet users. The researchers acknowledge that any extrapolation of the results to a larger population should be approached with caution. Equally, while it is arguable that the computer suite environment is a relatively natural setting for a sample of graduate students, there is no doubt that it is an artificial setting for a free and true assessment.

The choice of one Internet based system as the foundation of the data collected suggests that the strength or weaknesses of the dependent variables may be attributable only to the system studied. In answer to this, the researcher chose tasks which are characteristic of systems that support electronic commerce functionality so that the results can be used to further support and inform theory in electronic commerce success.

CONCLUSION AND FURTHER RESEARCH

Quality is a critical area for commercial enterprise. In terms of e-commerce success, businesses need to understand the impact usability has on potential customers, the satisfaction users experience with various aspects of the web site and crucially, what factors will influence users to transact on their web site. This study has found significant relationships between constructs representing information and system quality and success constructs measuring Intention to Use, User Satisfaction and Intention to Transact. Generally, this study found the following information and system quality constructs to be most important: ease of understanding, personalisation and reliability.

It can be concluded that the D&M IS success model (2003) for studying Information Systems success is equally useful in measuring the relative success of e-commerce systems. However, our sample is relatively small and there is a need to conduct further tests in other areas of e-commerce interaction.

This paper also proposes further research to confirm the specific findings of this research and to explore the interdependency of the factors of quality and success. Further studies should be developed to explore correlations between the quality factors to further understand the interplay between the constructs in their influence on Intention to Use, User Satisfaction and Intention to Transact.

REFERENCES

Ahuja, M., Gupta, B., & Raman, P. (2003). An empirical investigation of online consumer purchasing behavior. *Communications of the ACM*, *46*(12), 145–215. doi:10.1145/953460.953494

Aladwani, A. M., & Palvia, P. C. (2002). Developing and validating an instrument for measuring user-perceived web quality. *Information & Management*, *39*(6), 467–476. doi:10.1016/S0378-7206(01)00113-6

Bell, H., & Tang, N. K. H. (1998). The effectiveness of commercial Internet Web sites: A user's perspective. *Internet Research*, *8*(3), 219–228. doi:10.1108/10662249810217768

Benbunan-Fich, R. (2001). Using protocol analysis to evaluate the usability of a commercial web site. *Information & Management*, *39*(2), 151–163. doi:10.1016/S0378-7206(01)00085-4

Berthon, P., Pitt, L., & Prendergast, G. (1997). Visits, hits, caching and counting on the World Wide Web: Old wine in new bottles? *Internet Research: Electronic Networking Applications and Policy*, *7*(1), 5–8. doi:10.1108/10662249710159791

Chin, J. P., Norman, K. L., & Diehl, V. A. (1988). Development of a tool measuring user satisfaction of the human-computer interface. In *Proceedings of ACM SIGCHI 88* (pp. 213-218). ACM Publishing.

DeLone, W., & McLean, E. (1992). Information systems success: The quest for the dependent variable. *Information Systems Research*, *3*(1), 60–95. doi:10.1287/isre.3.1.60

DeLone, W., & McLean, E. (2003). The DeLone and McLean model of information systems success: A ten-year update. *Journal of Management Information Systems, 19*(4), 9–30.

Garrity, E. J., Glassberg, B., Kim, Y. J., Sanders, G. L., & Shin, S. K. (2005). An experimental investigation of web-based information systems success in the context of electronic commerce. *Decision Support Systems, 39*(3), 485–503. doi:10.1016/j.dss.2004.06.015

Gefen, D., & Straub, D. (2000). The relative importance of perceived ease of use in IS adoption: A study of e-commerce adoption. *Journal of the Association for Information Systems, 1*(8), 1–28.

Gill, J., & Johnson, P. (2002). *Research methods for managers* (3rd ed.). London: Paul Chapman Publishing. ISBN: 0-7619-4002-2.

Golden, W., Hughes, M., & Gallagher, P. (2003). Online retailing: What drives success? Evidence from Ireland. *Journal of End User Computing, 15*(3), 32–44.

Hagel, J., & Armstrong, A. G. (1997). Net gain: Expanding markets through virtual communities. *The McKinsey Quarterly, 1*, 141–153.

Hasan, H., & Tibbits, H. R. (2000). Strategic management of electronic commerce: An adaptation of the balanced scorecard. *Internet Research, 10*(5), 439–450. doi:10.1108/10662240010349453

Huizingh, E. K. R. E. (2000). The content and design of web sites: An empirical study. *Information & Management, 37*(3), 123–134. doi:10.1016/S0378-7206(99)00044-0

Ives, B., Olson, M. H., & Baroudi, J. J. (1983). The measurement of user information satisfaction. *Communications of the ACM, 26*(10), 785–793. doi:10.1145/358413.358430

Jeong, M., & Lambert, C. U. (2001). Adaptation of an information quality framework to measure customers' behavioral intentions to use lodging web sites. *Hospital Management, 20*, 129–146. doi:10.1016/S0278-4319(00)00041-4

Keevil, B. (1998). Measuring the usability index of your Web site. In *Proceedings of the 16th Annual International Conference on Computer Documentation* (pp. 271-277). ACM Publishing.

Koufaris, M. (2002). Applying the technology acceptance model and flow theory to online consumer behaviour. *Information Systems Research, 13*(2), 205–223. doi:10.1287/isre.13.2.205.83

Lala, V., Arnold, V., Sutton, S. G., & Guand, L. (2002). The impact of relative information quality of e-commerce assurance seals on Internet purchasing behavior. *International Journal of Accounting Information Systems, 3*, 237–253. doi:10.1016/S1467-0895(02)00069-6

Lee, Y., & Kozar, K. A. (2006). Investigating the effect of website quality on e-business success: An analytic hierarchy process (AHP) approach. *Decision Support Systems, 42*(3), 1383–1401. doi:10.1016/j.dss.2005.11.005

Lee, Y. W., Strong, D. M., Kahn, B. K., & Wang, R. Y. (2002). AIMQ: Methodology for information quality assessment. *Information & Management, 40*, 133–146. doi:10.1016/S0378-7206(02)00043-5

Lewis, J. R. (1994). Sample sizes for usability studies - additional considerations. *Human Factors, 36*(2), 368–378.

Lin, H. F., & Lee, G. G. (2006). Determinants of success for online communities: An empirical study. *Behaviour & Information Technology, 25*(6), 479–488. doi:10.1080/01449290500330422

Liu, C., & Arnett, K. P. (2000). Exploring the factors associated with web site success in the context of electronic commerce. *Information & Management*, *38*(1), 23–33. doi:10.1016/S0378-7206(00)00049-5

McKinney, V., Yoon, K., & Zahedi, F. M. (2002). The measurement of web-customer satisfaction: An expectation and disconfirmation approach. *Information Systems Research*, *13*(3), 296–315. doi:10.1287/isre.13.3.296.76

Molla, A., & Licker, P. S. (2001). E-commerce systems success: An attempt to extend and re-specify the DeLone and McLean model of IS success. *Journal of Electronic Commerce Success*, *2*(4), 1–11.

Negasha, S., Ryan, T., & Igbaria, M. (2003). Quality and effectiveness in web-based customer support systems. *Information & Management*, *40*, 757–768. doi:10.1016/S0378-7206(02)00101-5

Palmer, J. W. (2002). Web site usability, design, and performance metrics. *Information Systems Research*, *13*(2), 151–167. doi:10.1287/isre.13.2.151.88

Seddon, P. B. (1997). A respecification and extension of the DeLone and McLean model of IS success. *Information Systems Research*, *8*(3), 240–253. doi:10.1287/isre.8.3.240

Seddon, P. B., & Kiew, M.-Y. (1994, December). A partial test and development of the DeLone and McLean model of IS success. In *Proceedings of the International Conference on Information Systems*, Vancouver, British Columbia, Canada (pp. 99-110).

Taylor, S., & Todd, P. A. (1995). Understanding information technology usage: A test of competing models. *Information Systems Research*, *6*(2), 144–177. doi:10.1287/isre.6.2.144

van Iwaarden, J., van der Wiele, T., Ball, L., & Millen, R. (2004). Perceptions about the quality of web sites: A survey amongst students at Northeastern University and Erasmus University. *Information & Management*, *41*(8), 947–959. doi:10.1016/j.im.2003.10.002

Webb, B., & Sayer, R. (1998). Benchmarking small companies on the Internet. *Long Range Planning*, *31*(6), 815–827. doi:10.1016/S0024-6301(98)80018-6

Yang, Z., Cai, S., Zhou, Z., & Zhou, N. (2005). Development and validation of an instrument to measure user perceived service quality of information presenting web portals. *Information & Management*, *42*(4), 575–589. doi:10.1016/S0378-7206(04)00073-4

Young, D., & Benamati, J. (2000). Differences in public web sites: The current state of US firms. *Journal of Electronic Commerce Research*, *1*(3), 94–105.

Zhang, X., Keeling, K. B., & Pavur, R. J. (2000). Information quality of commercial web site home pages: An explorative analysis. In *Proceedings of the International Conference on Information Systems*, Brisbane, Australia (pp. 164-175).

APPENDIX A

Items adapted for the research instrument

Web Quality

Web Information Quality

- Completeness (Adapted from Yang, Cai, Zhou, & Zhou, 2005; Adapted from Lee et al., 2002)
- Ease of understanding (Adapted from Garrity et al., 2005; Palmer, 2002)
- Personalisation (McKinney et al., 2002; Aladwani & Palvia, 2002)
- Relevance (McKinney et al., 2002)
- Security (Aladwani & Palvia, 2002; Yang et al., 2005; Lee et al. 2002; van Iwaarden et al., 2004)

Web System Quality

- Usability (Adapted from Aladwani & Palvia, 2002; Gefen & Straub, 2000; Negash et al., 2003)
- Reliability (Adapted from McKinney et al., 2002)

Web Success

- Intention to use (Adapted from Koufaris, 2002; Garrity et al., 2005; Gefen & Straub, 2000)
- User Satisfaction (Adapted from Negash et al., 2003; Yang et al., 2005)
- Intention to transact (Adapted from Garrity et al., 2005; Gefen & Straub, 2000; Bell & Tang, 1998)

APPENDIX B

Suitability of previous instruments for use in this research (Figure 7).

Figure 7.

Web Based	Piloted	Reliability	Validation	Researcher(s)
No	No	Yes	No	Seddon & Kiew (1994)
Yes	No	No	No	Keevil (1998)
Yes	No	No	No	Bell and Tang (1998)
Yes	No	No	No	Benbunan-Fich (2001)
Yes	Yes	No	No	Young and Benamati (2000)
Yes	Yes	No	No	van Iwaarden et al, (2004)
Yes	No	Yes	Yes	Zhang, Keeling and Pavur (2000)
Yes	Yes	Yes	No	Jeong and Lambert (2001)
Yes	No	Yes	Yes	Palmer (2002)
No	Yes	Yes	Yes	Lee et al., (2002)
Yes	No	Yes	Yes	Koufaris (2002)
Yes	No	Yes	Yes	Aladwani and Palvia, (2002)
Yes	Yes	Yes	Yes	Gefen and Straub (2000)
Yes	Yes	Yes	Yes	McKinney, Yoon & Zahedi (2002)
Yes	Yes	Yes	Yes	Negash, Ryan and Igbaria (2003)
Yes	Yes	Yes	Yes	Yang et al, (2005)
Yes	Yes	Yes	Yes	Garrity et al, (2005)

This work was previously published in International Journal of E-Business Research Volume 6, Issue 1, edited by In Lee, pp. 68-84, copyright 2010 by IGI Publishing (an imprint of IGI Global).

Compilation of References

Aaij, S., & Brounen, D. (2002). High-tech IPOs: A tale of two continents. *Journal of Applied Corporate Finance*, *15*(1), 87–92. doi:10.1111/j.1745-6622.2002.tb00343.x

Aaker, D. A. (1996). *Building strong brands*. NY: Free Press.

Abdellaouia, M., Barrios, C., & Wakker, P. P. (2007). Reconciling introspective utility with revealed preference: Experimental arguments based on prospect theory. *Journal of Econometrics*, *138*(1), 356–378. doi:10.1016/j.jeconom.2006.05.025

Abellan-Perpiñan, J. M., Bleichrodt, H., & Pinto-Prades, J. L. (2009). *Journal of Health Economics*, *28*(6), 1039–1047. doi:10.1016/j.jhealeco.2009.09.002

Abraham, C., Ahuja, V., Ghosh, A. K., & Pakanati, P. (2002). *Inventory management using passive RFID tags: a survey*. Retrieved October 10, 2007, from http://www.it.iitb.ac.in/~it601/dep/paperlist/tag-identification-survey.pdf

ABS. (2002). *Small businesses in Australia 2001*. Retrieved October 10, 2008, from http://www.abs.gov.au/ausstats/abs@.nsf/e8ae5488b598839cca25682000131612/97452f3932f44031ca256c5b00027f19!OpenDocument

Achrol, R. S., & Kotler, P. (1999). Marketing in the network economy. *Journal of Marketing*, *63*, 146–163. doi:10.2307/1252108

ACNielsen. (2004). *Hong Kong online banking study*. Retrieved from http://acnielsen.com/news/20050714_ap.shtml

Ad Hamid, N. R. (2005). E-CRM: Are we there yet? *Journal of American Academy of Business*, *6*(1), 51–57.

AffStat. (2006). Affiliate program merchant facts & figures. *Shawn Collins Consulting*. Retrieved from http://www.shawncollinsconsulting.com

Agarwal, R., & Karahanna, E. (2000). Time flies when you're having fun: Cognitive absorption and beliefs about information technology usage. *MIS Quarterly*, *24*(4), 665–694. doi:10.2307/3250951

Ahuja, M., Gupta, B., & Raman, P. (2003). An empirical investigation of online consumer purchasing behavior. *Communications of the ACM*, *46*(12), 145–215. doi:10.1145/953460.953494

Ajzen, I., & Fishbein, M. (1980). *Understanding Attitudes and Predicting Social Behavior*. Englewood Cliffs, NJ: Prentice Hall.

Ajzen, I. (1991). The theory of planned behavior. *Organizational Behavior and Human Decision Processes*, *50*, 179–221. doi:10.1016/0749-5978(91)90020-T

Aladwani, A., & Palvia, P. (2002). Developing and validating an instrument for measuring user-perceived web quality. *Information & Management*, *39*(6), 467–476. doi:10.1016/S0378-7206(01)00113-6

Aladwani, A. (2001). Online banking: A field study of drivers, development challenges, and expectations. *International Journal of Information Management*, *21*, 213–225. doi:10.1016/S0268-4012(01)00011-1

Alavi, M., & Joachimsthalter, E. (1992). Revisiting DSS implementation research: A meta-analysis of the literature and suggestions for researchers. *MIS Quarterly*, *16*(1), 95–116. doi:10.2307/249703

Alba, J., Lynch, J., Weitz, B., Janiszewski, C., Lutz, R., Sawyer, A., & Wood, S. (1997). Interactive Home Shopping: Consumer, Retailer, and Manufacturer Incentives to Participate in Electronic Marketplaces. *Journal of Marketing*, *61*(3), 38–53. doi:10.2307/1251788

Albano, S., & Engels, D. (2002). *Auto-ID Center field trial: phase I summary.* Retrieved October 10, 2008, from http://autoid.mit.edu/whitepapers/MIT-AUTOID-TR006.PDF

Albers, S., Gehring, M., & Heuermann, C. (2003). A configurational approach to supply chain governance. In Seuring, S., Muller, M., Goldbach, M., & Schneidewind, U. (Eds.), *Strategy and organization in supply chains* (pp. 100–113). New York: Physica-Verlag.

Alberto, S. V. (2005). How potential conflict drives channel structure: Concurrent (direct and indirect) channels. *JMR, Journal of Marketing Research*, *42*(4), 507–515. doi:10.1509/jmkr.2005.42.4.507

Al-Gahtani, S. (2007). Information technology (IT) in Saudi Arabia: Culture and the acceptance and use of IT. *Information & Management*, *44*, 681–691. doi:10.1016/j.im.2007.09.002

AL-Harby. F., et al. (2008). The feasibility of biometrics authentication in e-commerce: User acceptance. In *Proceedings of the IADIS International Conference WWW/Internet*, Freiburg, Germany.

Al-Qirim, N. (2007). A research trilogy into E-Commerce adoption in small businesses in New Zealand. *Electronic Markets*, *17*, 263–285. doi:10.1080/10196780701635872

Alsajjan, B. (2008). *Internet banking acceptance model across cultures: The case of England and Saudi Arabia.* Unpublished doctoral dissertation, Brunel Business School, Brunel University, UK.

Al-Somali, S. A. (2008). *An investigation into the acceptance of online banking in Saudi Arabia.* Technovation.

Alwitt, L. F., & Prabhaker, P. R. (1992). Functional and Belief Dimensions of Attitudes Toward Television Advertising: Implications for Copytesting. *Journal of Advertising Research*, *32*(September/October), 30–42.

Anderson, R. E., & Srinivasan, S. (2003). E-satisfaction and e-loyalty: A contingency framework. *Psychology and Marketing*, *20*(3), 123–138. doi:10.1002/mar.10063

Anderson, J. C., & Gerbing, D. W. (1988). Structural equation modeling in practice: a review and recommended two-step approach. *Psychological Bulletin*, *103*(3), 411–423. doi:10.1037/0033-2909.103.3.411

Anderson, J. C., & Narus, J. A. (1990). A Model of Distributor Firm and Manufacturer Firm Working Partnerships. *Journal of Marketing*, *54*, 42–58. doi:10.2307/1252172

Anderson, E., Day, G. S., & Rangan, V. K. (1997). Strategic channel design. *Sloan Management Review*, *38*(4), 59–69.

Anderson, E., & Weitz, B. (1992). The use of pledges to build and sustain commitment in distribution channels. *JMR, Journal of Marketing Research*, *29*, 18–34. doi:10.2307/3172490

Andrews, L., Kiel, G., Drennan, J., Boyle, M. V., & Weerawardena, J. (2005). Gendered perceptions of experiential value in using web-based retail channels. *European Journal of Marketing*, *41*(5), 640–658. doi:10.1108/03090560710737660

Angeles, R. (2005). RFID technologies: supply-chain applications and implementation issues. *Information Systems Management*, *22*(1), 51–65. doi:10.1201/1078/44912.22.1.20051201/85739.7

Ariely, D., & Simonson, I. (2003). Buying, Bidding, Playing, or Competing? Value Assessment and Decision Dynamics in Online Auctions. *Journal of Consumer Psychology*, *13*(1&2), 113–123.

Armstrong, J. S., & Overton, T. S. (1977). Estimating Nonresponse Bias in Mail Surveys. *JMR, Journal of Marketing Research*, *14*, 396–402. doi:10.2307/3150783

Atkinson, M., & Kydd, C. (1997). Individual characteristics associated with world wide web use: An empirical study of playfulness and motivation. *The Data Base for Advances in Information Systems*, *28*(2).

Auger, P. (2005). The impact of interactivity and design sophistication on the performance of commercial websites for small businesses. *Journal of Small Business Management*, *43*(2), 119–137.

Baber, Z. (2005). *CyberAsia: The Internet and society in Asia.* Boston: Brill.

Bagozzi, R. P., & Yi, Y. (1991). Multi-trait multi-method matrices in consumer research. *The Journal of Consumer Research, 17*, 426–439. doi:10.1086/208568

Bailey, J., & Bakos, J. (1997). An exploratory study of the emerging role of electronic intermediaries. *International Journal of Electronic Commerce, 1*, 7–20.

Bailey, L. (1999, November). The triumph of the individual. *New Statesmen*, 1.

Bain & Company. (2002). *Catalog companies set multichannel standard.* Fighting Fire with Water -From Channel Conflict to Confluence.

Baker, E. W., Al-Gahtani, S. S., & Hubona, G. S. (2007). The effects of gender and age on new technology implementation in a developing country - testing the theory of planned behavior (TPB). *Information Technology & People, 20*(4), 352–375. doi:10.1108/09593840710839798

Bakos, J., & Brynjolfsson, E. (1993). From vendors to partners: Information technology and incomplete contracts in buyer–seller relationships. *Journal of Organizational Computing, 3*, 301–328. doi:10.1080/10919399309540206

Bakos, J., Lucas, H., Oh, W., Simon, G., Viswanathan, S., & Weber, B. (2005). The impact of e-commerce on competition in the retail brokerage industry. *Information Systems Research, 16*, 352–371. doi:10.1287/isre.1050.0064

Bakos, J. Y. (1997). Reducing buyer search costs: Implications for electronic marketplaces. *Management Science, 43*, 1676–1692. doi:10.1287/mnsc.43.12.1676

Balasubramanian, S., Prabhudev, K., & Nirup, M. M. (2003). Customer satisfaction in virtual environments: A study of online investing. *Management Science, 49*(7), 871–889. doi:10.1287/mnsc.49.7.871.16385

Bamford, J. D., Gomes-Casseres, B., & Robinson, M. S. (2003). *Mastering alliance strategy: a comprehensive guide to design, management, and organization.* San Francisco, CA: John Wiley & Sons.

Bandura, A. (1986). *Social Foundations of Thought and Action: A Social Cognitive Theory.* Englewood Cliffs, NJ: Prentice Hall.

Bapna, R., Goes, P., Gupta, A., & Yiwei Jin, Y. (2004). User Heterogeneity and its Impact on Electronic Auction Market Design: An Empirical Exploration. *Management Information Systems Quarterly, 28*(1), 21–43.

Barley, S. R., & Tolbert, P. S. (1997). Institutionalization and structuration: Studying the links between action and institution. *Organization Studies, 18*(1), 93–117. doi:10.1177/017084069701800106

Barner, R. (1996, March/April). The new millennium workplace: Seven changes that challenge managers-workers. *The Futurist*, 14–18.

Barnes, S. J., & Scornavacca, E. (2004). Mobile marketing: the role of permission and acceptance. *International Journal of Mobile Communications, 2*(2), 128–139. doi:10.1504/IJMC.2004.004663

Barney, J. (1991). Firm resources and sustained competitive advantage. *Journal of Management, 17*, 99–120. doi:10.1177/014920639101700108

Baron, R., & Kenny, D. (1986). The moderator-mediator variable distinction in social psychological research: Conceptual, strategic, and statistical considerations. *Journal of Personality and Social Psychology, 51*(6), 1173–1182. doi:10.1037/0022-3514.51.6.1173

Barua, A. C., & Whinston, A. B. (1996). Creating a collaboratory in cyberspace: Theoretical foundation and an implementation. *Journal of Organizational Computing, 5*(4), 417–442. doi:10.1080/10919399509540261

Batonda, G., & Perry, C. (2003). Approaches to relationship development processes in inter-firm networks. *European Journal of Marketing, 37*(10), 1457–1484. doi:10.1108/03090560310487194

Bauer, R. A., & Greyser, S. A. (1968). *Advertising in America: The Consumer View.* Boston: Harvard University Press.

Bauer, H. H., Falk, T., & Hammerschmidt, M. (2006). eTransQual: A transaction process-based approach for capturing service quality in online shopping. *Journal of Business Research, 59*(7), 866–875. doi:10.1016/j.jbusres.2006.01.021

Bauer, R. (1967). Consumer behavior as risk taking. In D. Cox, *Risk taking and information handling in consumer behavior*. Cambridge, MA: Harvard University Press.

Beatty, R. P., & Ritter, J. R. (1986). Investment banking, reputation and the underpricing of initial public offerings. *Journal of Financial Economics, 15*, 213–232. doi:10.1016/0304-405X(86)90055-3

Beck, R., & Weitzel, T. (2005). Some economics of vertical standards: Integrating SMEs in EDI supply chains. *Electronic Markets*, *15*, 313–322. doi:10.1080/10196780500302781

Beckinsale, M., Levy, M., & Powell, P. (2006). Exploring Internet adoption drivers in SMEs. *Electronic Markets*, *16*, 361–370. doi:10.1080/10196780600999841

Beer, S. (1984). The viable system model: its provenance, development, methodology and pathology. *The Journal of the Operational Research Society*, *35*(1), 7–25.

Beh, C. (2008). Friendster still tops but Facebook catching up. *Marketing-interactive.com*. Retrieved October 14, 2008, from http://www.marketing-interactive.com/news/5287

Belanger, F., Hiller, J. S., & Smith, W. J. (2002). Trustworthiness in electronic commerce: The role of privacy, security, and site attributes. *The Journal of Strategic Information Systems*, *11*, 245–270. doi:10.1016/S0963-8687(02)00018-5

Belenger, D., & Kargaonkar, P. (1980). Profiling the recreational shopper. *Journal of Retailing*, *56*, 77–92.

Bell, H., & Tang, N. K. H. (1998). The effectiveness of commercial Internet Web sites: A user's perspective. *Internet Research*, *8*(3), 219–228. doi:10.1108/10662249810217768

Bellenger, D. N., & Valencia, H. (1982). Understanding the Hispanic Market. *Business Horizons*, *25*(3), 47–50. doi:10.1016/0007-6813(82)90128-8

Benbunan-Fich, R. (2001). Using protocol analysis to evaluate the usability of a commercial web site. *Information & Management*, *39*(2), 151–163. doi:10.1016/S0378-7206(01)00085-4

Bendavid, Y., Wamba, S. F., & Lefebvre, L. (2006). *Proof of concept of an RFID-enabled supply chain in a B2B e-commerce environment.* Paper presented at the 8th International Conference on Electronic Commerce: The new e-commerce: Innovations for Conquering Current Barriers, Obstacles and Limitations to Conducting Successful Business on the Internet, Fredericton, Canada.

Bengtsson, M., Boter, H., & Vanyushyn, V. (2007). Integrating the Internet and marketing operations. A study of antecedents in firms of different size. *International Small Business Journal*, *25*, 27–48. doi:10.1177/0266242607071780

Benjamin, R., & Wigand, R. (1995). Electronic markets and virtual value chains on the information superhighway. *Sloan Management Review*, *36*, 62–72.

Bentler, P. M., & Hu, L. T. (1995). Evaluating model fit. In Hoyle, R. H. (Ed.), *Structural Equation Modeling* (pp. 76–99). Thousand Oaks, CA: Sage Publications.

Berg, B. L. (1995). *Qualitative Research Methods for the Social Sciences*. Boston: Allyn and Bacon.

Bergen, M., Kauffman, R., & Lee, D. (2005). Beyond the hype of frictionless markets: Evidence of heterogeneity in price and rigidity on the internet. *Journal of Management Information Systems*, *22*, 57–89.

Berkman, H. W., & Gilson, C. C. (1974). Consumer life styles and market segmentation. *Journal of the Academy of Marketing Science*, *2*(1), 189–200. doi:10.1007/BF02729512

Berry, L. L. (2000). Cultivating service brand equity. *Journal of the Academy of Marketing Science*, *28*, 128–137. doi:10.1177/0092070300281012

Berthon, P., Pitt, L., & Prendergast, G. (1997). Visits, hits, caching and counting on the World Wide Web: Old wine in new bottles? *Internet Research: Electronic Networking Applications and Policy*, *7*(1), 5–8. doi:10.1108/10662249710159791

Bestavros, A. (2000). Banking industry walks "tightrope" in personalization of Web services. *Bank Systems and Technology*, *37*, 54–56.

Bhargava, H., & Choudhary, V. (2004). Economics of an information intermediary with aggregation benefits. *Information Systems Research*, *15*, 22–36. doi:10.1287/isre.1040.0014

Bhatnagar, A., Misra, S., & Rao, H. (2000). On risk, convenience and internet shopping behavior. *Communications of the ACM*, *44*, 98–104. doi:10.1145/353360.353371

Bhattacherjee, A., & Premkumar, G. (2004). Understanding changes in belief and attitude toward information technology usage: a theoretical model and longitudinal test. *Management Information Systems Quarterly*, *28*(2), 229–254.

Bhuptani, M., & Moradpour, S. (2005). *RFID Field Guide: Deploying Radio Frequency Identification Systems*. Upper Saddle River, NJ: Prentice Hall.

Blackman, J., & Crompton, L. (1991). The usefulness of selected variables for predicting activity loyalty. *Leisure Sciences*, *13*(3), 205–220. doi:10.1080/01490409109513138

Blattberg, R. C., & Neslin, S. A. (1990). *Sales Promotion-Concepts, Methods, and Strategies*. Englewood Cliffs, NJ: Prentice-Hall.

Blau, P. M. (1964). *Exchange and Power in Social Life*. New York: John Wiley & Sons, Inc.

Bleecker, S. E. (1994, March/April). The virtual organization. *The Futurist*, *28*, 9–14.

Bollen, K. A. (1989). *Structural Equations with Latent Variables*. New York: Wiley.

Boudreau, M., Loch, K., Robey, D., & Straub, D. (1998). Going global: Using information technology to advance the competitiveness of the virtual transitional organization. *The Academy of Management Executive*, *12*(4), 120–128.

Bourgeois, V. W., & Pinder, C. C. (1983). Contrasting philosophical perspectives in administrative science: A reply to Morgan. *Administrative Science Quarterly*, *28*, 608–613. doi:10.2307/2393012

Bridis, T. (1998, August 27). New users battle Web Advertising. *Detroit News*, p. 6.

Broniarczyk, S., & Gershoff, A. (1997). Meaningless differentiation revisited. *Advances in Consumer Research. Association for Consumer Research (U. S.)*, *24*, 223–228.

Brown, G. H. (1952, June 9). Brand loyalty – fact or fiction? *Advertising Age*, *23*, 53–55.

Brown, D. H., Lockett, N., & Schubert, P. (2005). Preface to the focus theme section 'SMEs and E-Business. *Electronic Markets*, *15*, 76–78. doi:10.1080/10196780500083720

Brown, J. R., & Goolsbee, A. (2002). Does the Internet make markets more competitive? Evidence from the life insurance industry. *The Journal of Political Economy*, *110*, 481–507. doi:10.1086/339714

Brown, J. R., Chekitan, S. D., & Lee, D.-J. (2000). Managing marketing channel opportunism: the efficacy of alternative governance mechanisms. *Journal of Marketing*, *64*(2), 51–65. doi:10.1509/jmkg.64.2.51.17995

Brunso, K., & Grunert, K. G. (1998). Cross-cultural similarities and differences in shopping for food. *Journal of Business Research*, *42*(2), 145–150. doi:10.1016/S0148-2963(97)00114-8

Brunso, K., Scholderer, J., & Grunert, K. G. (2004). Closing the gap between values and behavior—A means-end theory of lifestyle. *Journal of Business Research*, *57*(6), 665–670. doi:10.1016/S0148-2963(02)00310-7

Brynjolfsson, E., & Smith, M. D. (2000). Frictionless Commerce? A comparison of Internet and conventional retailers. *Management Science*, *46*(4), 563–585. doi:10.1287/mnsc.46.4.563.12061

Bucklin, C. B., Thomas-Graham, P. A., & Webster, E. A. (1997). Channel conflict: When is it dangerous? *The McKinsey Quarterly*, *3*, 36–43.

Burgoon, J. K., Bonito, J. A., Bengtsson, B. Jr, Ramirez, A., Dunbar, N. E., & Miczo, N. (2000). Testing the Interactivity Model: Communication Processes, partner Assessments, and the Quality of Collaborative Work. *Journal of Management Information Systems*, *16*(3), 33–56.

Burke, K. (2005). The impact of firm size on Internet use in small businesses. *Electronic Markets*, *15*, 79–93. doi:10.1080/10196780500083738

Burke, R. R. (2002). Technology and the customer interface: What consumers want in the physical and virtual store. *Academy of Marketing Science Journal*, *30*(4), 411–432. doi:10.1177/009207002236914

Burns, E. (2005). *Hispanic Online Users Adopt Broadband.* Retrieved from http://www.clickz.com/3521301

Business Monitor International. (2008). *Hong Kong Telecommunications Report*. London: Business Monitor International.

Business Wire. (2000). What kind of dot-shopper are you? ebates.com survey first to uncover 'dot-shopping types' defining online purchasers. *BNET*. Retrieved March 13, 2008, from http://findarticles.com/p/articles/mi_m0EIN/is_2000_May_12/ai_62048317/print?tag=artBody;col1

Cabral, L. M. B. (2000). Stretching firm and brand reputation. *The Rand Journal of Economics*, *31*, 658–673. doi:10.2307/2696353

Cai, Y., & Cude, B. J. (2008). Online shopping. In Xiao, J. J. (Ed.), *Handbook of Consumer Finance Research*. New York: Springer Publishing. doi:10.1007/978-0-387-75734-6_9

Cao, M., & Shi, S. (2006). Signaling in the Internet craze of initial public offerings. *Journal of Corporate Finance*, *12*(4), 818–833. doi:10.1016/j.jcorpfin.2005.11.001

Carrie, A. J. (2004). *US eCommerce overview: 2004 to 2010: A six-year forecast of US online retail sales*. Business View Research Documents. Forrester Research.

Carroll, A., Barnes, S. J., Scornavacca, E., & Fletcher, K. (2007). Consumer perceptions and attitudes towards SMS advertising: recent evidence from New Zealand. *International Journal of Advertising*, *26*(1), 79–98.

Cartagena, C. (2005). *Latin Boom! Everything You Need to Know To Grow Your Business in the U.S. Hispanic Market*. New York: Ballantine Books.

Carter, R., & Hodgson, G. (2006). The impact of empirical tests of transaction cost economics on the debate on the nature of the firm. *Strategic Management Journal*, *27*, 461–476. doi:10.1002/smj.531

Carter, R. B., & Manaster, S. (1990). Initial public offerings and underwriter reputation. *The Journal of Finance*, *45*(4), 1045–1068. doi:10.2307/2328714

Carter, R. B., & Dark, F. H. (2006). *Underwriter reputation and IPO issuer alignment 1981-2000* (working paper). Ames, IA: Iowa State University.

Chan, S. C., & Lu, M. T. (2004). Understanding Internet banking adoption and use behavior: A Hong Kong perspective. *Journal of Global Information Management*, *12*, 2–43.

Chang, M. K., Cheung, W., & Lai, V. S. (2005). Literature derived reference models for the adoption of online shopping. *Information & Management*, *42*, 543–559. doi:10.1016/S0378-7206(04)00051-5

Chang, J., Torkzadeh, G., & Dhillon, G. (2004). Re-examining the measurement models of success for Internet commerce. *Information & Management*, *41*, 577–584. doi:10.1016/S0378-7206(03)00091-0

Chang, M., Cheung, W., & Lai, V. (2005). Literature derived reference models for the adoption of online shopping. *Information & Management*, *42*(4), 543–559. doi:10.1016/S0378-7206(04)00051-5

Charmaz, K. (2000). Grounded Theory: Objectivist and Constructivist Methods. In Denzin, N. K., & Lincoln, Y. S. (Eds.), *Handbook of qualitative research*. Thousand Oaks, CA: Sage Publications.

Chase, R. B., Jacobs, F. R., & Aquilano, N. J. (2006). *Operations Management for Competitive Advantage* (11th ed.). New York: McGraw-Hill.

Chatterjee, P., Hoffman, D. L., & Novak, T. P. (2003). Modeling the Clickstream: Implications for Web-based Advertising Efforts. *Marketing Science*, *22*(4), 520–541. doi:10.1287/mksc.22.4.520.24906

Chaudhuri, A., & Holbrook, M. B. (2001). The chain of effects from brand trust and brand affect to brand performance: the role of brand loyalty. *Journal of Marketing*, *65*(2), 81–93. doi:10.1509/jmkg.65.2.81.18255

Chen, S., & Dhillon, G. (2003). Interpreting dimensions of consumer trust in e-commerce. *Information Technology and Management*, *4*, 303–318. doi:10.1023/A:1022962631249

Chen, H. M., Chen, Q., & Kazman, R. (2007). The affective and cognitive impacts of perceived touch on online customers' intention to return in the Web-based eCRM environment. *Journal of Electronic Commerce in Organizations*, *5*(1), 69–91.

Chen, Q., & Chen, H. M. (2004). Exploring the success factors of eCRM strategies in practice. *Journal of Database Marketing & Customer Strategy Management*, *11*(4), 333–343. doi:10.1057/palgrave.dbm.3240232

Chen, C.-D. (2007). Predicting electronic toll collection service adoption: An integration of the technology acceptance model and the theory of planned behavior. *Transportation Research Part C, Emerging Technologies*, *15*(5), 300–311. doi:10.1016/j.trc.2007.04.004

Chen, L. (2002). Enticing online consumers: An extended technology acceptance perspective. *Information & Management*, *39*, 705–719. doi:10.1016/S0378-7206(01)00127-6

Cheng, J. M.-S. (2006). Consumer acceptance of the internet as a channel of distribution in Taiwan—a channel function perspective. *Technovation*, *26*(7), 856–864. doi:10.1016/j.technovation.2005.01.001

Cheng, T. C. E., Lai, L. C. F., & Yeung, A. C. L. (2008). The driving forces of customer loyalty: a study of Internet service providers in Hong Kong. *International Journal of E-Business Research*, *4*(4), 26–42.

Cheng, T. C. E., Lam, D. Y. C., & Yeung, A. C. L. (2006). Adoption of internet banking: an empirical study in Hong Kong. *Decision Support Systems*, *42*, 1558–1572. doi:10.1016/j.dss.2006.01.002

Chesbrough, H., & Teece, D. (1995). When is virtual virtuous? Organizing for innovation. *Harvard Business Review*, 65–72.

Cheung, C. M. K., Chan, G. W. W., & Limayem, M. (2005). A critical review of online consumer behavior: Empiric research. *Journal of Electronic Commerce in Organizations*, *3*(4), 1–19.

Chiang, K. P., & Dholakia, R. R. (2003). Factors driving consumer intention to shop online: an empirical investigation. *Journal of Consumer Psychology*, *13*(1), 177–183. doi:10.1207/153276603768344898

Childers, T. L., Carr, C. L., Peck, J., & Carson, S. (2001). Hedonic and utilitarian motivations for online retail shopping behavior. *Journal of Retailing*, *77*(4), 511–535. doi:10.1016/S0022-4359(01)00056-2

Chin, J. P., Norman, K. L., & Diehl, V. A. (1988). Development of a tool measuring user satisfaction of the human-computer interface. In *Proceedings of ACM SIGCHI 88* (pp. 213-218). ACM Publishing.

Chiou, J., & Cheng, C. (2003). Should a Company Have Message Boards on its Web Sites? *Journal of Interactive Marketing*, *17*(3), 50–61. doi:10.1002/dir.10059

Chircu, A., & Kauffman, R. (2000). Re-intermediation strategies in business-to-business electronic commerce. *International Journal of Electronic Commerce*, *4*, 7–42.

Cho, C. H., Lee, J. G., & Tharp, M. (2001). Different Forced-Exposure Levels to Banner Advertisements. *Journal of Advertising Research*, *41*(4), 45–56.

Chou, P., Chou, R., & Ko, K. (2009). Prospect theory and the risk-return paradox: some recent evidence. *Review of Quantitative Finance and Accounting*, *33*(3), 193–208. doi:10.1007/s11156-009-0109-z

Chou, W.-T. (2006). A forecast of the trends of Taiwan's online shopping development. Institute for Information Industry, Nov. 2, 2006. (in Chinese)

Chu, W., Choi, B., & Son, M. (2005). The role of on-line retailer brand and infomediary reputation in increasing consumer purchase intention. *International Journal of Electronic Commerce*, *9*, 115–127.

Chuang, M., & Shaw, W. (2005). *How RFID will impact supply chain networks?* Paper presented at the 2005 IEEE Engineering Management Conference, Newfoundland, Canada.

Churchill, G. (1979). A paradigm for developing better measures of marketing constructs. *JMR, Journal of Marketing Research*, *16*(1), 64–73. doi:10.2307/3150876

Citrin, A. V., Sprott, D. E., Silverman, S. N., & Stem, D. E. (2000). Adoption of internet shopping: the role of consumer innovativeness. *Industrial Management & Data Systems*, *100*(7), 294–300. doi:10.1108/02635570010304806

Clay, K., Krishnan, R., Wolff, E., & Fernandes, D. (2002). Retail strategies on the Web: Price and non-price competition in the online book industry. *The Journal of Industrial Economics*, *50*, 351–367.

Coase, R. H. (1937). The nature of the firm. *Economica*, *4*, 386–405. doi:10.1111/j.1468-0335.1937.tb00002.x

Coffey, A., & Atkinson, P. (1996). *Making sense of qualitative data: complementary research strategies*. Thousand Oaks, CA: Sage Publications.

Collins, D. J., & Whipple, N. N. (1994). *Using Bar Code: Why It's Taking Over*. Duxbury, MA: Data Capture Institute.

Compeau, D., & Higgins. (1995). Computer self-efficacy: Development of a measure and initial test. *MIS Quarterly*, *19*(2), 189–211. doi:10.2307/249688

Cox, J. (2004). Getting Connected. *Hispanic Business Magazine*. Retrieved December from http://www.hispanicbusiness.com/news/2004/11/22/getting_connected.htm

Cragg, P. B., & King, M. (1993). Small-firm computing: motivators and inhibitors. *Management Information Systems Quarterly*, *17*(1), 47–60. doi:10.2307/249509

Cronbach, L. J. (1951). Coefficient alpha and the internal structure of tests. *Psychometrika*, 31.

Cross, K. (1999). *Wither the Banner.* Retrieved from http://www.business2.com/b2/web/articles

Cuieford, J. P. (1965). *Fundamental Statistics in Psychology and Education* (4th ed.). New York: McGraw Hill.

Cunningham, R. M. (1956). Brand-Loyalty-What, Where, How Much. *Harvard Business Review*, *34*(1), 116–128.

Dahlstrom, R., McNeilly, K. M., & Speh, T. W. (1996). Buyer-seller relationships in the procurement of logistical services. *Journal of the Academy of Marketing Science*, *24*(2), 110–124. doi:10.1177/0092070396242002

Dave, D. (2003). *How to build and maintain a powerhouse brand: Branding basics*. Retrieved from http://www.DaveDolak.com

Davenport, D., & Pearlson, K. (1998). Two cheers for the virtual office. *Sloan Management Review*, 51–68.

David, R., & Han, S. (2004). A systematic assessment of the empirical support for transaction cost economics. *Strategic Management Journal*, *25*, 39–58. doi:10.1002/smj.359

Davidow, W. H., & Malone, M. S. (1992). *The virtual corporation: structuring and revitalizing the corporation of the 21st Century*. New York: Harper Collins.

Davis, F. D. (1989). Perceived usefulness, perceived ease of use and user acceptance of information technology. *Management Information Systems Quarterly*, *13*(3), 319–340. doi:10.2307/249008

Davis, E. W., & Spekman, R. E. (2004). *The Extended enterprise: gaining competitive advantage through collaborative supply chains*. Upper Saddle River, NJ: Prentice Hall.

Davis, F. D., Bagozzi, R. P., & Warshaw, P. R. (1989). User acceptance of computer technology: a comparison of two theoretical models. *Management Science*, *35*, 982–1002. doi:10.1287/mnsc.35.8.982

Davis, F. D. (1986). *Technology acceptance model for empirically testing new end-user information systems: Theory and results, Massachusetts Institute of Technology*. Unpublished doctoral dissertation.

Day, G. S. (1994). The capabilities of market-driven organizations. *Journal of Marketing*, *58*(4), 37–52. doi:10.2307/1251915

Day, G. S. (1969). A two-dimensional concept of brand loyalty. *JMR, Journal of Marketing Research*, *9*, 29–36.

Day, G. S., & Aaker, D. A. (1970). A guide to consumerism. *Journal of Marketing*, *34*, 12–19. doi:10.2307/1249814

de Chernatony, L., & Dall'Olmo, R. F. (1998). Defining a "brand": Beyond the literature with experts' interpretations. *Journal of Marketing Management*, *14*, 417–443. doi:10.1362/026725798784867798

Deignton, J. (1996). The future of interactive marketing. *Harvard Business Review*, *74*, 151–160.

Dekimpe, M. G., Pierre, F., Srinath, G., Gary, L. L., & Christophe, V. B. (1997). Generalizing about trade show effectiveness: A cross-national comparison. *Journal of Marketing*, *61*(4), 55–64. doi:10.2307/1252086

Deleersnyder, B., Geyskens, I., Gielens, K., & Dekimpe, M. G. (2002). How cannibalistic is the Internet channel? *International Journal of Research in Marketing*, *19*(4), 337–348. doi:10.1016/S0167-8116(02)00099-X

DeLone, W., & McLean, E. (1992). Information systems success: The quest for the dependent variable. *Information Systems Research*, *3*(1), 60–95. doi:10.1287/isre.3.1.60

DeLone, W., & McLean, E. (2003). The DeLone and McLean model of information systems success: A ten-year update. *Journal of Management Information Systems*, *19*(4), 9–30.

den Poel, D., & Leunis, J. (1999). Consumer acceptance of the Internet as a channel of distribution. *Journal of Business Research, 45*(3), 249–256. doi:10.1016/S0148-2963(97)00236-1

Denzin, N. K., & Lincoln, Y. S. (2005). *The Sage Handbook of Qualitative Research* (3rd ed.). Thousand Oaks, CA: Sage Publications.

der Heijden, H., Verhagen, T., & Creemers, M. (2003). Understanding online purchase intentions: Contributions from technology and trust perspectives. *European Journal of Information Systems, 12*(1), 41–48. doi:10.1057/palgrave.ejis.3000445

Deshpande, R., Hoyer, W. D., & Donthu, N. (1986). The Intensity of Ethnic Affiliation: A Study of the Sociology of Hispanic Consumption. *The Journal of Consumer Research, 13*(September), 214–220. doi:10.1086/209061

Deshpande, R. (2002). Performance companies. *International Journal of Medical Marketing, 2*(3), 225–232. doi:10.1057/palgrave.jmm.5040080

Devaraj, S., Fan, M., & Kohli, R. (2002). Antecedents of B2C channel satisfaction and preference: Validating e-commerce metrics. *Information Systems Research, 13*(3), 316–333. doi:10.1287/isre.13.3.316.77

Dholakia, U. M. (2005). The usefulness of bidders' reputation ratings to sellers in online auctions. *Journal of Interactive Marketing, 19*(1), 31–40. doi:10.1002/dir.20029

Dholakia, R. R. (1999). Going shopping: key determinants of shopping behaviors and motivations. *International Journal of Retail and Distribution Management, 27*(4), 154–165. doi:10.1108/09590559910268499

Dholakia, R. R., & Chiang, K. P. (2003). Shoppers in cyberspace: are they from Venus or Mars and does it matter? *Journal of Consumer Psychology, 13*(1), 171–176. doi:10.1207/153276603768344889

Diamond, J., & Pintel, G. (1996). *Retailing* (6th ed.). Upper Saddle River, NJ: Prentice Hall.

Dick, A. S., & Basu, K. (1994). Customer Loyalty: Toward an Integrated Conceptual Framework. *Journal of the Academy of Marketing Science, 22*(2), 99–113. doi:10.1177/0092070394222001

Dickson, M. W., Den Hartog, D. N., & Mitchelson, J. K. (2003). Research on leadership in a cross-cultural context: Making progress, and raising new questions. *The Leadership Quarterly, 14*(6), 729–768. doi:10.1016/j.leaqua.2003.09.002

Dillon, T., & Reif, H. (2006). Identifying purchase perceptions that promote frequent e-commerce buying. *International Journal of Electronic Marketing and Retailing, 1*, 48–66. doi:10.1504/IJEMR.2006.010095

DMD. (2006). RFID Application in Inventory Control. *The DecisionMakers' Direct: new insights, new tools, new solutions*. Retrieved November 21, 2008, from http://www.decisioncraft.com/dmdirect/pdf/rfidapplications.pdf

Dodds, W., Monroe, K., & Grewal, D. (1991). Effects of price, brand and store information on buyers' product evaluation. *JMR, Journal of Marketing Research, 28*, 307–319. doi:10.2307/3172866

Dodourova, M. (2009). Alliances as strategic tools: a cross-industry study of partnership planning, formation and success. *Management Decision, 47*(5). doi:10.1108/00251740910960150

Doney, P., & Cannon, J. (1997). An examination of the nature of trust in buyer seller relationships. *Journal of Marketing, 61*, 35–51. doi:10.2307/1251829

Donthu, N., & Garcia, A. (1999). The internet shopper. *Journal of Advertising Research, 39*(3), 52–58.

Doolin, B., Dillon, S., Thompson, F., & Corner, J. L. (2005). Perceived risk, the Internet shopping experience and online purchasing behavior: a New Zealand perspective. *Journal of Global Information Management, 13*(2), 66–68.

Ducoffe, R. H. (1996). Advertising Value and Advertising on the Web. *Journal of Advertising Research, 36*(5), 21–35.

Duffy, D. L. (2005). Affiliate marketing and its impact on e-commerce. *Journal of Consumer Marketing, 22*(2/3), 161–163. doi:10.1108/07363760510595986

Duncan, O. D. (1986). Path analysis: Sociological example. *American Journal of Sociology, 72*(1), 1–16. doi:10.1086/224256

Dwivedi, Y., Kiang, M., Williams, M., & Lal, B. (2008). Profiling research published in the Journal of Electronic Commerce Research. *Journal of Electronic Commerce Research*, *9*, 77–91.

Dyer, J. (1997). Effective collaboration: How firms minimize transaction costs and maximize transaction value. *Strategic Management Journal*, *18*, 535–556. doi:10.1002/(SICI)1097-0266(199708)18:7<535::AID-SMJ885>3.0.CO;2-Z

Easingwood, C., & Coelho, F. (2003). Single versus multiple channel strategies: Typologies and drivers. *Service Industries Journal*, *23*, 31–46. doi:10.1080/02642060412331300872

Economist Intelligence Unit. (2007). *The 2007 e-readiness rankings - Raising the bar*. Retrieved October 29, 2008 from http://a330.g.akamai.net/7/330/25828/20070420195432/graphics.eiu.com/files/ad_pdfs/2007Ereadiness_Ranking_WP.pdf

Eid, R., & Trueman, M. (2002). The Internet: New international marketing issues. *Management Research News*, *25*(12), 54–67. doi:10.1108/01409170210783278

Eisenhardt, K. M. (1989). Building theories from case study research. *Academy of Management Review*, *14*, 532–550. doi:10.2307/258557

Ellram, L., Londe, B. L., & Weber, M. (1999). Retail logistics. *International Journal of Physical Distribution and Logistics Management*, *29*(7), 477–494. doi:10.1108/09600039910291993

Eng, T. (2008). The effects of the internet on network structures and business performance. *Electronic Markets*, *18*, 91–101. doi:10.1080/10196780701797706

Ernst & Young. (2007). *Doing Business in Hong Kong*. Hong Kong: Ernst & Young.

Eyuboglu, N., & Kahadayi, S. (2005). Dealer-manufacturer alienation in a multiple channel system the moderating effect of structural variables. *Journal of Marketing Channels*, *12*(3), 5–27. doi:10.1300/J049v12n03_02

Falk, T., Schepers, J., Hammerschmidt, M., & Bauer, H. H. (2007). Identifying cross-channel dissynergies for multichannel service providers. *Journal of Service Research*, *10*(2), 143–160. doi:10.1177/1094670507306683

Faucheux, C. (1997). How virtual organizing is transforming management science. *Communications of the ACM*, *40*(9), 50–55. doi:10.1145/260750.260766

Feather, F. (2002). *Future consumer.com*. Toronto: Warwick Publishing.

Featherman, M., & Pavlou, P. (2003). Predicting e-services adoption: A perceived risk facets perspective. *International Journal of Human-Computer Studies*, *59*(4), 451–474. doi:10.1016/S1071-5819(03)00111-3

Fillis, I., & Wagner, B. (2005). E-business development. An exploratory investigation of the small firm. *International Small Business Journal*, *23*, 604–634. doi:10.1177/0266242605057655

Filson, D. (2004). The impact of e-commerce strategies on firm value: lessons from Amazon.com and its early competitors. *The Journal of Business*, *77*, S135–S154. doi:10.1086/381640

Fingleton, J. (1997). Competition between intermediated and direct trade and the timing of disintermediation. *Oxford Economic Papers*, *49*, 543–556.

Finkenzeller, K. (2003). *RFID Handbook: Fundamentals and Applications in Contactless Smart Cards and Identification*. West Sussex, UK: John Wiley & Sons.

Fishbein, M., & Ajzen, I. (1975). *Belief, attitude, intention, and behavior: An introduction to theory and research*. Reading, MA: Addison-Wesley.

Fisher, J., Craig, A., & Bentley, J. (2007). Moving from a web presence to e-Commerce: The importance of a business--web strategy for small-business owners. *Electronic Markets*, *17*(4), 253–262. doi:10.1080/10196780701635864

Fleisch, E., & Tellkamp, C. (2003). *The impact of inventory inaccuracy on retail supply chain performance: a simulation study*. Retrieved from http://autoid.mit.edu/whitepapers/STG-AUTOID-WH003.pdf

Fones, M. (1981, April 6). Two Worlds Together: Towns Along the US-Mexico Border Could Provide Marketing Edge. *Advertising Age*, 8.

Fornell, C., & Larcker, D. F. (1981). Evaluating Structural Equations Models with Unobservable Variables and Measurement Error. *JMR, Journal of Marketing Research*, *18*, 39–50. doi:10.2307/3151312

Fornell, C., & Larcker, D. (1981). Evaluating structural equation models with unobservable variables and measurement error. *JMR, Journal of Marketing Research*, *18*, 39–50. doi:10.2307/3151312

Fox, L. (2000) Affiliate marketing makes headway *Upside, 12*, 4, 176.

Frazier, G. (1999). Organizing and managing channels of distribution. *Journal of the Academy of Marketing Science*, *27*(2), 226–240. doi:10.1177/0092070399272007

Friedman, B., Kahn, P. H. Jr, & Howe, D. C. (2000). Trust online. *Communications of the ACM*, *43*, 34–40. doi:10.1145/355112.355120

Furnell, S., & Evangelatos, K. (2007). Public awareness and perceptions of biometrics. *Computer Fraud & Security*, *1*, 8–13. doi:10.1016/S1361-3723(07)70006-4

Furnell, S., & Karweni, T. (1999). Security implications of electronic commerce: A survey of consumers and business. *Internet Research. Electronic Networking Applications and Policy*, *9*, 372–382. doi:10.1108/10662249910297778

Garbarino, E., & Strahilevitz, M. (2004). Gender differences in the perceived risk of buying online and the effects of receiving a site recommendation. *Journal of Business Research*, *57*, 768–775. doi:10.1016/S0148-2963(02)00363-6

Garfinkel, S., & Rosenberg, B. (2005). *RFID: Applications, Security, and Privacy*. Upper Saddle River, NJ: Addison-Wesley.

Garrity, E. J., Glassberg, B., Kim, Y. J., Sanders, G. L., & Shin, S. K. (2005). An experimental investigation of web-based information systems success in the context of electronic commerce. *Decision Support Systems*, *39*(3), 485–503. doi:10.1016/j.dss.2004.06.015

Garud, R., & Kotha, S. (1993). Using the brain as metaphor to model flexible production systems. *Academy of Management Review*, *19*(4), 671–698. doi:10.2307/258741

Gefen, D., & Straub, D. (2000). The relative importance of perceived ease of use in IS adoption: A study of e-commerce adoption. *Journal of the Association for Information Systems*, 1–8. Retrieved from http://jais.isworld.org?articles/1-8/article.htm.

Gefen, D. (2003). TAM or just plain habit: A look at experienced online shoppers. *Journal of End User Computing*, *15*(3), 1–13.

Gefen, D., & Straub, D. (1997). Gender differences in the perception and use of e-mail: An extension to the Technology Acceptance Model. *Management Information Systems Quarterly*, *21*(4), 389–400. doi:10.2307/249720

Gefen, D., Karahanna, E., & Straub, D. W. (2003). Trust and TAM in online shopping: An integrated model. *MIS Quarterly*, *27*, 51–90.

Gefen, D., & Straub, D. (2000). The relative importance of perceived ease of use in IS adoption: A study of e-commerce adoption. *Journal of the Association for Information Systems*, *1*(8), 1–28.

Gehrig, T. (1993). Intermediation in search markets. *Journal of Economics & Management Strategy*, *2*, 97–120. doi:10.1111/j.1430-9134.1993.00097.x

Gephart, R. (2004). Qualitative Research and the Academy of Management Journal. *Academy of Management*, *47*, 454–462.

Ghose, A., & Dou, W. (1998). Interactive Functions and Their Impacts on the Appeal of Internet Presence Sites. *Journal of Advertising Research*, *8*, 29–43.

Giaglis, G., Klein, S., & O'Keefe, R. (2002). The role of intermediaries in electronic marketplaces: Developing a contingency model. *Information Systems Journal*, *12*, 231–246. doi:10.1046/j.1365-2575.2002.00123.x

Gibson, C., & Zellmer-Bruhn, M. (2001). Metaphors and meaning: an intercultural analysis of the concept of teamwork. *Administrative Science Quarterly*, *46*(2), 274–303. doi:10.2307/2667088

Giddens, A. (1984). *The constitution of society*. Berkeley, CA: University of California Press.

Giesing, I. (2003). *User perceptions related to identification through biometrics within electronic business*. Unpublished master's thesis, University of Pretoria, Pretoria, South Africa.

Gill, J., & Johnson, P. (2002). *Research methods for managers* (3rd ed.). London: Paul Chapman Publishing. ISBN: 0-7619-4002-2.

Gioi, D., & Poole, P. (1984). Scripts in organizational behavior. *Academy of Management Review*, *9*(4), 449–459. doi:10.2307/258285

Girard, T., Korgaonkar, P., & Silverblatt, R. (2003). Relationship of type of product, shopping orientations, and demographics with preference for shopping on the Internet. *Journal of Business and Psychology*, *18*(1), 101–120. doi:10.1023/A:1025087021768

Glaser, B. G. (1978). *Theoretic Sensitivity*. Mill Valley, CA: The Sociology Press.

Glaser, B. G., & Strauss, A. L. (1967). *The discovery of grounded theory: strategies for qualitative research*. New York: Aldine Publishing Company.

Glesne, C. (2006). *Becoming qualitative researchers* (3rd ed.). Boston: Pearson Education, Inc.

Godin, S. (1999). *Permission Marketing: Turning Strangers into Friends, and Friends into Customers*. New York: Simon and Schuster.

Goethals, F., & Vandenbulcke, J. (2005). Two basic types of business-to-business integration. *International Journal of E-Business Research*, *1*(1), 1–15.

Golden, W., Hughes, M., & Gallagher, P. (2003). Online retailing: What drives success? Evidence from Ireland. *Journal of End User Computing*, *15*(3), 32–44.

Gommans, M., Krishan, K. S., & Scheddold, K. B. (2001). From Brand loyalty to e-loyalty: A conceptual framework. *Journal of Economic and Social Research*, *3*(1), 43–58.

Gordon, M. (1981). Models of Pluralism: The New American Dilemma. *The Annals of the American Academy of Political and Social Science*, *4*(54), 178–188. doi:10.1177/000271628145400115

Gouldner, A. W. (1960). The norm of reciprocity: A preliminary statement. *American Sociological Review*, *25*(2), 161–178. doi:10.2307/2092623

Grace, D., & O'Cass, A. (2005). Service branding: Consumer verdicts on service brands. *Journal of Retailing and Consumer Services*, *12*, 125–139. doi:10.1016/j.jretconser.2004.05.002

Grandon, E. E., & Pearson, J. M. (2004). Electronic commerce adoption: An empirical study of small and medium US businesses. *Information & Management*, *42*, 19–216.

Grover, V., & Teng, J. (2001). E-commerce and the information market. *Communications of the ACM*, *44*, 79–86. doi:10.1145/367211.367272

Gudonavičienė, R., Alijošienė, S., & Aukščionis, L. (2008). The peculiarities of conflicts management in marketing channels. *Economics & Management*, 261-271.

Guernica, A., & Kasperuk, I. (1982). *Reaching the Hispanic Market Effectively*. New York: McGraw-Hill.

Gunasekaran, A., Forker, L., & Kobu, B. (2000). Improving operations performance in a small company: a case study. *International Journal of Operations & Production Management*, *20*(3), 316–335. doi:10.1108/01443570010308077

Gunn, R., & Burroughs, M. (1996). Work spaces that work: Designing high-performance offices. *The Futurist*, 19–24.

Gupta, A., Su, B., & Walter, Z. (2004). An empirical study of consumer switching from traditional to electronic channels: A purchase-decision process perspective. *International Journal of Electronic Commerce*, *8*, 131–161.

Gutman, J. (1982). A means-end chain model based on consumer categorization processes. *Journal of Marketing*, *46*(Spring), 60–72. doi:10.2307/3203341

Hagel, J., & Armstrong, A. G. (1997). Net gain: Expanding markets through virtual communities. *The McKinsey Quarterly*, *1*, 141–153.

Hair, J. F., Anderson, R. E., Tatam, R. L., & Black, W. C. (1998). *Multivariate Data Analysis* (5th ed.). Upper Saddle River, NJ: Prentice-Hall.

Hair, J. (2006). *Multivariate data analysis*. Upper Saddle River, NJ: Pearson Education, Inc.

Håkansson, H., & Snehota, I. (1992). Analyzing business relationships. In *Developing Relationships in Business Networks*. London: Routledge.

Halldorsson, A., Skjott-Larsen, T., & Kotzab, H. (2003). Inter-organizational theories behind supply chain management – discussion and applications. In Seuring, S., Muller, M., Goldbach, M., & Schneidewind, U. (Eds.), *Strategy and organization in supply chains* (pp. 32–46). New York: Physica-Verlag.

Hamill, J. (1997). The Internet and international marketing. *International Marketing Review, 14*(5), 300–323. doi:10.1108/02651339710184280

Hanley, K. W., Kumar, A. A., & Seguin, P. J. (1993). Price stabilization in the market for new issues. *Journal of Financial Economics, 34*(2), 177–197. doi:10.1016/0304-405X(93)90017-6

Hardwick, M., & Bolton, R. (1997). The industrial virtual enterprise. *Communications of the ACM, 40*(9), 59–60. doi:10.1145/260750.260770

Harrington, L. (2002). *Winning supply chain strategies: adversity breeds creativity.* Retrieved October 10, 2007, from http://www.inboundlogistics.com/articles/features/1202_feature02.shtml

Harrison, T., Waite, K., & Hunter, G. L. (2006). The internet, information and empowerment. *European Journal of Marketing, 40*(9/10), 972–993. doi:10.1108/03090560610680961

Harrison-Walker, L. J. (2001). E-Complaining: A content analysis of an Internet complaint forum. *Journal of Services Marketing, 15*(5), 397–412. doi:10.1108/EUM0000000005657

Hart, C. W. (1996). Made to Order. *Marketing Management, 5*(2), 12–22.

Harter, D. E., & Slaughter, S. A. (2003). Quality improvement and infrastructure activity costs in software development: a longitudinal analysis. *Management Science, 49*(6), 784–800. doi:10.1287/mnsc.49.6.784.16023

Hasan, H., & Tibbits, H. R. (2000). Strategic management of electronic commerce: An adaptation of the balanced scorecard. *Internet Research, 10*(5), 439–450. doi:10.1108/10662240010349453

Hashim, N. A. (2009). E-commerce and SMEs – The need for caution. *Prometheus, 27*, 125–140. doi:10.1080/08109020902895268

Hawkins, D. I., Best, R. J., & Coney, K. A. (2004). *Consumer Behavior: Building Marketing Strategy* (9th ed.). Richard D. Irwin, Inc.

Heide, J. B., Wathne, K. H., & Rokkan, A. I. (2007). Interfirm monitoring, social contracts, and relationship outcomes. *JMR, Journal of Marketing Research, XLIV*, 425–433. doi:10.1509/jmkr.44.3.425

Heinrich, C. (2005). *RFID and Beyond: Growing Your Business Through Real World Awareness*. Indianapolis, IN: Wiley Publishing Incorporated.

Hellard, R. B. (1995). *Project partnering: principle and practice*. London: Thomas Telford Publications. doi:10.1680/pppap.20436

Hennig-Thurau, T., Gwinner, K. P., Walsh, G., & Gremler, D. D. (2004). Electronic word-of-mouth platforms: What motivates consumers to articulate themselves on the Internet. *Journal of Interactive Marketing, 18*(1), 38–52. doi:10.1002/dir.10073

Herman, A. (2002). Major bank signs up for digital signature verification technology. *Biometric Technology Today, 10*.

Hernandez, S. A., & Newman, L. M. (1992). Choice of English vs. Spanish Language in Advertising to Hispanics. *Journal of Current Issues and Research in Advertising, 14*(2), 1–11.

Hoffman, D., & Novak, T. (1996). Marketing in Hypermedia Computer Mediated Environments: Conceptual Foundations. *Journal of Marketing, 80*(4), 50–68. doi:10.2307/1251841

Hoffman, D. L., & Novak, T. P. (1997). A new marketing paradigm for electronic commerce. *The Information Society, 13*(1), 43–54. doi:10.1080/019722497129278

Hoffman, D. L., & Novak, T. P. (2000). How to acquire customers on the web. *Harvard Business Review*, (May-June): 2000.

Hoffman, D. L., Novak, T. P., & Peralta, M. (1999). Building consumer trust online. *Communications of the ACM, 42*, 80–85. doi:10.1145/299157.299175

Hofstede, G. J., Pedersen, P., & Hofstede, G. H. (2002). *Exploring Culture: Exercises, Stories, and Synthetic Cultures*. Yarmouth, ME: Intercultural Press.

Hofstede, G. (2001). *Culture's consequences: Comparing values, behaviors, institutions, and organizations across nations* (2nd ed.). Thousand Oaks, CA: Sage.

Holzwarth, M., Janiszewski, C., & Neumann, M. M. (2006). The Influence of Avatars on Online Consumer Shopping Behavior. *Journal of Marketing, 70*(4), 19–36. doi:10.1509/jmkg.70.4.19

Hong Kong Monetary Authority. (2008). *Banking policy and supervision*. Retrieved from http://www.info.gov.hk/hkma/eng/bank/index.htm

Hsu, M. H., Yen, C. H., Chiu, C. M., & Chang, C. M. (2006). A longitudinal investigation of continued online shopping behavior: an extension of the theory of planned behavior. *International Journal of Human-Computer Studies, 64*, 889–904. doi:10.1016/j.ijhcs.2006.04.004

Hu, P., Chau, P., Sheng, O., & Tam, K. (1999). Examining the technology acceptance model using physician acceptance of telemedicine technology. *Journal of Management Information Systems, 16*(2), 91–112.

Hu, Y., Sundara, S., Chorma, T., & Srinivasan, J. (2005). *Supporting RFID-based item tracking applications in Oracle DBMS using a bitmap datatype.* Paper presented at the 31st International Conference on Very Large Data Bases, Trondheim, Norway.

Huber, N., Michael, K., & McCathie, L. (2006). *Barriers to RFID in the Supply Chain.* Paper presented at the RFID Eurasia, Turkey.

Hubona, G., & Blanton, J. (1996). Evaluating system design features. *International Journal of Human-Computer Studies, 44*(1), 93–118. doi:10.1006/ijhc.1996.0005

Huffman, C., & Kahn, B. E. (1998). Variety for Sale: Mass Customization or Mass Confusion? *Journal of Retailing, 74*, 491–513. doi:10.1016/S0022-4359(99)80105-5

Huizingh, E. K. R. E. (2000). The content and design of web sites: An empirical study. *Information & Management, 37*(3), 123–134. doi:10.1016/S0378-7206(99)00044-0

Hunsberger, B. (2008). Nike tests Facebook. *The Oregonian.* Retrieved September 17, 2008, from http://www.oregonlive.com/business/index.ssf/2008/06/nike_tests_facebook.html

IBM Learning Services. (2001). *e-Business Selling Workshop: Opportunity Identification (Pain Chains).*

Igbaria, M., & Parsuraman, S. (1989). A path analytic study of individual characteristics, computer anxiety and attitudes toward microcomputers. *Journal of Management, 15*(3), 373–388. doi:10.1177/014920638901500302

Interactive Advertising Bureau. (2006). *Reach U.S. Hispanics Through Online Marketing.* Retrieved from http://www.iab.net/insights_research/iab_research/1666

Internet World Stats. (2009). *Internet world stats: Usage and population statistics.* Retrieved January 24, 2009, from http://www.internetworldstats.com/stats.htm

Ives, B., Olson, M. H., & Baroudi, J. J. (1983). The measurement of user information satisfaction. *Communications of the ACM, 26*(10), 785–793. doi:10.1145/358413.358430

J.C. Williams Group. (2001). *Multi-channel retail report 2001.* Retrieved March 2005, from http://www.shop.org

Jacoby, J., & Kyner, D. B. (1973). Brand loyalty vs. repeat purchasing behavior. *JMR, Journal of Marketing Research, 10*(1), 1–9. doi:10.2307/3149402

Jaggia, S., & Thosar, S. (2004). The medium-term aftermarket in high-tech IPOs: patterns and implications. *Journal of Banking & Finance, 28*(5), 931–950. doi:10.1016/S0378-4266(03)00040-2

Jain, A. (2000). Biometric identification. *Communications of the ACM, 43*(2), 91–98. doi:10.1145/328236.328110

Jain, A. (2004). Biometric recognition: How do I know who you are? In *Proceedings of the IEEE 12th Signal Processing and Communications Applications Conference.*

James, T. (2006). Determining the itention to use biometric devices: An application and extension of the Technology Acceptance Model. *Journal of Organizational and End User Computing, 18*(3), 1–24.

James, C. (1992). Relationship specific assets and the pricing of underwriter services. *The Journal of Finance, 47*(5), 1865–1885. doi:10.2307/2328999

Jarvenpaa, S. L., Tractinsky, N., & Vitale, M. (2000). Consumer trust in internet stores. *Information Technology and Management, 1*, 45–71. doi:10.1023/A:1019104520776

Jarvenpaa, S. L., & Todd, P. A. (1997). Is There a Future for Retailing on the Internet? In Peterson, R. A. (Ed.), *Electronic Marketing and the Consumer* (pp. 139–154). Thousand Oaks, CA: Sage.

Jayatilaka, B., Klein, H. K., & Lee, J. (2007). On Categorizing the IS Research literature - User Oriented Perspective. *Informing Science Journal, 10*, 237–271.

Jensen, M., & Meckling, W. (1976). Theory of the firm: managerial behavior, agency costs, and ownership structure. *Journal of Financial Economics, 3*, 305–360. doi:10.1016/0304-405X(76)90026-X

Jeong, M., & Lambert, C. U. (2001). Adaptation of an information quality framework to measure customers' behavioral intentions to use lodging web sites. *Hospital Management, 20*, 129–146. doi:10.1016/S0278-4319(00)00041-4

Jih, W. J., Lee, S. F., & Fang, S. R. (2006). Investigating the Impact of Customer Relationship Management Practices of E-Commerce on Online Customers' Website Satisfaction: A Model-Building Approach. *International Journal of E-Business Research, 2*(4), 61–77.

Jih, W. J., Wong, S., & Chang, C. B. (2004). Effect of Perceived Risk on Adoption of Internet Banking: An Empirical Investigation. *International Journal of Electronic Business Research, 1*(1), 91–105.

Jih, W. J., & Lee, S. F. (2004). Relationship between Online Shoppers' Motivation and Life Style Indicators. *Journal of Computer Information Systems, 44*(2), 65–73.

Johnson, G., Bruner, G. C. II, & Kumar, A. (2006). Interactivity and its facts revisited. *Journal of Advertising, 35*(4), 35–52. doi:10.2753/JOA0091-3367350403

Johnson, E. J., Moe, W., Fader, P., Bellmann, S., & Lohse, J. (2004). On the depth and dynamics of online search behavior. *Management Science, 50*, 299–308. doi:10.1287/mnsc.1040.0194

Johnson, J. M., & Miller, R. E. (1988). Investment banker prestige and the underpricing of initial public offerings. *Financial Management, 17*(2), 19–29. doi:10.2307/3665523

Johnston, H. R., & Carrico, S. R. (1988). Developing capabilities to use information strategically. *Management Information Systems Quarterly, 12*(1), 37–47. doi:10.2307/248801

Johnston, J., & Madura, J. (2002). The performance of Internet firms following their initial public offering. *Financial Review, 37*(4), 525–550. doi:10.1111/1540-6288.00027

Jones, P., Clarke-Hill, C., Shears, P., Comfort, D., & Hillier, D. (2004). Radio frequency identification in the UK: opportunities and challenges. *International Journal of Retail & Distribution Management, 32*(3), 164–171. doi:10.1108/09590550410524957

Kahneman, D., & Tversky, A. (1979). Prospect theory: An analysis of decision under risk. *Econometrica, 47*(2), 263–292. doi:10.2307/1914185

Karahanna, E., Ahuja, M., Srite, M., & Galvin, J. (2002). Individual differences and relative advantage: The case of GSS. *Decision Support Systems, 32*(4), 327–341. doi:10.1016/S0167-9236(01)00124-5

Karakaya, F., & Khalil, O. (2004). Determinants of Internet adoption in small and medium-sized enterprises. *International Journal of Internet and Enterprise Management, 2*, 341–365. doi:10.1504/IJIEM.2004.005568

Kärkkäinen, M. (2003). Increasing efficiency in the supply chain for short shelf life goods using RFID tagging. *International Journal of Retail & Distribution Management, 31*(10), 529–536. doi:10.1108/09590550310497058

Kauffman, R., & Walden, E. (2001). Economics and electronic commerce: survey and directions for research. *International Journal of Electronic Commerce, 5*, 5–116.

Kauffmann, R. J., & Wang, B. (2001). New buyers' arrival under dynamic pricing market microstructure: The case of group-buying discounts on the Internet. *Journal of Management Information Systems, 18*, 157–188.

Kaufman, C., & Hernandez, S. (1990, October/November). Barriers to Coupon Use: A View from the Bodega. *Journal of Advertising Research, 30*, 18–25.

Keeney, R. L. (1999). The value of Internet commerce to the customer. *Management Science, 45*(4), 533–542. doi:10.1287/mnsc.45.4.533

Keevil, B. (1998). Measuring the usability index of your Web site. In *Proceedings of the 16th Annual International Conference on Computer Documentation* (pp. 271-277). ACM Publishing.

Keller, K. L. (1998). *Strategic brand management: Building, measuring, and managing brand equity*. Upper Saddle River, NJ: Prentice Hall.

Kennedy, A. (2006). Electronic customer relationship management (eCRM): Opportunities and challenges in a digital world. *Irish Marketing Review*, *18*(1-2), 58–68.

Kenny, D. A. (1979). *Correlation and causality*. New York: Wiley.

Kerem, T. (2005). *Advances in the Economics of Information Systems*. Hershey, PA: IGI Global.

Kimery, K., & McCord, M. (2002). Third party assurances: Mapping the road to trust in e-retailing. *Journal of Information Technology Theory and Application*, *4*(2), 63–81.

Kincad, J. W. (2003). *Customer Relationship Management: Getting It Right!*Upper Saddle River, NJ: Prentice-Hall.

Klein, L. R. (1998). Evaluating the potential of interactive media through a new lens: search versus experience goods. *Journal of Business Research*, *41*(3), 195–203. doi:10.1016/S0148-2963(97)00062-3

Knight, P. (2002). *Small-Scale Research*. London: Sage.

Korgaonkar, P., Silverblatt, R., & Girard, T. (2006). Online retailing, product classifications, and consumer preferences. *Internet Research*, *16*(3), 267–288. doi:10.1108/10662240610673691

Korgaonkar, P. K., Karson, E. J., & Lund, D. (2000). Hispanics and Direct Marketing Advertising. *Journal of Consumer Marketing*, *17*(2), 137–157. doi:10.1108/07363760010317204

Korgaonkar, P. K., Silverblatt, R., & O'Leary, B. (2001). Web Advertising and Hispanics. *Journal of Consumer Marketing*, *18*(2), 134–150. doi:10.1108/07363760110386009

Koufaris, M. (2002). Applying the technology acceptance model and flow theory to online consumer behavior. *Information Systems Research*, *13*(2), 205–223. doi:10.1287/isre.13.2.205.83

Koufaris, M. (2002). Applying the technology acceptance model and flow theory to online consumer behaviour. *Information Systems Research*, *13*(2), 205–223. doi:10.1287/isre.13.2.205.83

Kripanont, N. (2007). *Examining a Technology Acceptance Model of Internet usage by academics within Thai business schools. School of Information Systems*. Unpublished doctoral dissertation, Victoria University, Melbourne, Australia.

Krippendorff, K. (2004). *Content Analysis: An Introduction to Its Methodology*. Thousand Oaks, CA: Sage.

Krishnamurthy, S. (2001). A comprehensive analysis of permission marketing. *Journal of Computer-Mediated Communication*, *6*(2).

Kuehn, A. (1962). Consumer brand choice as a learning process. *Journal of Advertising Research*, *2*, 10–17.

Kuk, G. (2004). Effectiveness of vendor-managed inventory in the electronics industry: Determinants and outcomes. *Information & Management*, *41*(5), 645–654. doi:10.1016/j.im.2003.08.002

Kula, V., & Tatoglu, E. (2003). An exploratory study of Internet adoption by SMEs in an emerging market economy. *European Business Review*, *15*(5), 324–333. doi:10.1108/09555340310493045

Kumar, V., & Shah, D. (2004). Building and Sustaining Profitable Customer Loyalty for the 21st Century. *Journal of Retailing*, *1*(80), 317–330. doi:10.1016/j.jretai.2004.10.007

Kusev, P., van Schaik, P., Ayton, P., Dent, J., & Chater, N. (2009). Exaggerated risk: Prospect theory and probability weighting in risky choice. *Journal of Experimental Psychology. Learning, Memory, and Cognition*, *35*(6), 1487–1505. doi:10.1037/a0017039

Kwak, H., Fox, R. J., & Zinkhan, G. M. (2002). What products can be successfully promoted and sold via the Internet. *Journal of Advertising Research*, *42*(1), 23–38.

Lakoff, G., & Johnson, M. (1980). *Metaphors we live by*. Chicago: University of Chicago Press.

Lakoff, G., & Nunez, R. E. (2001). *Where Mathematics Comes From: How the Embodied Mind Brings Mathematics into Being*. New York: Basic Books.

Lal, R., & Sarvary, M. (1999). When and how is the Internet likely to decrease price competition? *Marketing Science*, *18*, 485–503. doi:10.1287/mksc.18.4.485

Lala, V., Arnold, V., Sutton, S. G., & Guand, L. (2002). The impact of relative information quality of e-commerce assurance seals on Internet purchasing behavior. *International Journal of Accounting Information Systems, 3*, 237–253. doi:10.1016/S1467-0895(02)00069-6

Land, K. C. (1969). *Principles of path analysis*. San Francisco, CA: Jossey-Bass.

Larzelere, R. E., & Huston, T. L. (1980). The dyadic trust scale: toward understanding interpersonal trust in close relationships. *Journal of Marriage and the Family, 42*, 595–604. doi:10.2307/351903

Lau, A. (2008). Facebook tops social networks in Hong Kong. *Marketing-interactive.com*. Retrieved October 14, 2008, from http://www.marketing-interactive.com/news/5328

Laurie, A. J., et al. (2007). Towards understanding user perceptions of authentication technologies. In *Proceedings of the 2007 ACM workshop on Privacy in electronic society*, Alexandria, VA. New York: ACM.

Laux, D., et al. (2007). *Customer acceptance of biometric technology in the hotel industry*.

Lederer, A., Maupin, D., Sena, M., & Zhuang, Y. (2000). The technology acceptance model and the World Wide Web. *Decision Support Systems, 29*(3), 269–282. doi:10.1016/S0167-9236(00)00076-2

Lee, S. F., Jih, W. J., & Fang, S. R. (2006). Investigating the impact of customer relationship management practices of e-commerce on online customers' website satisfaction: A model-building approach. *International Journal of E-Business Research, 2*(4), 61–77.

Lee, H. Y., Qu, H. L., & Kim, Y. S. (2007). A study of the impact of personal innovativeness on online travel shopping behavior – a case study of Korean travelers. *Tourism Management, 28*(3), 886–897. doi:10.1016/j.tourman.2006.04.013

Lee, Z., & Gosain, S. (2002). A longitudinal price comparison for music cds in electronics and brick-and-mortar markets: Pricing strategies in emergent electronic commerce. *The Journal of Business Strategy, 19*, 55–71.

Lee, J. (2004). Discriminant analysis of technology adoption behavior: A case of Internet technologies in small businesses. *Journal of Computer Information Systems, 44*(4), 57–66.

Lee, Y., Lee, Z., & Larsen, K. R. T. (2003). Coping with Internet channel conflict. *Communications of the ACM, 46*(7), 137–142. doi:10.1145/792704.792712

Lee, M., & Turban, E. (2001). A trust model for consumer Internet shopping. *International Journal of Electronic Commerce, 6*, 75–91.

Lee, Y., & Kozar, K. A. (2006). Investigating the effect of website quality on e-business success: An analytic hierarchy process (AHP) approach. *Decision Support Systems, 42*(3), 1383–1401. doi:10.1016/j.dss.2005.11.005

Lee, Y. W., Strong, D. M., Kahn, B. K., & Wang, R. Y. (2002). AIMQ: Methodology for information quality assessment. *Information & Management, 40*, 133–146. doi:10.1016/S0378-7206(02)00043-5

Lee, J., Jayatilaka, B., & Kwok, R. *(2002). Virtual Organizations: Duality of Human Identities in Consciousness and Entity. In* Proceedings of the Information Resources Management Association International Conference *(IRMA).*

Lee, Y., Cheng, F., & Leung, Y. (2004). *Exploring the impact of RFID on supply chain dynamics*. Paper presented at the 2004 Winter Simulation Conference, Yorktown Heights, NY.

Lefebvre, L., Lefebvre, E., Bendavid, Y., Wamba, S. F., & Boeck, H. (2006). *RFID as an enabler of B-to-B e-Commerce and its impact on business processes: a pilot study of a supply chain in the retail industry.* Paper presented at the 39th Annual Hawaii International Conference on System Sciences, Hawaii, USA.

Leggatt, H. (2007). Online Spending Power of Hispanic Population Increasing. *Bizreport*. Retrieved from http://www.bizreport.com/2007/07/onlinespendingpowerofHispanicpopulationincreasing.html

Legris, P., Ingham, J., & Collerette, P. (2003). Why do people use information technology? A critical review of the technology acceptance model. *Information & Management, 40*(3), 191–204. doi:10.1016/S0378-7206(01)00143-4

Lehmann, D. (1999). Consumer behavior and Y2K. *Journal of Marketing Channels*, *63*(5), 14–18.

Lehmann, E. E. (2003). Pricing behaviour on the Web: Evidence from online travel agencies. *Empirica*, *30*, 379–396. doi:10.1023/B:EMPI.0000005240.01953.e0

Klein, L. R. (1998). Evaluating the potential of interactive media through a new lens; search versus experience goods. *Journal of Business Research*, *41*(3), 195–203. doi:10.2307/1914185

Leland, H. E., & Pyle, D. H. (1977). Information asymmetries, financial structure, and financial intermediation. *The Journal of Finance*, *32*(2), 371–387. doi:10.2307/2326770

Lennie, I. (1999). Managing metaphorically. *Studies in Cultures, Organizations and Societies*, *5*, 46–59. doi:10.1080/10245289908523520

Lesser, J. A., & Marie, A. H. (1986). The generalizability of psychographic market segments across geographic location. *Journal of Marketing*, *50*(1), 18–27. doi:10.2307/1251275

Levin, A. M., Levin, I. P., & Weller, J. A. (2005). A multi-Attribute Analysis of Preferences for Online and Offline Shopping: Differences across Products, Consumers, and Shopping Stages. *Journal of Electronic Commerce Research*, *6*(4), 281–290.

Levy, M., & Powell, P. (2003). Exploring SME Internet adoption: Towards a contingent model. *Electronic Markets*, *13*, 173–181. doi:10.1080/1019678032000067163

Levy, M., Powell, P., & Worrall, L. (2005). Strategic intent and e-Business in SMEs: Enablers and inhibitors. *Information Resources Management Journal*, *18*, 1–20.

Lewis, M., & Grimes, A. (1999). Metatriangulation: Building theory from multiple paradigms. *Academy of Management Review*, *24*, 672–690. doi:10.2307/259348

Lewis, J. R. (1994). Sample sizes for usability studies - additional considerations. *Human Factors*, *36*(2), 368–378.

Lewis-Beck, M. S. (1994). *Factor Analysis and Related Techniques*. London: Sage Publications.

Li, C. C. (1975). *Path analysis: A primer*. Pacific Grove, CA: The Boxwood Press.

Lian, J. W., & Lin, T. M. (2008). Effects of consumer characteristics on their acceptance of online shopping: comparison among different product types. *Computers in Human Behavior*, *24*, 48–65. doi:10.1016/j.chb.2007.01.002

Liang, T. P., & Huang, J. S. (1998). An empiric study of consumer acceptance of products in electronic markets: A Transactional Cost Model. *Decision Support Systems*, *24*(1), 29–43. doi:10.1016/S0167-9236(98)00061-X

Liao, S. (1999). The adoption of virtual banking: An empirical study. *International Journal of Information Management*, *19*, 63–74. doi:10.1016/S0268-4012(98)00047-4

Libai, B., Biyalogorsky, E., & Gerstner, E. (2003). Setting referral fees in affiliate marketing. *Journal of Service Research*, *5*(4), 303–315. doi:10.1177/1094670503005004003

Lichtenstein, S., & Williamson, K. (2006). Understanding consumer adoption of internet banking: An interpretive study in the Australian banking context. *Journal of Electronic Commerce Research*, *7*, 50–66.

Lightner, N. J. (2003). What users want in e-commerce design: effects of age, education and income. *Ergonomics*, *46*(1-3), 153–168. doi:10.1080/00140130303530

Lim, K. S., & Razzaque, M. A. (1997). Brand loyalty and situational effects: An interactionist perspective. *Journal of International Consumer Marketing*, *9*(4), 95–115. doi:10.1300/J046v09n04_06

Lim, K. H., Leung, K., Sia, C. L., & Lee, M. K. (2004). Is e-commerce boundary-less? Effects of individualism-collectivism and uncertainty avoidance on Internet shopping. *Journal of International Business Studies*, *35*, 545–559. doi:10.1057/palgrave.jibs.8400104

Limayem, M., Khalifa, M., & Frini, A. (2000). What makes consumers buy from internet? A longitudinal study of online shopping. *IEEE Transactions on Systems, Man, and Cybernetics*, *30*(4), 421–432. doi:10.1109/3468.852436

Lin, J. C. C., & Lu, H. P. (2000). Towards an understanding of the behavioural intention to use a web site. *International Journal of Information Management*, *20*, 197–208. doi:10.1016/S0268-4012(00)00005-0

Lin, H. F., & Lee, G. G. (2006). Determinants of success for online communities: An empirical study. *Behaviour & Information Technology*, *25*(6), 479–488. doi:10.1080/01449290500330422

Lipstein, B. (1959). The dynamics of brand loyalty and brand switching. In *Proceedings of the Fifth Annual Conference of the Advertising Research Foundation* (pp. 101-108). New York: Advertising Research Foundation.

Liu, S., & Silverman, M. (2001). A practical guide to biometric security technology. *IT Professional*, 27–32. doi:10.1109/6294.899930

Liu, C., & Arnett, K. P. (2000). Exploring the factors associated with web site success in the context of electronic commerce. *Information & Management*, *38*(1), 23–33. doi:10.1016/S0378-7206(00)00049-5

Ljungqvist, A., & Wilhelm, W. J. (2003). IPO pricing in the dot-com bubble. *The Journal of Finance*, *58*(2), 723–752. doi:10.1111/1540-6261.00543

Ljungqvist, A., & Wilhelm, W. J. (2005). Does prospect theory explain IPO market behavior? *The Journal of Finance*, *60*(4), 1759–1790. doi:10.1111/j.1540-6261.2005.00779.x

Loughran, T., & Ritter, J. R. (2004). Why has IPO underpricing changed over time? *Financial Management*, *33*(3), 5–37.

Lu, Y., Zhou, T., & Wang, B. (2009). Exploring Chinese users' acceptance of instant messaging using the theory of planned behavior, the technology acceptance model, and the flow theory. *Computers in Human Behavior*, *25*(1), 1–252. doi:10.1016/j.chb.2008.06.002

Lu, D., & Choy, T. K. L. (2004). Competitive supply chain development through strategic alliance and dynamic relationship. In *Proceedings of Supply Chain Management and Information System 2004,* Hong Kong (pp. 287-296).

Lumsden, J. (2006). Online-questionnaire design guidelines. In Reynolds, R. A., Woods, R., & Baker, J. D. (Eds.), *Handbook of Research on Electronic Surveys and Measurements*. Hershey, PA: IGI Global.

Magazine, C. (2007). *Online U.S. Hispanics are Internet Trend Setters.* Retrieved from http://www.contactomagazine.com

Malone, T. W., Yates, J., & Benjamin, R. I. (1989). The logic of electronic markets. *Harvard Business Review*, *30*, 484–497.

Malone, T. W., Yates, J., & Benjamin, R. I. (1987). Electronic market and electronic hierarchies. *Communications of the ACM*, *30*(6), 484–497. doi:10.1145/214762.214766

Manchanda, P., Dube, J. P., Goh, K. Y., & Chintagunta, P. K. (2006). The Effect of Banner Advertising on Internet Purchasing. *JMR, Journal of Marketing Research*, *43*(February), 98–108. doi:10.1509/jmkr.43.1.98

Marasini, R., Ions, K., & Ahmad, M. (2008). Assessment of e-business adoption in SMEs. A study of manufacturing industry in the UK North East region. *Journal of Manufacturing Technology Management*, *19*, 627–643. doi:10.1108/17410380810877294

Margulis, S. (1977). Conceptions of privacy: Current status and next steps. *The Journal of Social Issues*, *33*, 5–12.

Marin, G. (1989, December). Barbara VanOss Marin, Regina Otero-Sabogal, Fabio Sabogal, and Eliseo J. Perez-Stable, The Role of Acculturation in the Attitudes, Norms, and Expectancies of Hispanic Smolers. *Journal of Cross-Cultural Psychology*, *20*, 399–415. doi:10.1177/0022022189204005

Marin, G., & Marin, B. V. (1991). *Research with Hispanic Populations*. Thousand Oaks, CA: SAGE.

Marshall, B. B., Crutchley, C. E., & Lending, D. (2004). Early Internet IPOs versus subsequent entrants. *Journal of Economics & Finance*, *28*(1), 104–116. doi:10.1007/BF02761458

Martin, B. A. S., Van Durme, J., Raulas, M., & Merisavo, M. (2003). Email Advertising: Exploratory insights from Finland. *Journal of Advertising Research*, *43*(September), 293–300.

Mathwick, C. (2002). Understanding the Online Consumer: A Typology of Online Relational Norms and Behavior. *Journal of Interactive Marketing*, *16*(1), 40–55. doi:10.1002/dir.10003

Matthew, L. M., Ostrom, A. L., Roundtree, R. I., & Bitner, M. J. (2000). Self-service technologies: Understanding customer satisfaction with technology based service customers. *Journal of Marketing*, *64*, 50–65.

McCathie, L., & Michael, K. (2005). *Is it the end of barcodes in supply chain management?* Paper presented at the Proceedings of the Collaborative Electronic Commerce Technology and Research Conference LatAm, Talca, Chile.

McCoy, S., Everard, A., Polak, P., & Galletta, D. F. (2007). The Effects of Online Advertising. *Communications of the ACM, 50*(3), 84–88. doi:10.1145/1226736.1226740

McFarland, D. J., & Hamilton, D. (2006). Adding contextual specificity to the technology acceptance model. *Computers in Human Behavior, 22*, 427–447. doi:10.1016/j.chb.2004.09.009

McGowan, P., & Durkin, M. G. (2002). Toward an understanding of Internet adoption at the marketing/entrepreneurship interface. *Journal of Marketing Management, 18*, 361–377. doi:10.1362/0267257022872451

McGowan, P., Durkin, M. G., Allen, L., Dougan, C., & Nixon, S. (2001). Developing competencies in the entrepreneurial small firm for use of the Internet in the management of customer relationships. *Journal of European Industrial Training, 25*, 126–136. doi:10.1108/EUM0000000005443

McKinney, V., Yoon, K., & Zahedi, F. M. (2002). The measurement of web-customer satisfaction: An expectation and disconfirmation approach. *Information Systems Research, 13*(3), 296–315. doi:10.1287/isre.13.3.296.76

McWilliam, G. (2000). Building Stronger Brands through Online Communities. *Sloan Management Review, 41*(3), 43–54.

Mehrtens, J., Cragg, P. B., & Mills, A. M. (2001). A model of Internet adoption by SMEs. *Information & Management, 39*, 165–176. doi:10.1016/S0378-7206(01)00086-6

Melewar, T. C., & Smith, N. (2003). The Internet revolution: some global marketing implications. *Marketing Intelligence & Planning, 21*(6), 363–369. doi:10.1108/02634500310499220

Merton, R. C. (1973). Theory of rational option pricing. *The Bell Journal of Economics and Management Science, 4*(1), 141–183. doi:10.2307/3003143

Meyers-Levy, J., & Tybout, A. (1989). Schema congruity as a basis for product evaluation. *The Journal of Consumer Research, 16*(1), 39–54. doi:10.1086/209192

Michael, K., & McCathie, L. (2005). *The pros and cons of RFID in supply chain management.* Paper presented at the International Conference on Mobile Business, Sydney, Australia.

Michman, R. D., Mazze, E. M., & Greco, A. J. (2003). *Lifestyle Marketing: Reaching the New American Consumer*. Westport, CT: Praeger Publishers.

Miller, D. (1991). *Handbook of research design and social measurement*. Newbury Park, CA: Sage Publications.

Mills-Harris, M., Soylemezoglu, A., & Saygin, C. (2005). *RFID data-based inventory management of time-sensitive materials.* Paper presented at the Industrial Electronics Society (IECON 2005), Raleigh, NC.

Min, S., Song, S., & Keebler, J. S. (2002). An Internet-mediated market orientation (IMO): Building a theory. *Journal of Marketing Theory and Practice, 10*(2), 1–11.

Mintzberg, H., Quinn, J. B., & Voyer, J. (1995). *The strategy process*. Englewood Cliffs, NJ: Prentice-Hall.

Miroshnik, V. (2002). Culture and international management: a review. *Journal of Management Development, 21*(7), 521–544. doi:10.1108/02621710210434647

Mittal, B. (1994). Public Assessment of TV Advertising: Faint Praise and Harsh Criticism. *Journal of Advertising Research, 34*(February), 35–53.

Modahl, M. (2000). *Now or never: How companies must change Today to Win The Battle for Internet Consumers* (pp. 169–174). New York: HarperCollins.

Molander, J. (2005). The evolution of affiliate marketing. *AffilateTip.com*, Oct 20, 2005. Retrieved from http://affiliatetip.com/news/

Molla, A., & Licker, P. S. (2001). E-commerce systems success: An attempt to extend and re-specify the DeLone and McLean model of IS success. *Journal of Electronic Commerce Success, 2*(4), 1–11.

Monroe, K. B., & Guiltinan, J. P. (1975). A path-analytic exploration of retail patronage influences. *The Journal of Consumer Research, 2*, 19–28. doi:10.1086/208612

Moon, J., & Kim, Y. (2001). Extending the TAM for a world-wide-web context. *Information & Management, 38*, 217–230. doi:10.1016/S0378-7206(00)00061-6

Moon, Y. (2000). *BizRate.com* (Case 9-501-024, 1-22). Cambridge, MA: Harvard Business School.

Moore, G. C., & Benbasat, I. (1991). Development of an instrument to measure the perceptions of adopting an information technology innovation. *Information Systems Research, 2*, 192–222. doi:10.1287/isre.2.3.192

Morgan, G. (2006). *Images of organization*. Thousand Oaks, CA: Sage.

Morton, S. F., Zettelmeyer, F., & Silva-Risso, J. (2001). Internet car retailing. *The Journal of Industrial Economics, 49*, 501–519.

Mowshowitz, A. (1994). Virtual organization: A vision of management in the information age. *The Information Society, 10*(4), 267–288. doi:10.1080/01972243.1994.9960172

Muehling, D. D. (1987). An Investigation of Factors Underlying Attitude-Toward-Advertising In-General. *Journal of Advertising, 16*(1), 32–40.

Mukherjee, A., & Hoyer, W. (2001). The effect of novel attributes on product evaluation. *The Journal of Consumer Research, 28*(3), 462–472. doi:10.1086/323733

Mukhopadhyay, S. K., Yao, D., & Yue, X. (2008). Information sharing of value-adding retailer in a mixed channel hi-tech supply chain. *Journal of Business Research, 61*(9), 950–958. doi:10.1016/j.jbusres.2006.10.027

Muller, M. (2003). *Essentials of Inventory Management*. New York: American Management Association.

Muniz, A., & O'Guinn, T. C. (2001). Brand Community. *The Journal of Consumer Research, 58*, 20–38.

Murphy, S. (2008). The view from the top. *Chain Store Age, 84*(9), 58.

Murphy, T. (2000). *Web rules: How the Internet is changing the way consumers make choices*. Chicago: Dearborn.

Nadilo, R. (1998). On-Line Research Taps Consumers Who Spend. *Marketing Research, 8*(June), 12.

Nair, G., & Pleasance, D. (2005). Mitigating channel conflict. *The McKinsey Quarterly, 3*, 16–17.

Negasha, S., Ryan, T., & Igbaria, M. (2003). Quality and effectiveness in web-based customer support systems. *Information & Management, 40*, 757–768. doi:10.1016/S0378-7206(02)00101-5

Nelson, P. J. (1970). Information and consumer behavior. *The Journal of Political Economy, 78*(2), 311–329. doi:10.1086/259630

Nemeth, P., Toth, L., & Hartvanyi, T. (2006). *Adopting RFID in supply chains*. Paper presented at the IEEE International Conference on Mechatronics.

Neslin, S. A., Grewal, D., Leghorn, R., Shankar, V., Teerling, M. L., Thomas, J. S., & Verhoef, P. C. (2006). Opportunities and challenges in multichannel customer management. *Journal of Service Research, 9*(2), 95–112. doi:10.1177/1094670506293559

Newman, J. W., & Werbel, R. A. (1973). Multivariate analysis of brand loyalty for major household appliances. *JMR, Journal of Marketing Research, 10*(4), 404–409. doi:10.2307/3149388

News, B. B. C. (2008, December 16). Legal papers served via Facebook. *BBC* (Online). Retrieved December 17, 2008, from http://news.bbc.co.uk/2/hi/asia-pacific/7785004.stm

Ng, I. (2007). Establishing a service channel: A transaction cost analysis of a channel contract between a cruise line and a tour operator. *Journal of Services Marketing, 2*, 4–14. doi:10.1108/08876040710726257

Ngai, E. W. T. (2005). Customer relationship management research (1992-2002): An academic literature and classification. *Marketing Intelligence & Planning, 23*(6-7), 582–605. doi:10.1108/02634500510624147

Nielsen, J. F., Høst, V., & Mols, N. P. (2005). Adoption of Internet-based marketing channels by small and medium-sized manufacturers. *International Journal of E-Business Research, 1*(2), 1–23.

Nielson Company. (2009). *Global online shopping report, February 2008*. Retrieved January 24, 2009 from http://www.nielsen.com/solutions/GlobalOnlineShoppingReportFeb08.pdf

Nokia. (2008). *Nokia and Google to Offer Google Search to Nokia customers worldwide*. Retrieved April 14, 2008, from http://www.nokia.com/A4136001?newsid=1190585

Norton, S. W., & Norton, W. Jr. (1988). An Economic perspective on the information content of magazine advertisements. *International Journal of Advertising*, *7*(2), 138–148.

Nosek, J., & Shephard, G. (1995). View point: Making groupware payoff-British model. *Communications of the ACM*, *38*(6), 11–13. doi:10.1145/213859.214784

Nowlis, S., & Simonson, I. (1996). The effect of new product features on brand choice. *JMR, Journal of Marketing Research*, *32*, 36–46. doi:10.2307/3152011

NRF/Forrester Research. (2002). *December 2002 NRF/Forrester Online Retail Index*.

Numberger, S., & Rennhak, C. (2005). The future of B2C e-commerce. *Electronic Markets*, *15*, 269–282. doi:10.1080/10196780500209077

Nunnally, J. C. (1978). *Psychometric theory* (2nd ed.). NY: McGraw-Hill.

O'Cass, A., & Fenech, T. (2003). Web retailing adoption: exploring the nature of Internet users web retailing behaviour. *Journal of Retailing and Consumer Services*, *10*(2), 81–94. doi:10.1016/S0969-6989(02)00004-8

O'Gorman, L. (2003). Comparing passwords, tokens, and biometrics for user authentication. *Proceedings of the IEEE*, *91*, 2021–2040. doi:10.1109/JPROC.2003.819611

O'Reilly, K., & Paper, D. (2009). The role of vendor in the development of eCRM tools. *Qualitative Market Research -. International Journal (Toronto, Ont.)*, *12*(4), 404–427.

Oberthur Technologies; Zain Group Selects Oberthur Technologies' End-to-End Security Platform for Africa's Biggest Mobile Commerce Service. (2009, March 3). *Science Letter*, 3882.

O'Leary, D., Kuokka, D., & Plant, R. (1997). Artificial intelligence and virtual organizations. *Communications of the ACM*, *40*(1), 52–69. doi:10.1145/242857.242871

Oliver, R. L., & Anderson, E. (1995). Behavior- and outcome-based sales control systems: Evidence and consequences of pure-form and hybrid governance. *Journal of Personal Selling & Sales Management*, *15*(4).

Olmedo, E. L., & Padilla, A. M. (1978). Empirical and Construct Validation of a Measure of Acculturation for Mexican Americans. *The Journal of Social Psychology*, *105*, 179–187.

Olson, J. C., & Reynolds, T. J. (1984). Understanding consumers' cognitive structures: Implications for advertising strategy. In Percy, L., & Woodside, A. G. (Eds.), *Advertising and consumer psychology* (pp. 77–90). Lexington, MA: Lexington Books.

Olson, J. C. (1977). Price as an informational cue: Effects on product evaluations. In A. G. Woodside et al. (Eds.), *Consumer and industrial buying behavior* (pp. 267-286). NY: North Holland.

OMG UK. (2008). *Affiliate contract*. Retrieved February 2008, from https://omgadmin.co.uk/en/clientarea/affiliates/welcome.asp

Orlikowski, W. J., & Robey, D. (1991). Information technology and the structuring of organizations. *Information Systems Research*, *2*(2), 143–169. doi:10.1287/isre.2.2.143

Osmonbekov, T., Bello, D. C., & Gilliland, D. I. (2009). The impact of e-business infusion on channel coordination, conflict and reseller performance. *Industrial Marketing Management*, *38*(7), 778–784. doi:10.1016/j.indmarman.2008.03.005

Oxoby, R. J., & Finnigan, H. (2007). Developing heuristic-based quality judgments: Blocking in consumer choice. *Psychology and Marketing*, *24*, 295–393. doi:10.1002/mar.20162

Pallant, J. (2005). *SPSS survival manual. Milton, Keynes*. UK: Open University Press.

Palmer, R. C. (1995). *The Bar Code Book: Revised and Expanded*. Dublin, NH: Helmers Publishing, Inc.

Palmer, I., & Dunford, R. (1996). Conflicting uses of methapors: reconceptualizing their use in the field of organizational change. *Academy of Management Review*, *21*(3), 691–717. doi:10.2307/258999

Palmer, J. W., & Griffith, D. A. (1998). An emerging role of web site design for marketing. *Communications of the ACM*, *41*(3). doi:10.1145/272287.272296

Palmer, J. W. (2002). Web site usability, design, and performance metrics. *Information Systems Research, 13*(2), 151–167. doi:10.1287/isre.13.2.151.88

Palmer, J. W., & Speier, C. (1997). *A Typology of Virtual Organizations: An Empirical Study*. Retrieved from http://hsb.baylor.edu/ramsower/ais.ac.97/papers/palm_spe.htm

Pan, X., Ratchford, B. T., & Shankar, V. (2003). The evolution of price dispersion in Internet retail markets. *Advances in Applied Microeconomics, 12*, 85–105. doi:10.1016/S0278-0984(03)12004-4

Papatla, P., & Bhatnagar, A. (2002). Choosing the right mix of on-line affiliates: How do you select the best? *Journal of Advertising, 31*(3), 69–81.

Parasuraman, S., & Igbaria, M. (1990). An examination of gender differences in the determinants of computer anxiety and attitudes toward microcomputers among managers. *International Journal of Man-Machine Studies, 32*(3), 327–340. doi:10.1016/S0020-7373(08)80006-5

Park, C., & Jun, J. K. (2003). A cross-cultural comparison of Internet buying behavior – effects of Internet usage, perceived risks, and innovativeness. *International Marketing Review, 20*(5), 534–553. doi:10.1108/02651330310498771

PartnerCentric. (2006) *Affiliate survey*. Retrieved February 2008, from http://www.marketingsherpa.com/cs/amscr2006/study.html

Parvatiyar, A., & Sheth, J. N. (2001). Customer relationship management: Emerging practice, process, and discipline. *Journal of Economic & Social Research, 3*(2), 1–34.

Patton, M. Q. (2002). *Qualitative research and evaluaion methods*. Thousand Oaks, CA: Sage Publication, Inc.

Paul, A. P. (2003). *Consumer acceptance of electronic commerce: Integrating trust and risk with the Technology Acceptance Model* (*Vol. 7*, pp. 101–134). Armonk, NY: M. E. Sharpe, Inc.

Pauwels, K., & Neslin, S. A. (2006). *Building with bricks and mortar: The revenue impact of opening physical stores in a multichannel environment*. Hanover, NH: Dartmouth College.

Pavlou, P. A., & Chai, L. (2002). What drives electronic commerce across cultures? A cross-cultural empirical investigation of the theory of planned behavior. *Journal of Electronic Commerce Research, 3*(4), 240–253.

Pavlou, P. A., & Fygenson, M. (2006). Understandings and predicting electronic commerce adoption: An extension of the theory of planned behavior. *Management Information Systems Quarterly, 30*(1), 115–143.

Pentland, B. (1995). Grammatical models of organizational processes. *Organization Science, 6*(5), 541–556. doi:10.1287/orsc.6.5.541

Peppard, J. (2000). Customer Relationship Management (CRM) in Financial Services. *European Management Journal, 18*(3), 312–326. doi:10.1016/S0263-2373(00)00013-X

Peppers, D., Rogers, M., & Dorf, B. (1999). Is Your Company Ready for One-to-One Marketing? *Harvard Business Review, 77*, 151–160.

Perea y Monsuwe, T., Dellaert, B., & Ko, D. (2004). What drives consumers to shop online? A literature review. *International Journal of Service Industry Management, 15*, 102–121. doi:10.1108/09564230410523358

Peter, J. P. (1979). Reliability: A review of psychometric basics and recent marketing practices. *JMR, Journal of Marketing Research, 16*, 6–7. doi:10.2307/3150868

Peterson, R. A., Balasubramanian, S., & Bronnenberg, B. J. (1997). Exploring the Implications of the Internet for Consumer Marketing. *Journal of the Academy of Marketing Science, 25*(4), 329–346. doi:10.1177/0092070397254005

Pew Internet & American Life Project. (2005). *CAN-SPAM a year later*. Retrieved February 2008, from http://www.pewinternet.org/report_display.asp?r=102

Phau, I., & Poon, S. M. (2000). Factors influencing types of products and services purchased over the Internet. *Journal of Internet Research, 10*(2), 102–113. doi:10.1108/10662240010322894

Pieters, R., Rosbergen, E., & Wedel, M. (1999). Visual attention to repeated print advertising: A test of scanpath theory. *JMR, Journal of Marketing Research, 36*, 424–438. doi:10.2307/3151998

Pikkarainen, T. (2004). Consumer acceptance of online banking: An extension of the technology acceptance model. *Internet Research, 14*(3), 224–235. doi:10.1108/10662240410542652

Pine, B. J. (1993). *Mass Customization: The New Frontier in Business Competition*. Boston: Harvard Business School Press.

Pires, G. D., Stanton, J., & Rita, P. (2006). The internet, consumer empowerment and marketing strategies. *European Journal of Marketing, 40*(9/10), 936–949. doi:10.1108/03090560610680943

Pitknow, J., & Kehoe, C. (1996). Emerging trends in the WWW user population. *Communications of the ACM, 39*(6), 106–108. doi:10.1145/228503.228525

Pitt, L. F., Berthon, P., Watson, R. T., & Ewing, M. (2001). Pricing strategy and the Net. *Business Horizons, 44*(2), 45–54. doi:10.1016/S0007-6813(01)80022-7

Pitt, L. F., Berthon, P., Watson, R. T., & Zinkhan, G. M. (2002). The Internet and the birth of real consumer power. *Business Horizons, 45*(4), 7–14. doi:10.1016/S0007-6813(02)00220-3

Plouffe, C. R. (2001). Research report: Richness versus parsimony in modelling technology adoption decisions-understanding merchant adoption of a smart card-based payment system. *Information Systems Research, 12*(2), 208–222. doi:10.1287/isre.12.2.208.9697

Png, I. (2001). Dimensions of national culture and corporate adoption of IT infrastructure. *IEEE Transactions on Engineering Management*, 48.

Podsakoff, P. M., Mackenzie, S. B., Lee, J. Y., & Podsakoff, N. P. (2003). Common method biases in behavioral research: A critical review of the literature and recommended remedies. *The Journal of Applied Psychology, 88*, 879–903. doi:10.1037/0021-9010.88.5.879

Pollay, R. W., & Mittal, B. (1993). Here's the Beef: Factors, Determinants and Segments of Consumer Criticism of Advertising. *Journal of Marketing, 57*(July), 99–114. doi:10.2307/1251857

Poon, S., & Swatman, P. M. C. (1997). Small business use of the Internet. Findings from Australian case studies. *International Marketing Review, 14*, 385–402. doi:10.1108/02651339710184343

Porter, M. E. (2001, March). Strategy and the Internet. *Harvard Business Review*, 63.

Prahalad, C. K., & Ramaswamy, V. (2004). Co-creation Experiences: The Next Practice in Value Creation. *Journal of Interactive Marketing, 18*(3), 5–14. doi:10.1002/dir.20015

Rajgopal, S., Venkatachalam, M., & Kotha, S. (2002). Managerial actions, stock returns, and earnings: the case of business-to-business Internet firms. *Journal of Accounting Research, 40*(2), 529–556. doi:10.1111/1475-679X.00060

Rao, A. R., & Monroe, K. B. (1989). The effect of price, brand name, and store name on buyers' perception of product quality: an integrative review. *JMR, Journal of Marketing Research, 26*, 351–357. doi:10.2307/3172907

Ray, A. (2001). Affiliate schemes prove their worth. *Marketing London, 2001*, 29–30.

Rayport, J., & Sviokla, J. (1995). Exploiting virtual value chain. *Harvard Business Review*, 75–85.

Read, B. (2008). Mobile CRM: "Are we there yet? *Customer Interaction Solutions, 27*(5), 5.

Reichenbach, H. (1958). *The philosophy of space & time*. New York: Dover Publications, Inc.

Reichheld, A., & Sasser, W. E. Jr. (1990). Zero defection: quality comes to services. *Harvard Business Review, 68*, 105–111.

Reinartz, W. J., & Kumar, V. (2002). The mismanagement of customer loyalty. *Harvard Business Review, 80*(7), 86.

Reisch, L. A. (2001). The Internet and sustainable consumption: Perspectives on a Janus face. *Journal of Consumer Policy, 24*, 251–286. doi:10.1023/A:1013977509623

Reynolds, F. D., Darden, W. R., & Martin, W. (1974). Developing an image of the store-loyal customer. *Journal of Retailing, 50*(4), 73–84.

Rezabakhsh, B., Bornemann, D., Hansen, U., & Schrader, U. (2006). Consumer Power: A Comparison of the Old Economy and the Internet Economy. *Journal of Consumer Policy, 29*, 3–36. doi:10.1007/s10603-005-3307-7

Rha, J.-Y., & Widdows, R. (2002). The Internet and the consumer: Countervailing power revisited. *Prometheus, 20*, 107–118. doi:10.1080/08109020210137493

Rhee, H., Riggins, F. J., & Kim, C. (2009). The impact of product type and perceived characteristics of the web on multifaceted online shopping behavior. *Journal of Organizational Computing and Electronic Commerce*, *19*, 1–29. doi:10.1080/10919390802198907

Rindfleisch, A., & Heide, J. B. (1997). Transaction cost analysis: Past, present and future applications. *Journal of Marketing*, *61*(October), 30–54. doi:10.2307/1252085

Ritter, J. R., & Welch, I. (2002). A review of IPO activity, pricing, and allocations. *The Journal of Finance*, *57*(4), 1795–1828. doi:10.1111/1540-6261.00478

Robinson, P. (1991a). *Criteria for scale selection and evaluation. Measures of personality and social psychological attitudes*. San Diego, CA: Academic Press.

Rock, K. (1986). Why new issues are underpriced. *Journal of Financial Economics*, *15*, 187–212. doi:10.1016/0304-405X(86)90054-1

Rodgers, S., & Harris, M. A. (2003). Gender and e-commerce: an exploratory study. *Journal of Advertising Research*, *43*(3), 322–329.

Rogers, J. L. (1989/1990, December/January). Consumer Response to Advertising Mail. *Journal of Advertising Research*, *29*, 18–24.

Rogers, E. M. (1995). *Diffusion of Innovations* (4th ed.). New York: The Free Press.

Rohm, A. J., & Swaminathan, V. (2004). A typology of online shoppers based on shopping motivations. *Journal of Business Research*, *57*(7), 748–757. doi:10.1016/S0148-2963(02)00351-X

Rosenberg, L. J. (1974). A new approach to distribution conflict management. *Business Horizons*, *17*(5), 67–74. doi:10.1016/0007-6813(74)90042-1

Rosenbloom, B. (2007). Multi-channel strategy in business-to-business markets: Prospects and problems. *Industrial Marketing Management*, *36*(1), 4–9. doi:10.1016/j.indmarman.2006.06.010

Roslow, P., & Roslow, S. (1980, June). How U.S. Latins Use Radio. *Journal of Advertising Research*, *20*, 19–22.

Ross, D. (2005). E-CRM from a study chain management perspective. *Information Systems Management*, *22*(1), 37–44. doi:10.1201/1078/44912.22.1.20051201/85737.5

Ross, S. (1973). The economic theory of agency: The principal's problem. *The American Economic Review*, *63*, 134–139.

Rossman, M. (1994). Multicultural Marketing: Selling to a Diverse America. *American Management Association*, *1994*, 46–80.

Rowley, J. E. (2000). Understanding loyalty and loyal: The customer perspective. In *Proceedings of the Academy of Marketing*.

Rubinstein, A., & Wolinksky, A. (1987). Middlemen. *Quarterly Journal of Economics,* 581- 593.

Sahadev, S. (2005). Exploring the role of expert power in channel management: An empirical study. *Industrial Marketing Management*, *34*(5), 487–494. doi:10.1016/j.indmarman.2004.11.002

Salkin, S. (1999). Fear of buying. *Logistics Management and Distribution Report*, *8*(5), 101–104.

Samiee, S. (1998). The Internet and international marketing: is there a fit? *Journal of Interactive Marketing*, *12*(4), 5–21. doi:10.1002/(SICI)1520-6653(199823)12:4<5::AID-DIR2>3.0.CO;2-5

Sanchez-Franco, M. J. (2006). Exploring the influence of gender on the web usage via partial least squares. *Behaviour & Information Technology*, *25*(1), 19–36. doi:10.1080/01449290500124536

Sanders, P. (1982). Phenomenology: A new way of viewing organizational research. *Academy of Management Review*, *7*(3), 353–360. doi:10.2307/257327

Sands, M. (2003). Integrating the Web and e-mail into a push-pull strategy. *Qualitative Market Research –. International Journal (Toronto, Ont.)*, *6*, 27–37.

SAP AG. (2002). *Adaptive supply chain networks*. Retrieved October 10, 2007, from http://www.sap.com/solutions/business-suite/scm/pdf/50056466.pdf

Sarkar, M. B., Butler, B., & Steinfield, C. (1995). Intermediaries and cybermediaries: A continuing role for mediating players in the electronic marketplace. *Journal of Computer-Mediated Communication, 1*, 111–133.

Sarma, S. (2004). Integrating RFID. *Queue, 2*(7), 50–57. doi:10.1145/1035594.1035620

Saunders, J. (1993). Branding financial services. *International Journal of Bank Marketing, 11*(6), 32–38. doi:10.1108/02652329310045747

Sawhney, M., & Kotler, P. (2001). Marketing in the age of information democracy. In Iacobucci, D. (Ed.), *Kellogg on Marketing* (pp. 386–409). New York: Wiley.

Schiffman, L. G., & Kanuk, L. L. (1997). *Consumer Behavior* (6th ed., pp. 238–240). Upper Saddle River, NJ: Prentice-Hall.

Schlosser, A. E., White, T. B., & Lloyd, S. M. (2006). Converting Web Site Visitors into Buyers: How Web Site Investment Increases Consumer Trusting Beliefs and Online Purchase Intentions. *Journal of Marketing, 70*(2), 133–148. doi:10.1509/jmkg.70.2.133

Schniederjans, M. J., & Cao, Q. (2002). *e-Commerce Operations Management*. River Edge, NJ: World Scientific.

Schoemaker, P. J. H. (1982). The expected utility model: Its variants, purposes, evidence and limitations. *Journal of Economic Literature, 20*, 529–563.

Schramm, W. (1971). *Notes on case studies of instructional media projects*. Washington, DC: Academy for Educational Development.

Schultz, D. (2000). Customer/Brand Loyalty in an Interactive Marketplace. *Journal of Advertising Research, 40*(3), 41–53.

Schultze, U., & Orlikowski, W. J. (2001). Metaphors of virtuality: shaping an emergent reality. *Information and Organization, 11*(1), 45–77. doi:10.1016/S1471-7727(00)00003-8

Scullin, S. S., Fjermestad, & Romano, N. C., Jr. (2004). E-relationship maketing: changes in traditional marketing as an outcome of electronic customer relationship management. *Journal of Enterprise Information Systems, 17*(6), 410–415.

Seddon, P. B. (1997). A respecification and extension of the DeLone and McLean model of IS success. *Information Systems Research, 8*(3), 240–253. doi:10.1287/isre.8.3.240

Seddon, P. B., & Kiew, M.-Y. (1994, December). A partial test and development of the DeLone and McLean model of IS success. In *Proceedings of the International Conference on Information Systems*, Vancouver, British Columbia, Canada (pp. 99-110).

Segal, M. N., & Sosa, L. (1983). Marketing to the Hispanic Community. *California Management Review, 26*(1), 120–134.

Sekaran, U. (2000). *Research methods for business: A skill-building approach*. New York: John Wiley and Sons.

Sen, R., & King, R. (2003). Revisit the debate on intermediation, disintermediation and re-intermediation due to e-commerce. *Electronic Markets, 13*, 51–60. doi:10.1080/1019678032000067181

Shang, R., Chen, Y., & Shen, L. (2005). Extrinsic versus intrinsic motivations for consumers to shop online. *Information & Management, 42*(3), 401–413. doi:10.1016/j.im.2004.01.009

Shani, D., & Chalasani, S. (1992). Exploiting niches using relationship marketing. *Journal of Consumer Marketing, 9*(3), 33–42. doi:10.1108/07363769210035215

Shankar, V., Urban, G. L., & Sultan, F. (2002). Online trust: A stakeholder perspective, concepts, implications, and future directions. *The Journal of Strategic Information Systems, 11*, 325–344. doi:10.1016/S0963-8687(02)00022-7

Shapiro, C., & Varian, H. R. (1999). *Information rules: A strategic guide to the network economy*. Boston: Harvard Business School Press.

Sharma, S. K., Wickramasinghe, N., & Singh, M. (2005). Building systems around reverse value chain: A new system development approach. *International Journal of Management and Enterprise Development, 2*, 93–105. doi:10.1504/IJMED.2005.006032

Shergill, G. S., Sarrafzadeh, A., Diegel, O., & Shekar, A. (2008). Computerized sales assistants: The application of computer technology to measure consumer interest – A conceptual framework. *Journal of Electronic Commerce Research, 9*, 176–191.

Sheth, J. N., & Sisodia, R. S. (1999). Revisiting Marketing's Lawlike Generalizations. *Journal of the Academy of Marketing Science*, *27*(1), 71–87. doi:10.1177/0092070399271006

Shih, H. P. (2008). Continued use of a Chinese online portal: An empirical study. *Behaviour & Information Technology*, *27*(3), 201–209. doi:10.1080/01449290600802403

Shih, H. (2004). An empirical study on predicting user acceptance of e-shopping on the Web. *Information & Management*, *41*(3), 351–368. doi:10.1016/S0378-7206(03)00079-X

Shrimp, T. A. (2003). *Advertising promotion: Supplementary aspects of integrated marketing communications* (5th ed.). Orlando, FL: Dryden Press.

Silk, A. J., Klein, L. R., & Bernt, E. R. (2001). The Emerging Position of the Internet as an Advertising Medium. *NETNOMICS: Economic Research and Electronic Networking*, *3*(2), 129–148. doi:10.1023/A:1011455329830

Simmons, G., Armstrong, G. A., & Durkin, M. G. (2008). A conceptualization of the determinants of small business website adoption. *International Small Business Journal*, *26*, 351–389. doi:10.1177/0266242608088743

Sin, L., & Tse, A. (2002). Profiling internet shoppers in Hong Kong: demographic, psychographic, attitudinal and experiential factors. *Journal of International Consumer Marketing*, *15*(1), 7–28. doi:10.1300/J046v15n01_02

Sipior, J. C., Ward, B. T., & Bonner, P. G. (2004). Should spam be on the menu? *Communications of the ACM*, *47*(6), 59–63. doi:10.1145/990680.990681

Siu, N. Y., & Cheng, M. M. (2001). A study of the expected adoption of online shopping - the case of Hong Kong. *Journal of International Consumer Marketing*, *13*(3), 87–106. doi:10.1300/J046v13n03_06

Slater, S. F. (1997). Developing a customer value-based theory of the firm. *Journal of the Academy of Marketing Science*, *25*(2), 162–167. doi:10.1007/BF02894352

Smits, M., & Janssen, R. (2008). Impact of electronic auctions on health care markets. *Electronic Markets*, *18*, 19–29. doi:10.1080/10196780701797607

So, W. C., Wong, T. N., & Sculli, D. (2005). Factors affecting intentions to purchase via the Internet. *Industrial Management & Data Systems*, *105*(9), 1225–1244. doi:10.1108/02635570510633275

Solomon, M. (2007). *Consumer Behavior* (7th ed.). Upper Saddle River, New Jersey: Prentice-Hall, Inc.

Solomon, M. R. (1999). *Consumer Behavior* (4th ed., pp. 208–209). Upper Saddle River, NJ: Prentice-Hall.

Solomon, M. (1999). The value of status and the status of value. In Holbrook, M. B. (Ed.), *Consumer Value: A Framework for Analysis and Research*. New York: Routledge. doi:10.4324/9780203010679.ch3

South, S. J., & Spitze, G. (1994). Housework in martial and non-marital households. *American Sociological Review*, *59*, 327–347. doi:10.2307/2095937

Spann, M., & Tellis, G. J. (2006). Does the Internet Promote Better Consumer Decisions? The Case of Name-Your-Own-Price Auctions. *Journal of Marketing*, *70*(1), 65–78. doi:10.1509/jmkg.2006.70.1.65

Spink, A., & Jansen, B. (2008). Trends in searching for commerce related information on web search engines. *Journal of Electronic Commerce Research*, *9*, 154–161.

Srinivasan, S. S., Anderson, R., & Ponnavolu, K. (2002). Customer loyalty in E-commerce: An exploration of its antecedents and consequences. *Journal of Retailing*, *78*(1), 41–50. doi:10.1016/S0022-4359(01)00065-3

Steel, E. (2009, March 18). WPP, Google to Fund Web-Ad Research. *Wall Street Journal*, B6.

Stein, G. (2004). Trademark bidding on Google. *Clickz.com*. Retrieved February 2008, from http://www.clickz.com/showPage.html?page=3341321

Steinfield, C. (2003). *New directions in research on e-commerce*. West Lafayette, IN: Purdue University Press.

Steinfield, C. (2001). Pillars of virtual commerce: Leveraging physical and virtual presence in the new economy. *Info*, *3*(3), 203–213. doi:10.1108/14636690110801923

Steinfield, C., Bouwman, H., & Adelaar, T. (2002). The dynamics of click-and-mortar electronic commerce: Opportunities and management strategies. *International Journal of Electronic Commerce*, *7*(1), 93–119.

Stevens, J. (1996). *Applied Multivariate Statistics for the Social Sciences* (3rd ed.). Mahwah, NJ: Lawrence Elbaum Associates.

Stone, G. P. (1996). City shoppers and urban identification: Observation on the social psychology of the city life. *American Journal of Sociology*, *60*, 36–55. doi:10.1086/221483

Storey, V., Straub, D., Stewart, K., & Welke, R. (2000). A conceptual investigation of the E-commerce industry. *Communications of the ACM*, *43*(7), 117–124. doi:10.1145/341852.341871

Strader, T. J., & Shaw, M. J. (1999). Consumer cost differences for traditional and Internet markets. *Internet Research: Electronic Networking Applications and Policy*, *9*, 82–92. doi:10.1108/10662249910264819

Strader, T. J., & Shaw, M. J. (1997). Characteristics of electronic markets. *Decision Support Systems*, *21*(3), 185–198. doi:10.1016/S0167-9236(97)00028-6

Straub, D. (1998). Testing the technology acceptance model across cultures: A three country study. *Information & Management*, *33*(1), 1–11. doi:10.1016/S0378-7206(97)00026-8

Straughn, R., & Albers-Miller, N. (2001). An international investigation of cultural and demographic effects of domestic retail loyalty. *International Marketing Review*, *18*(5), 521–541. doi:10.1108/EUM0000000006044

Strauss, A., & Corbin, J. (1990). *Basics of qualitative research: grounded theory procedures and techniques*. Newbury Park, CA: Sage Publications.

Stump, R. L., & Heide, J. B. (1996). Controlling supplier opportunism in industrial relationships. *JMR, Journal of Marketing Research*, *33*, 431–441. doi:10.2307/3152214

Suh, B., & Han, I. (2002). Effect of trust on customer acceptance of Internet banking. *Electronic Commerce Research and Applications*, *1*, 247–263. doi:10.1016/S1567-4223(02)00017-0

Sun, T., & Johnson, A. (2009, March 4). *In down times, Hispanic Market is booming*. Retrieved from http://www.msnbc.msn.com/id/29470143

Suri, R., Long, M., & Monroe, K. B. (2003). The impact of the Internet and consumer motivation on evaluation of prices. *Journal of Business Research*, *56*, 379–390. doi:10.1016/S0148-2963(01)00228-4

Swift, R. S. (2001). *Accelerating Customer Relationships Using CRM and Relationship Technologies*. Upper Saddle River, NJ: Prentice-Hall.

Swinyard, W. R., & Smith, S. M. (2003). Why people (don't) shop online: A lifestyle study of the internet consumer. *Psychology and Marketing*, *20*(7), 567–597. doi:10.1002/mar.10087

Szajna, B. (1994). Software evaluation and choice: Predictive evaluation of the technology acceptance instrument. *Management Information Systems Quarterly*, *18*(3), 319–324. doi:10.2307/249621

Szymanski, D. M., & Hise, R. T. (2000). E-satisfaction: An initial examination. *Journal of Retailing*, *76*(3), 309–322. doi:10.1016/S0022-4359(00)00035-X

Tamilia, R. D. (2002). Conventional channels of distribution and electronic intermediaries: A functional analysis. *Journal of Marketing Channels*, *9*(3-4), 27–35. doi:10.1300/J049v09n03_03

Tamura, H. (2005). Behavioral models for complex decision analysis. *European Journal of Operational Research*, *166*, 655–665. doi:10.1016/j.ejor.2004.03.038

Tan, S. J. (1999). Strategies for reducing consumers' risk aversion in Internet shopping. *Journal of Consumer Marketing*, *16*(2), 163–180. doi:10.1108/07363769910260515

Tan, M., & Teo, T. (2000). Factors influencing the adoption of Internet banking. *Journal of the Association for Information Systems*, *1*(1).

Taylor, S., & Todd, P. A. (1995). Understanding information technology usage: A test of competing models. *Information Systems Research*, *6*(2), 144–177. doi:10.1287/isre.6.2.144

Tellis, E., & Ortiz, V. (2008). *Generations of Exclusion: Mexican Americans, Assimilation, and Race*. New York: Russell Sage Foundation.

Tellkamp, C., Angerer, A., Fleisch, E., & Corsten, D. (2005). *From pallet to shelf: improving data quality in retail supply chains using RFID.* Retrieved October 10, 2007, from http://www.autoidlabs.org/uploads/media/AUTOIDLABS-WP-BIZAPP-019.pdf

Telser, L. G. (1980). A theory of self-enforcing agreements. *The Journal of Business*, *53*(1), 27–44. doi:10.1086/296069

Teltzrow, M., Meyer, B., & Lenz, H. (2007). Multi-channel consumer perceptions. *Journal of Electronic Commerce Research*, *8*, 18–31.

Teo, T. S. H., & Lim, V. K. G. (2000). Gender differences in internet usage and task preferences. *Behaviour & Information Technology*, *19*(4), 283–295. doi:10.1080/01449290050086390

Tersine, R. (1998). *Principles of Inventory and Material Management* (4th ed.). Upper Saddle River, NJ: Prentice Hall.

Terziovski, M. (2003). The relationship between networking practices and business excellence: A study of small to medium enterprises (SMEs). *Measuring Business Excellence*, *7*, 78–92. doi:10.1108/13683040310478011

Tinker, T. (1986). Metaphor or reification: are radical humanists really libertarian anarchists? *Journal of Management Studies*, *23*(4), 363–384. doi:10.1111/j.1467-6486.1986.tb00423.x

Toker, A., & Varnali, K. (2009). A Framework for E-Potential Assessment in Online Markets. *Journal of Internet Business*, *6*, 81–110.

Torkzadeh, G., & Dhillon, G. (2002). Measuring factors that influence the success of Internet commerce. *Information Systems Research*, *13*(2), 187–204. doi:10.1287/isre.13.2.187.87

Tradedoubler. (2007). Affiliate contract. Retrieved February 2008, from http://www.tradedoubler.com/pan/aInfo Center.?textKey=INFORMATION_AFFILIATE_INFO-CENTER_AFFILIATE_AGREEMENT

Trevino, L. K., & Webster, J. (1992). Flow in Computer-Mediated Communication. *Communication Research*, *19*(5), 539–573. doi:10.1177/009365092019005001

Trusov, M., Bucklin, E. B., & Pauwels, K. (2009). Effects of word-of-mouth versus traditional marketing: findings from an Internet social networking site. *Journal of Marketing*, *73*, 90–102. doi:10.1509/jmkg.73.5.90

Tsay, A., & Agrawal, N. (2004). Channel conflict and coordination in the e-commerce age. *Production and Operations Management*, *13*, 93–110.

Tseng, F.-C. (2008). Network effects and multiple adoption: Two fundamental factors in the competition of e-commerce. *Proceedings of the 4th IEEE International Conference on Management of Innovation and Technology*, 498-502.

Tsoukas, H. (1991). The missing link: a transformational view of metaphors in organizational science. *Academy of Management Review*, *16*(3), 566–585. doi:10.2307/258918

Tung, F. C. (2007). Using the e-CRM information system in the hi-tech industry: Predicting salesperson intentions. *Journal of American Academy of Business*, *11*(2), 131–136.

Turban, E., King, D., Lee, J., & Viehland, D. (2004). *Electronic Commerce: A Managerial Perspective*. Upper Saddle River, NJ: Prentice-Hall.

U. S. Census Bureau. (2004). *2002 economic census: Software publishers: 2002*. Retrieved from http://www.census.gov/prod/ec02/ec0251i06t.pdf

U.S. Census Bureau. (2006). *U.S. Hispanic Population: 2006.* Retrieved from http://www.census.gov/population/socdemo/hispanic/cps2006/CPS_Powerpoint_2006.pdf

U.S. Census Bureau. (2008). *Hispanic Heritage Month 2008: Sept. 15-Oct. 15.* Retrieved from http://www.census.gov/Press-Release/www/releases/archive/cb08ff-15.pdf

U.S. Census Bureau. (2009). *Service sector statistics quarterly retail e-commerce sales*. Retrieved January 24, 2009, from http://www.census.gov/mrts/www/ecomm.html

U.S. Public Interest Research Group. (2000, April 25). *Public comment on barriers to electronic commerce, response to call by U.S. Department of Commerce* (65 [). Boston: USDC.]. *Federal Register*, 15898.

Urban, L. G. (2004). The emerging era of customer advocacy. *Sloan Management Review*, *45*(2), 77–82.

Valdes, I. (1991, October). Understanding Hispanic Consumers. *American Demographics*, S1–S4.

Valdes, I. M., & Sadane, M. H. (1995). *Hispanic Market Handbook*. Detroit, MI: Gale Research Inc.

Valencia, H. (1989). Hispanic Values and Subcultural Research. *Journal of the Academy of Marketing Science, 17*(winter), 23-28.

ValueClick. (2006). *Investor relations overview*. Los Angeles: Author.

van Iwaarden, J., van der Wiele, T., Ball, L., & Millen, R. (2004). Perceptions about the quality of web sites: A survey amongst students at Northeastern University and Erasmus University. *Information & Management, 41*(8), 947–959. doi:10.1016/j.im.2003.10.002

Van Osselaer, S. M. J., & Alba, J. W. (2000). Consumer learning and brand equity. *The Journal of Consumer Research, 27*, 1–16. doi:10.1086/314305

Van Raaij, W. F. (1998). Interactive communication: Consumer power and initiative. *Journal of Marketing Communications, 4*, 1–8. doi:10.1080/135272698345843

Van Slyke, C., Comunale, C. L., & Belanger, F. (2002). Gender differences in perceptions of web-based shopping. *Communications of the ACM, 45*(8), 82–86. doi:10.1145/545151.545155

Varshney, U. (2008, November/December). Business models for mobile commerce service: requirements, design, and the future. *IT Pro*, 48-55.

Veal, A. J. (2005). *Business research methods: A managerial approach*. Upper Saddle River, NJ: Pearson Education Australia.

Venkatesh, V. (2003). User acceptance of information technology: Toward a unified view. *Management Information Systems Quarterly, 27*(3), 425–478.

Venkatesh, V., & Davis, F. D. (1996). A model of the antecedents of perceived ease of use: Development and test. *Decision Sciences, 27*(3), 181–451. doi:10.1111/j.1540-5915.1996.tb01822.x

Venkatesh, V., & Davis, F. D. (2000). A theoretical extension of the technology acceptance model: four longitudinal field studies. *Management Science, 46*, 186–204. doi:10.1287/mnsc.46.2.186.11926

Venkatesh, V., & Morris, M. G. (2000). Why don't men ever stop to ask for directions? Gender, social influence, and their role in technology acceptance and usage behavior. *MIS Quarterly, 24*, 115–139. doi:10.2307/3250981

Venyon, Thailand bank cooperate on contactless mobile payments. (2009, February 10). *Contactless News*. Retrieved March 10, 2009, from http://www.contactlessnews.com/2009/02/10/venyon-thailand-bank-cooperate-on-contactless-mobile-payments

Viale, D. (1996). *Basics of Inventory Management: From Warehouse to Distribution Center*. Menlo Park, CA: Crisp Publications Incorporated.

Vijayasarathy, L. R. (2004). Predicting consumer intentions to use on-line shopping: the case for an augmented technology acceptance model. *Information & Management, 41*, 747–762. doi:10.1016/j.im.2003.08.011

Vijayasarathy, L. R. (2002). Product characteristics and Internet shopping intentions. *Journal of Internet Research, 12*(5), 441–426.

Voss, H. (1996). Virtual organizations: The future is now. *Strategy and Leadership, 24*(4), 12–16. doi:10.1108/eb054559

Walthieu, L. (2000). *Yesmail.com* (9-500-092). Cambridge, MA: Harvard Business School.

Wang, E. T.-G., Yeh, H.-Y., & Jiang, J. J. (2006). The relative weights of Internet shopping fundamental objectives: Effect of lifestyle differences. *Psychology and Marketing, 23*(5), 356–367. doi:10.1002/mar.20116

Wang, W., & Benbasat, I. (2007). Recommendation agents for electronic commerce: Effects of explanation facilities on trusting beliefs. *Journal of Management Information Systems, 23*, 217–246. doi:10.2753/MIS0742-1222230410

Wang, M. S., Chen, C. C., Chang, S. C., & Yang, Y. H. (2007). Effects of online shopping attitudes, subjective norms and control beliefs on online shopping intentions: a test of the Theory of Planned Behaviour. *International Journal of Management, 24*(2), 296–303.

Wang, Y. S., Wang, M. Y., Lin, H. H., & Tang, T. I. (2003). Determinants of user acceptance of Internet banking: An empirical study. *International Journal of Service Industry Management, 14*(5), 501–518. doi:10.1108/09564230310500192

Want, R. (2004). The magic of RFID. *Queue*, *2*(7), 40–48. doi:10.1145/1035594.1035619

Ward, M. R. (2001). Will online shopping compete more with traditional retailing or catalog shopping? *NETNOMICS: Economic Research and Electronic Networking*, *3*(2), 103–117. doi:10.1023/A:1011451228921

Warrington, T. B., Abgrab, N. J., & Caldwell, H. M. (2000). Building trust to develop competitive advantage in e-business relationships. *Competitiveness Review*, *10*, 160–168.

Wathne, K. H., & Heide, J. B. (2000). Opportunism in interfirm relationships: Forms, outcomes and solutions. *Journal of Marketing*, *64*(October), 36–51. doi:10.1509/jmkg.64.4.36.18070

Wayman, J. L., & Alyea, L. (2000). *Picking the best biometric for your applications. National biometric test center collected works* (pp. 269–275). San Jose, CA: National Biometric Test Center.

Webb, K. L. (2002). Managing channels of distribution in the age of electronic Commerce. *Industrial Marketing Management*, *31*, 95–102. doi:10.1016/S0019-8501(01)00181-X

Webb, B., & Sayer, R. (1998). Benchmarking small companies on the Internet. *Long Range Planning*, *31*(6), 815–827. doi:10.1016/S0024-6301(98)80018-6

Webster, C. (1992). The Effects of Hispanic Subcultural Identification on Information Search Behavior. *Journal of Advertising Research*, *32*(September/October), 54–62.

Weiner, M., Nohria, N. E. R. G., Hickman, A., & Smith, H. (1997). Value networks: The future of the U.S. electric utility industry. *Sloan Management Review*, 21–34.

Werther, W. (1999, March/April). Structure-driven strategy and virtual organization design. *Business Horizon*, 13-18.

Werts, C. E., & Linn, R. L. (1960). Path analysis: Psychological examples. *Psychological Bulletin*, *74*(2), 194–212.

Wigand, R. T., Steinfield, C. W., & Markus, M. L. (2005). Information technology standards choices and industry structure outcomes: The case of the U.S. home mortgage industry. *Journal of Management Information Systems*, *22*, 165–191.

Wikstrom, S. (1996). The customer as co-producer. *European Journal of Marketing*, *30*(4), 6. doi:10.1108/03090569610118803

Wilbon, A. D. (2002). Predicting survival of high-technology initial public offering firms. *The Journal of High Technology Management Research*, *13*(1), 127–141. doi:10.1016/S1047-8310(01)00052-9

Wild, T. (1997). *Best Practice in Inventory Management*. New York: John Wiley & Sons.

Williamson, O. E. (1991). Comparative economic organization: The analysis of discrete structural alternatives. *Administrative Science Quarterly*, *36*, 269–296. doi:10.2307/2393356

Williamson, O. E. (1992). *Markets and hierarchies*. New York: Free Press.

Williamson, O. (1985). *The economic institutions of capitalism*. New York: The Free Press.

Wilson, R. (1998, May 15). Manufacturer's dilemma: To sell or not to sell directly.

Wind, Y. (2008). A Plan to Invent the Marketing We Need Today. *MIT Sloan Management Review*, *49*(4), 21–28.

Wind, Y., & Mahajan, V. (2002). Convergence marketing. *Journal of Interactive Marketing*, *16*(2), 64–79. doi:10.1002/dir.10009

Wind, Y., & Rangaswamy, A. (2001). Customerization: The next revolution in mass customization. *Journal of Interactive Marketing*, *15*(1), 13–32. doi:10.1002/1520-6653(200124)15:1<13::AID-DIR1001>3.0.CO;2-#

Wolf, A. (2009). Surviving 2009: cost –cutting, e-commerce considered critical. *Twice*, *24*(1), 74.

Woodward, J. D. (2001). *Army biometric applications*. RAND Corporation.

Wrench, J. S., McCroskey, J. C., & Richmond, V. P. (2008). *Human Communication in Everyday Life*. Boston: Pearson Education.

Wu, S.-I. (2006). A comparison of the behavior of different customer clusters towards Internet bookstores. *Information & Management*, *43*(8), 986–1001. doi:10.1016/j.im.2006.09.002

Wu, J. (2006). Core capabilities for exploring electronic banking. *Journal of Electronic Commerce Research, 7*, 111–122.

Wulf, K. D., Schillewaert, N., Muylle, S., & Rangarajan, D. (2006). The role of pleasure in web site success. *Information & Management, 43*, 434–446. doi:10.1016/j.im.2005.10.005

Wymer, S. A., & Regan, E. A. (2005). Factors influencing e-commerce adoption and use by small and medium businesses. *Electronic Markets, 15*, 436–453. doi:10.1080/10196780500303151

Xu, M., Ferrand, B. M., & Roberts, M. (2008). The last mile of e-commerce: Unattended delivery from the consumers and retailers' perspectives. *International Journal of Electronic Marketing and Retailing, 2*, 20–38. doi:10.1504/IJEMR.2008.016815

Yahoo. (2008). *Yahoo! oneSearch™ Launches as the Mobile Search Service on T-Mobile USA's New Web-2goSM Service*. Retrieved January 11, 2009, from http://yhoo.client.shareholder.com/press/releasedetail-mobile.cfm?ReleaseID=350074

Yan, R., & Pei, Z. (2009). Retail services and firm profit in a dual-channel market. *Journal of Retailing and Consumer Services, 16*(4), 306–314. doi:10.1016/j.jretconser.2009.02.006

Yang, K. (2004). A comparison of attitudes towards Internet advertising among lifestyle segments in Taiwan. *Journal of Marketing Communications, 10*(3), 195–212. doi:10.1080/1352726042000181657

Yang, M. H., Lin, W. S., Liu, S. C., Chao, H. Y., & Chen, S. H. (2008). Developing the partner relationship management system for franchised electronic stores. *International Journal of Technology Management, 43*, 176–193. doi:10.1504/IJTM.2008.019414

Yang, B., & Lester, B. (2005). Gender differences in e-commerce. *Applied Economics, 37*, 2077–2089. doi:10.1080/00036840500293292

Yang, Z., Cai, S., Zhou, Z., & Zhou, N. (2005). Development and validation of an instrument to measure user perceived service quality of information presenting web portals. *Information & Management, 42*(4), 575–589. doi:10.1016/S0378-7206(04)00073-4

Yang, C.-H. (2007). Analysis of the consumer behavior of Taiwan's Internet users. Institute for Information Industry. Oct. 8, 2007. (in Chinese)

Yankelovich, S. (1981). *Spanish USA: A Study of the Hispanic Market in the United States*. New York: White, Inc.

Yin, R. (1994). *Case Study Research: Design and Methods* (2nd ed.). London: Sage Publications.

Yin, R. (2003). *Case study research: Design and methods*. London: Sage Publications.

Yogesh, M., & Malhotra, D. (1999). *Extending the Technology Acceptance Model to account for social influence*. Theoretical Bases and Empirical Validation.

Young, D., & Benamati, J. (2000). Differences in public web sites: The current state of US firms. *Journal of Electronic Commerce Research, 1*(3), 94–105.

Yu, C. (2007). What drives enterprises to trading via B2B e-marketplaces? *Journal of Electronic Commerce Research, 8*, 84–100.

Yu, T. K., & Wu, G. S. (2007). Determinants of Internet shopping behavior: an application of Reasoned Behaviour Theory. *International Journal of Management, 24*(4), 744–764.

Zajac, E. J., & Olsen, C. P. (1993). From transaction cost to transaction value analysis: Implications for the study of inter-organizational strategies. *Journal of Management Studies, 30*, 131–145. doi:10.1111/j.1467-6486.1993.tb00298.x

Zajtchuk, R., & Satava, R. (1997). Medical application of virtual reality. *Communications of the ACM, 40*(9), 63–64. doi:10.1145/260750.260768

Zanna, M. P., & Rempel, J. K. (1988). *Attitudes: A new look at an old concept*. New York: Cambridge University Press.

Zebra Technologies. (2007). *Bar coding 101... What you need to know*. Retrieved October 8, 2007, from http://www.zebra.com/id/zebra/naen/documentlibrary/whitepapers/bar_coding_101.DownloadFile.File.tmp/WP1144Barcode101_R5_1.07.pdf?dvar1=White%20Paper&dvar2=Bar%0Coding%20101...What%20You%20Need%20to%20Know

Zeithaml, V. A. (1981). How consumer evaluation processes differ between goods and services. In Donnelly, J. H., & George, W. R. (Eds.), *Marketing of Services* (pp. 186–190). American Marketing Association.

Zhang, X., Keeling, K. B., & Pavur, R. J. (2000). Information quality of commercial web site home pages: An explorative analysis. In *Proceedings of the International Conference on Information Systems*, Brisbane, Australia (pp. 164-175).

Zhao, F. (2006). *Maximize Business Profits through E-partnerships*. Hershey, PA: IGI Global.

Zhou, M., Dresber, M., & Windle, R. (2007). Online reputation systems: Design and strategic practices. *Decision Support Systems*, *44*, 785–797. doi:10.1016/j.dss.2007.10.001

Zhou, N., Zhuang, G., & Yip, L. S. (2007). Perceptual difference of dependence and its impact on conflict in marketing channels in China: An empirical study with two-sided data. *Industrial Marketing Management*, *36*(3), 309–321. doi:10.1016/j.indmarman.2005.10.001

Zikmund, W. G., & Catalanello, R. F. (1976). Managing channel conflict through channel development. *Journal of the Academy of Marketing Science*, *4*(4), 801–813. doi:10.1007/BF02729839

Zipkin, P. (2000). *Foundations of Inventory Management*. Boston: McGraw-Hill.

About the Contributors

In Lee is a professor in the School of Computer Sciences at the College of Business and Technology at Western Illinois University. He is a two-time winner of the Research Excellence Award in the College of Business and Technology at WIU. He is the founding editor-in-chief of the International Journal of E-Business Research. He has published his research in such journals as *Communications of the ACM, IEEE Transactions on Systems, Man and Cybernetics, IEEE Transactions on Engineering Management, International Journal of Production Research, Decision Support Systems, Computers and Operations Research, International Journal of Production Economics, Computers and Education, International Small Business Journal, Computers in Human Behavior, Computers and Industrial Engineering*, and others. His current research interests include Web 2.0 technology development and management, investment strategies for computing technologies, and mobile services. He has also served as a reviewer for a number of premier journals. He received his PhD from University of Illinois at Urbana-Champaign.

***

Thomas Acton is a lecturer in Business Information Systems at the J.E. Cairnes School of Business & Economics, National University of Ireland, Galway, Ireland. His main research interests lie in decision support systems, usability and acceptance, and mobile devices. He holds a PhD degree on Decision Support for Small-Screen Information Systems. He is a graduate of the National University of Ireland, with other qualifications in mathematics, education, software design, and commerce. He has a number of journal publications, book chapters and conference papers, and is Associate Editor on the European Journal of Information Systems.

Milam Aiken, PhD, is currently professor of MIS and Chair of the MIS/POM Department in the School of Business Administration at the University of Mississippi. He received his PhD degree in MIS from the University of Arizona in 1991. Since 1990, he has published more than 120 journal articles on group decision support systems and artificial intelligence.

Fahad AL-Harby is a PhD researcher at the University of Bradford in UK. He received his BSc. degree in computer science from King Saud University, and M.Sc. degree in software systems for e-business and WWW from the University of Bradford in 2006. He has many years industrial experience as a computer programmer, systems analyst and an IT specialist at the Ministry of Finance in Saudi Arabia. His research interests include information system security within organisations, biometrics authentication and e-commerce/e-government.

Enrique P. Becerra (PhD. Florida Atlantic University) is an assistant professor in the department of marketing at Texas State University-San Marcos, specializing in international marketing, and ethnic and niche marketing. Dr. Becerra has professional experience in retailing, banking, and international business. His research emphasis is in the area of consumer behavior, ethnic and cross-cultural research, and personality influences. His work has previously been published in the *Journal of Computer Mediated Communications*, the *Journal of Virtual Worlds Research*, the *International Journal of Electronic Marketing and Retailing*, and the *Journal of Electronic Commerce Research.*

Rick Carter is professor of finance at Iowa State University. Carter currently teaches real estate, investments and security analysis courses at both the undergraduate and graduate levels. He received his MBA in 1985 and his PhD in Finance in 1987, both from the University of Utah. Prior to receiving his graduate degrees, Carter was Supervisor of Data Processing and of Policy-owner Service at Mutual of New York and Terminal Manager at Stott and Davis Motor Express at their Syracuse office. His research interests include investment banking, security analysis, and the effects of e-commerce on financial services and capital acquisition. He has published in such journals as the *Journal of Financial and Quantitative Analysis, Electronic Commerce Research, the Journal of the Association for Information Systems, the Journal of Finance, the Journal of Financial Research, the Financial Review, the Journal of Portfolio Management, and the Journal of Accounting and Public Policy and Financial Management.*

T. C. Edwin Cheng is chair professor of Management in the Department of Logistics and Maritime Studies of The Hong Kong Polytechnic University. He obtained a BSc[Eng](Hons) from the University of Hong Kong, an MSc from the University of Birmingham, and a PhD and an ScD from the University of Cambridge. He has previously taught in Canada, England and Singapore. Cheng's research interests are in Operations Management. He has published over 430 papers in such journals as *Management Science*, *Operations Research* and *MIS Quarterly*, and co-authored five books. He received the Outstanding Young Engineer of the Year Award from the Institute of Industrial Engineers, U.S.A., in 1992, and the Croucher Senior Research Fellowship (the top science award in Hong Kong) in 2001. He was named one of the "most cited scientists" in All Fields, in Computer Science, and in Engineering over the period 1999-2009 by the *ISI Web of Knowledge* in 2009. According to the *ISI Web of Science*, he has attained an *h*-index of 23 (i.e., having produced 23 publications each attracting 23 or more citations). A paper in the *International Journal of Production Economics* in 2009 ranked him as the most productive researcher and one of the top *h*-index researchers in Production/Operations Management in the world in the past half-century.

Man Wai Chung completed her M.Sc. (Operations Management) with distinction at The Hong Kong Polytechnic University in 2009. Before entering the Operations Management field, she worked for Ernst & Young on Event Management and Adidas Sourcing Limited as a management trainee in Knowledge Management. She had lived in Canada and completed her B.B.A. (majoring jointly in Economics and Management & Organization Studies) at Simon Fraser University. A member of the Junior Team Canada, she visited Beijing and Hong Kong as an ambassador representing the Canadian education industry in 2004. She is also an active member in the society and has participated in various community activities, such as the school rebuilding projects with Xuan Yuen Education Fund Association, the Equestrian Events of the 2008 Beijing Olympic Games, as well as the 2009 East Asian Games.

Fergle J. D'Aubeterre is the Application Services & Architecture Team Lead at Flint Transfield Services Ltd (FT-SERVICES), Canada. He obtained his PhD in Information Systems from The University of North Carolina at Greensboro and his MBA from Central Michigan University. His research interests include electronic commerce, business processes, Semantic Web, IT security and privacy, and global IT management. Fergle has published papers in journals such as *Journal of the Association for Information Systems, European Journal of Information Systems, Information Systems Journal*; *Electronic Government an International Journal; the International Journal of Electronic Commerce Research*; *Encyclopedia of E-Commerce; E-Government*, and *Mobile Commerce*; the *Proceedings of the International Conference on Information Systems*; the *Proceedings of Americas Conference on Information Systems*; the *Proceedings of Global Information Technology Management*; the *Proceedings of the Design Science Research in Information Systems and Technology*; and the *Proceedings of the Decision Sciences Institute*.

Paul Fox is an assistant professor in the Department of Management and Technology at La Salle Business Engineering School – Ramon Llull University in Barcelona, Spain. His research focuses on technology-driven innovation, Internet business models, and Internet marketing.

Linwu Gu, PhD, is currently associate professor of MIS at Indiana University of Pennsylvania. She earned her Master's degree in Computer Science in 2001 and her PhD in MIS in 2005, both from the University of Mississippi. Her research areas include group decision support systems, strategic information systems, and e-business.

Dane Hamilton graduated from the School of Information Systems and Technology, at the University of Wollongong with first class honors in the Bachelor of Information and Communication Technology (eBusiness major) in 2007. Dane received the Price Water House Coopers award for the highest grade awarded to a thesis in this program. He has worked as a consultant for Ernst and Young, and more recently as an area manager for a large retail company in Australia.

Soongoo Hong is an associate professor at the Department of Management Information Systems at Dong-A University in Korea. He received his Ph.D. in Management Information Systems in 2000 and Mater of Arts in Management in 1995 from the University of Nebraska – Lincoln. Prior to joining Dong-A University he was an assistant professor at Texas A&M International University. His research interests include SME's informatization, knowledge management, data warehousing, and IT impacts on organizations.

Bandula Jayatilaka was an Assistant Professor at Binghamton University, in New York and currently he is at University of Nizwa as an Associate Professor. His current research interests are: IT sourcing, IT and organizational change including E-Business organizations, knowledge management, epistemological foundations, globalization and IT and IS development. He has published in journals such as European Journal of Information System, Journal of AIS, and Database for Advancement of IS. He has provided editorial and reviewing services to ICIS, MISQ, EJIS and several other journals and conferences. Before joining the academic community, he worked for General Electric Company at Johnson Space Center NASA in Houston. He earned his PhD in MIS from the University of Houston.

Wen-Jang (Kenny) Jih is a professor of computer information systems at the Jennings A. Jones College, Middle Tennessee State University. He recently has conducted research in the areas of knowledge management, information strategy, and e-commerce. His papers have been published in International Journal of E-Business Research, Information Resource Management Journal, International Journal of Information Management, International Journal of Knowledge Management Studies, Journal of Information Systems Education, Journal of Computer Information Systems, Computers in Industry, International Journal of Enterprise Information Systems, and others. He received his doctorate degree in Business Computer Information Systems from the University of North Texas in 1985.

Mumtaz Kamala A Lecturer, Computer Systems Administration Course Leader and Admission Tutor at the University of Bradford, School of Computing, Informatics and Media, Department of Computing. He graduated as an engineer from the University of Technology/Baghdad in 1981 then finished his PhD degree at the University of Sheffield in 1992 under the supervision of Prof. Bob Boucher. The research was to study the flow behaviour by visualising the fluid using laser sheet then capture images aiming to analyse them in real time. Dr. Kamala moved to Bradford University in 1995 when he started developing his interest and research activity in Information Management and Security supervising many PhD students in this field and collaborate with the School of Management doing research in Information Security Management, and Total Quality Management. Dr. Kamala has many successful PhD studies and publications.

Pairin Katerattanakul is an associate professor in the Department of Business Information Systems, Western Michigan University. He received his Ph.D. in Management Information Systems and Master of Arts in Marketing from the University of Nebraska – Lincoln. His research and teaching interests are in information systems discipline, business modeling, electronic business, and marketing aspects of electronic commerce. He has published his research in several academic journals, including Communications of the ACM, European Journal of Information Systems, Communications of the AIS, INFOR, Journal of Electronic Commerce in Organizations, and International Journal of Information Technology & Decision Making.

Ben B. Kim is a Chair of the Department of Management and also a Director of the Center for Electronic Commerce and Information Systems at the Seattle University. Before joining Seattle University, he taught at the University of Nebraska - Lincoln for ten years. He holds Ph.D., MBA and BA degrees from University of Minnesota, University of Washington and Seoul National University, respectively. His teaching and research interests are in building and managing corporate information systems, including web applications, database systems, and object-oriented systems.

Ruth C. King is an Associate Professor at the Bryan School of Business and Economics, University of North Carolina at Greensboro. She received her PhD from the University of Texas at Austin in the Information Systems area. Her research interests include IT governance and strategy, electronic commerce, IS design and development, and IT human capital management. She has published in leading journals such as *Information Systems Research, Journal of Management Information Systems, Decision Sciences, Communications of ACM, International Journal of Electronic Commerce, Information and*

Management, European Journal of Information Systems, Electronic Market, and *International Journal of Human Resource Management*.

Pradeep K. Korgaonkar is the InternetCoast Institute Adams Professor of Marketing at Florida Atlantic University. His research interests are in the areas of e-commerce, advertising, and retailing. His research has been published in numerous journals including the *Journal of Advertising*, the *Journal of Advertising Research*, the *Journal of Business Research*, the *Journal of Computer Mediated Communications*, the *Journal of Consumer Marketing*, the *Journal of Current Issues and Research in Advertising*, the *Journal of Electronic Commerce Research*, the *Journal of Internet Research*, the *Journal of Retailing*, and *the Journal of Marketing Research* among others.

Euntae "Ted" Lee received his Ph.D. in Management Information Systems from University of Nebraska and MS in Computer Science from The Pennsylvania State University. Dr. Lee is currently an associate professor of MIS in the Fogelman College of Business and Economics at the University of Memphis. His primary research interests are knowledge management and sourcing, BRMS (Business Rules Management Systems), database systems, ethical decision making in IT use, data warehousing/mining, and strategic use of organizational information systems.

James J. Lee is a Genevieve Albers Professor of Albers School of Business and Economics. Prior to that he had served as an assistant professor at State University of New York at Binghamton. He teachs primarily in the areas of E-Commerce & Information Systems, E-Business, and Web Applications. His research is in the areas of Enterprise Integration in Social Context, Information Technology Conceptualization, and Communication Structure. Dr. Lee's publications have appeared in such journals as the Communications of ACM, Journal of Information Technology, Industrial Management & Data Systems, and many more.

Su-Fang Lee is an associate professor in the Department of Business Administration at the Overseas Chinese Institute of Technology in Taichung, Taiwan. She received her doctorate degree from the College of Management of DaYeh University in Taiwan. Her recent research interests primarily focus on Internet applications in marketing and knowledge management. She has published in Information Resources Management Journal, International Journal of E-Business Research, International Journal of Business Studies, Journal of Computer Information Systems, and many academic journals in Taiwan.

Katina Michael (katina@uow.edu.au) is an Associate Professor in the School of Information Systems and Technology in the Faculty of Informatics at the University of Wollongong. She received a Bachelor of Information Technology degree from the University of Technology, Sydney (UTS) in 1996, a PhD degree in information technology and communications in 2003, and a Masters of Transnational Crime Prevention in the Faculty of Law from the University of Wollongong in 2009. Before joining the University of Wollongong in 2002 to teach and research in eBusiness, she worked as a senior network and business planner at Nortel Networks. Dr. Michael is a senior member of the Institute of Electrical and Electronics Engineers. Most recently she has guest edited several special issues in the *Proceedings of the IEEE* (PIEEE) and the *Journal of Theoretical and Applied Electronic Commerce Research* (JTAER) on the theme of RFID Innovation and Supply Chain Management.

Kelley O'Reilly is a PhD Candidate in Management Information Systems (MIS) at Utah State University. She has more than twenty years of executive level experience in franchise retail and service businesses and has held positions including Senior Vice President of Strategy and Brand Development, Vice President of Sales & Marketing, and Vice President of Advertising. Her research interests are electronic marketing, customer relationship management (CRM), and improving customer-company interactions. Her work has been published in Qualitative Market Research: An International Journal, and Informing Sciences: The International Journal of an Emerging Transdiscipline. Her work is forthcoming in the Journal of Information Technology Case and Application Research, and the Encyclopedia of E-Business Development and Management in the Global Economy.

Rami Qahwaji was born in 1972. He got his BSc in electrical engineering from the University of Mustansiriyah (Baghdad) in 1994 and his MSc in control and computer engineering from the same university in 1997. He got his PhD in computer vision systems in 2002 from Bradford University (UK). He is currently a Reader in Visual Computing at the School of Computing, Informatics and Media at Bradford University (UK). He is the principal investigator for two major EPSRC grants, total value of around 425K GBP. He is also involved in 3 small research grants. His publications include around 85 refereed papers, over 30 conference presentations and 7 PhD completions. He was invited to give talks in Belgium, Italy, France, USA, the Netherlands and Jordan. He is the conference Co-Chair for the International Conference on Cyberworlds 2009 (UK) and CSAA09 - Egypt) and had been a member of the technical committees for more than 16 international conferences. Dr Qahwaji is a Charted Engineer (CEng– IET), a Fellow of the Higher Education academy (HEA) and a member of AGU, IET and IEEE. Dr Qahwaji has refereed research proposals for different funding bodies and he is a reviewer for many journals.

Murray Scott is a lecturer in Business Information Systems at the J.E. Cairnes School of Business & Economics, National University of Ireland, Galway, Ireland. His main research interests lie in IS success, e-government and innovation in public sector information systems. He has Masters degrees from the National University of Ireland and the University of Dundee in the areas of Commerce and Philosophy respectively. He has a number of journal publications, book chapters and conference papers.

Ravi Sen joined Texas A&M University in 2003 as an Assistant Professor in the Department of Information and Operations Management at the Mays Business School. He received his PhD in Business Administration from the University of Illinois at Urbana–Champaign in 2003. Prior to that, he completed his MBA from Manchester Business School (UK) in 1995, and his BE (Computer Science) from Devi Ahilya Vishwavidyalaya (Indore, India) in 1993. His research interests include economics of electronic commerce, open source software, and software security. He has published in the *Journal of Management Information Systems, Decision Support Systems, International Journal of Electronic Commerce, Communications of the AIS, Electronic Markets, Journal of Electronic Commerce Research*, and others.

Vikram Sethi, Director, Institute of Defense Studies and Education; Advisor to the Dean, Corporate Programs and Relations, Raj Soin College of Business and Director of the Institute of Defense Studies and Education (IDSE). Vikram received his PhD from the University of Pittsburgh. He previously held the position of Chair, Department of Information Systems and Operations Management, Raj Soin College of Business; Associate Professor and PhD Coordinator, University of Texas, Arlington. He also serves as Director, Data Intensive Supply Chain Research Center, Wright State University. He has published more than 50 articles in refereed journals and conferences, and produced several white papers in the areas of global information systems, human capital management, organizational transformation and information system project risk assessment. His book entitled, *Organizational Transformation Through Business Process Reengineering*, was published by Prentice Hall in 1993.

Ultan Sharkey is a PhD student at the J.E. Cairnes School of Business & Economics, National University of Ireland, Galway, Ireland. He is a graduate of National University of Ireland, University College Dublin and National University of Ireland, Galway with qualifications in commerce, law and information systems. He is currently pursuing a PhD in decision support systems for electronic commerce.

Ronnie Silverblatt is an Associate Professor of Management and International Business, College of Business Administration, at Florida International University. She holds a PhD in Business Administration from Georgia State University. Some of her prior works have been published in *Industrial Relations, Journal of Labor Studies, Journal of Business and Psychology, Journal of Business Research, Journal of Computer Mediated Communications, Journal of Internet Research* and the *Journal of Applied Business Research.* Her areas of interest include human resource management, Hispanics and e-commerce.

Troy Strader is professor of Information Systems in the Drake University College of Business and Public Administration. Strader received his PhD in Business Administration (Information Systems) from the University of Illinois at Urbana-Champaign in 1997. His research interests include online consumer behavior, information technology adoption, mobile commerce, and impact of the Internet and e-business on initial public offerings. He has published in the International Journal of E-Commerce, Communications of the ACM, the *European Journal of Information Systems, the Journal of the Association of Information Systems, Decision Support Systems, Electronic Markets* and other academic and practitioner journals and books. He has co-edited two books, the *Handbook on Electronic Commerce, and Mobile Commerce: Technology, Theory and Applications.*

Yuan-Cheng Tsai received his BS degree in mechanical engineering from Chung Cheng Institute of Technology (CCIT) in Taiwan, and MS degree in mechanical engineering from Da-Yeh University in Taiwan. After obtaining his doctorate degree in business (marketing concentration) from Da-Yeh University, he has been teaching in Department of Marketing and Distribution Management at Overseas Chinese University located in Taichung, Taiwan. His research interests are marketing, knowledge management, new product development, human factor engineering, logistics management, and e-business.

Fan-Chen Tseng is an Associate Professor in the Department of Information and Electronic Commerce at Kainan University (KNU), Taiwan. He received his BSEE and MSEE degrees from the department of electrical engineering at Taiwan University in 1984 and 1986, respectively. After that, he was with the Swire Group (Taiwan), United Microelectronics Corporation (UMC), and National Yi-Lan University. He received his PhD degree in Information Engineering at Taiwan University in 2002. His current research and teaching interests are in the area of e-commerce, online consumer behavior, and Internet marketing. He has papers published in Technovation, International Journal of E-Business Research,

International Journal of Electronic Business Management, and Electronic Commerce Studies, and has papers presented in IEEE international conferences.

Zsolt Ugray is an Assistant Professor in Management Information Systems at the Jon M. Huntsman School of Business of Utah State University. He received his PhD in Management Science and Information Systems from The University of Texas at Austin. His research interests include issues related to the efficient use of information systems by end-users in business and education, and the successful development of information systems and optimization solutions. He published in the INFORMS Journal on Computing, Interacting with Computers, International Journal of Mobile Learning and Organization and other journals.

Kaan Varnali, Ph.D. is an Assistant Professor of Advertising at the School of Communications, Istanbul Bilgi University, Turkey. His current research interests focus on marketing through new media, and the impact of technology on the universe of marketing. He has a particular interest on mobile consumer behavior, especially on the factors that drive consumers' value creation process through mobile experiences. His prior research have appeared on several scholarly outlets as journal articles, conference proceedings, books, and book chapters. He also has professional experience in the areas of e-ventures consultancy and web-based strategy development.

Samuel Fosso Wamba is a senior lecturer in Information Technology at the University of Wollongong, Australia. His current research focuses on business value of IT, inter-organisational system (e.g., RFID technology) adoption and use, supply chain management, electronic commerce and mobile commerce. He published in *European Journal of Information Systems, International Journal of Production Economics, Information Systems Frontiers, Journal of Theoretical and Applied Electronic Commerce Research, Hawaii International Conference on Systems Science* (HICSS) and *Americas Conference on Information Systems* (AMCIS). He is co-organizing special issues on RFID for the *Business Process Management Journal* and *Pacific Asia Journal of the Association for Information Systems*. He has been served as mini-track chair on the same topic for the 15th AMCIS, The 6th, 7th and 8th Workshops on e-Business, A Pre-ICIS Workshop Sponsored by AIS SIGeBIZ. Dr Wamba is CompTIA RFID+ Certified Professional, Academic Co-Founder of RFID Academia and Founder and CEO of e-m-RFID.

Jianfeng Wang, PhD, is currently associate professor of MIS at Indiana University of Pennsylvania. He earned his Bachelor's degree in Economics from Fudan University in 1991 and his PhD in MIS from the University of Mississippi in January 2003. His research areas include economics of information technology, group decision support systems, database technology, and organizational impacts of information technology.

Jonathan Wareham is an associate professor of Information Systems at ESADE - Ramon Llull University . His research focuses on the intersection of information technology, economics and strategy. Wareham's research has been published or forthcoming in over 60 refereed journals and proceedings such as *MIS Quarterly, Decision Support Systems, IEEE Transactions on Engineering Management, IEEE Transactions on Professional Communication, IEEE Computer, Journal of Medical Internet Research, Communications of the ACM, Telecommunications Policy, Information and Organization, Information*

Systems Journal, Journal of Information Technology. He currently serves as Associate Editor for *Information Systems Research* as well as Director of Research at ESADE.

W. H. Yeung is a Senior Manager and Zone Manager of Retail Banking Division of Dah Sing Bank Ltd. He obtained a doctor of Business Administration from The Hong Kong Polytechnic University, a Master of Business Administration and a Master of Science in Banking degree from the City University of Hong Kong. He is a Fellow Member of the Association of Chartered Certified Accountants, an Associate Member of the Hong Kong Institute of Certified Public Accountants and an Associate Member of the Hong Kong Institute of Bankers. He has over 23 years of banking experience.

Fang Zhao (PhD) is an Associate Professor of Management in the American University of Sharjah, the UAE. She has authored/edited and published 3 research books in the areas of e-business and IT management over the past few years. She established the international refereed journal – *International Journal of e-Business Management* (ISSN: 1835-5412) in 2007 and has been the Editor-in-Chief ever since. She has also published over 60 research works in peer-reviewed journals, books and conferences internationally in the areas of entrepreneurship and innovation, research commercialization, knowledge management, TQM, etc. Her current research interests are in m-commerce and e-government practices in the Middle East.

Index

E

F

G

H

I

Q

R

S

T

CPSIA information can be obtained at www.ICGtesting.com
Printed in the USA
BVOW050226231111
276395BV00008B/3/P

9 781613 504628